HANDBOOK OF THERMAL INSULATION APPLICATIONS

HANDBOOK OF THERMAL INSULATION APPLICATIONS

by

D.E. Croy **D.A. Dougherty**

EMC Engineers, Inc.
Denver, Colorado

np

NOYES PUBLICATIONS
Park Ridge, New Jersey, USA

Library of Congress Catalog Card Number: 83-22118
ISBN: 0-8155-0968-5
ISSN: 0270-9155
Printed in the United States

Published in the United States of America by
Noyes Publications
Mill Road, Park Ridge, New Jersey 07656

10 9 8 7 6 5 4 3 2 1

Library of Congress Cataloging in Publication Data

Croy, D.E.
Handbook of thermal insulation applications.

(Energy technology review, ISSN 0270-9155 ; no. 89)
Bibliography: p.
Includes index.
1. Insulation (Heat)- -Handbooks, manuals, etc.
I. Dougherty, D.A. II. EMC Engineers, Inc. III. Title.
IV. Series.
TH1715.C74 1984 621.402'4 83-22118
ISBN 0-8155-0968-5

Foreword

This handbook of thermal insulation applications is a comprehensive compilation of design information and data on thermal insulation materials and assemblies for building envelopes and mechanical systems. It will provide readily accessible information for those involved with thermal insulation for energy conservation.

Energy costs during the 1980s are expected to remain at an elevated level, whether or not energy shortages continue; therefore energy conservation techniques will continue to remain important. Thermal insulation will have a major role in energy conserving efforts during the decade. Three billion pounds of insulation were produced in 1980; the amount is expected to more than double by 1990.

The handbook has been organized to provide the architect, designer, and engineer with a powerful decision-making tool for the design of new buildings and mechanical systems, as well as for the retrofit of existing buildings and mechanical systems. If the information presented is used judiciously, the overall result will be to conserve energy.

The handbook is divided into six basic sections, as follows:

Section 1.0 includes the objective and scope of the manual.

Section 2.0 discusses heat transfer fundamentals, and gives examples of heat transfer through composite slabs and cylinders. Methods of determining surface and interface temperatures and the thickness of insulation required to prevent condensation on chilled surfaces are also presented. Finally, the physics of heat transfer inside thermal insulation is discussed.

Section 3.0 provides a description of common generic insulating materials. Thermophysical properties, manufacturing processes, and typical applications for each generic insulation are discussed. A table of thermophysical properties for each material is presented, followed by a comparative table summarizing the major advantages, disadvantages, and limitations for each generic insulation material. This section also provides a sequence of tables showing the thermal properties of building materials and insulation. For industrial and piping insulation, the temperature dependency of thermal conductivity, the maximum service temperature, and the material density are tabulated. Finally, this section includes a discussion of vapor barriers, including types and material properties.

Section 4.0 is a graphic presentation of approximately 150 wall, roof, window, door, and gasketing sections, as well as piping, tank, and duct insulation sections. This graphics section is indicative of existing and new methods in design and construction, as well as in retrofitting for energy conservation.

Section 5.0 discusses basis economic parameters and methods of determining optimum insulation thickness.

Section 6.0 includes a description of the FORTRAN and BASIC Computer Programs to compute various characteristics of composite slabs and insulated piping, including overall heat transfer coefficient, unit heat transmission loss, unit density, and unit heat capacity.

The information in the book is from *Handbook of Thermal Insulation Applications,* prepared by D.E. Croy and D.A. Dougherty of EMC Engineers, Inc., for the Naval Civil Engineering Laboratory, June 1982.

The table of contents is organized in such a way as to serve as a subject index and provides easy access to the information contained in the book.

Advanced composition and production methods developed by Noyes Data are employed to bring this durably bound book to you in a minimum of time. Special techniques are used to close the gap between "manuscript" and "completed book." In order to keep the price of the book to a reasonable level, it has been partially reproduced by photo-offset directly from the original report and the cost saving passed on to the reader. Due to this method of publishing, certain portions of the book may be less legible than desired.

Preface

During the 1980s, energy shortages will continue to increase, led by a decline in oil production by the United States of at least 1 million barrels per day [Trias, Priscilla F. "Industry News," *ASHRAE Journal,* Vol. 23, No. 3, p. 13 (March 1981)] . This conclusion, contained in a report to the President and Congress by the Panel on Energy, Natural Resources, and the Environment of the President's Commission for a National Agenda for the Eighties, serves to reinforce an opinion held by most energy professionals: the energy problem in the United States is far from being solved. The cost of energy purchased from foreign sources is now tens of billions of dollars every year, fueling our inflationary spiral. With domestic production declining, the Panel also concludes that United States energy policy in the 1980s should be aimed at achieving higher efficiency levels in the use of raw energy resources, and that conservation is the best energy policy.

Clearly, thermal insulation will have a major role in energy conservation during the next decade. According to industry experts, 3 billion pounds of insulation were produced in 1980 and this figure may more than double by 1990 (see Goutte, 1981, and McDuff, 1980, in bibliography). A recent study conducted by the Lawrence Berkeley Laboratory concluded that the projected energy requirement by buildings in this country in the year 2000—some 32 quads (quadrillion Btu)—could be cut in half using only energy conservation techniques known today. Every $10 spent on conservation would save at least 1 barrel of oil—equivalent to buying oil at $10 a barrel.

In recent years, energy conservation programs have emphasized reducing the energy consumption of both existing and newly constructed buildings and mechanical systems; these programs will continue to focus on these areas in the future. Such programs have created a need for information on thermal insulation and building material properties, as well as on assemblies of these materials.

In January 1980, the first edition of the *Building Insulation Materials Compilation* was assembled. This previous manual, published as report number CR80.001 represented results of an investigation conducted by Dynatech R/D Company, a division of Dynatech Corporation, under contract with the Naval Civil Engineering Laboratory, sponsored by the Naval Material Command. The contents of this previous report were limited primarily to discussions of building insulation thermophysical properties. In the past few years, rapid advances have taken place in the area of insulation assemblies, both for new construction and retrofit applications. The Naval Civil Engineering Laboratory (NCEL), Port Hueneme, California, recognized the need to document these advances; consequently, EMC Engineers, Inc. was retained to generate a revision and expansion of the previous manual.

An attempt was made during the creation of the revised manual to provide information in a form that could be easily used and interpreted by practicing architects and engineers. This newly revised and expanded edition, presented in the following pages, which includes information on insulation assemblies for buildings, mechanical equipment, piping systems, and other industrial systems, has been retitled *Handbook of Thermal Insulation Applications.*

ACKNOWLEDGEMENTS

The staff of EMC Engineers, Inc. would like to express its appreciation for the support and assistance obtained from Dr. Robert Alumbaugh, P.E., and Mr. Spencer R. Conklin, P.E., of NCEL.

NOTICE

Contents and Subject Index

1.0

Introduction

1.1 Objective

The objective of this handbook is to provide a comprehensive compilation of design information and data on thermal insulation materials and assemblies for building envelopes and mechanical systems. This information is intended to be useful to architects, designers, and engineers, both in designing new buildings and in altering existing buildings to conserve energy.

This handbook was prepared under the authority of Contract No. N62474-81-C-9395 issued by the Naval Civil Engineering Laboratory (NCEL), Port Hueneme, California, to EMC Engineers, Inc., Denver, Colorado.

1.2 Scope

This handbook provides up-to-date design information and data on thermal insulation materials and assemblies for building envelopes including:

- o Insulation components.
- o Building materials.
- o Combinations of insulation components and building materials.

Information and data on mechanical insulation systems are also presented. Such systems include:

- o Piping and industrial insulation components.
- o Piping insulation systems.
- o Ducting insulation systems.
- o Tank and vessel insulation systems.
- o Other special applications.

The improved work also includes the expansion, modification, and updating of the existing Building Insulation Materials Compilation including additional information and illustrations concerning:

- o Basics of heat transfer.
- o Vapor barriers.
- o Economics of insulation.

- Building envelope assemblies.
- Mechanical system assemblies.
- Material specifications (ASTM standards).
- List of manufacturers and associations.
- Useful formulas and data tables.
- Conversion factors.

1.3 General Comments

This handbook is designed for use by those with experience in building and mechanical systems design. The intent throughout the manual is to present information in a useful and easily understood form.

The insulation system designer will, of course, want to consult federal, state, and local standards and building codes applicable to any particular project [1-6]*.

Though much is known about insulation, the field of knowledge is expected to expand considerably within the next several years. Agencies and associations having a high interest in energy conservation and thermal insulation include the U.S. Department of Energy (DOE), the National Bureau of Standards (NBS), the U.S. Department of Housing and Urban Development (HUD), the American Society for Testing and Materials (ASTM), the American Society of Heating, Refrigerating and Air-Conditioning Engineers (ASHRAE), and a host of federal, state, professional, and trade associations too numerous to list individually. Two of these agencies, DOE and NBS, have prepared a National Program Plan for Building Thermal Envelope Systems and Insulating Materials [7-10] identifying critical research and standards needed in energy conservation technology.

Areas of concern regarding thermal insulation include standard test procedures for measuring performance, the effects of high thermal capacity in building elements, the effects of temperature, humidity, settling, and shrinkage of insulation on performance, research on smoldering characteristics and fire retardants, fire performance data, analytical models, and field test methods for evaluating the performance of components, systems, and entire buildings (see also [11]). Because of the rather rapid changes occurring in the field of thermal insulation, it is recommended that periodic updates to this manual be made when the amount of new information available is enough to warrant a revised edition. Users of this handbook are encouraged to provide feedback to the authors and the Naval Civil Engineering Laboratory as to suggested changes or additions.

This handbook does not present a detailed method for determining building loads. Such methods (see, for example, References [12-17]) depend on local weather conditions. It is noted, however, that the Federal Housing Authority (FHA) has established minimum insulation standards for new one- and two-family dwellings which are based on the average number of degree-days at the dwelling location. A summary of these standards is presented in Table 1-1 on the following page; a degree-day map is given in

*Numbers in brackets indicate references listed at the end of the section.

Figure 1-1 on page 4. The Department of Defense and ASHRAE have also established similar criteria [18,19].

Table 1-1

FHA MINIMUM INSULATION STANDARDS FOR NEW ONE- AND TWO-FAMILY DWELLINGS

	R-Values*		
	GAS OR OIL		
Degree - Days	Ceilings	Walls	Floors
Above 7000	38	17	19
6001-7000	30	12	11
4501-6000	30	12	11
3501-4500	30	12	11
2501-3500	22	11	0
1001-2500	19	11	0
1000 and under	19	11	0
	ELECTRIC RESISTANCE		
Above 7000	38	17	19
6001-7000	38	17	19
4501-6000	30	17	19
3501-4500	30	17	19
2501-3500	30	17	11
1001-2500	22	12	0
1000 and under	19	11	0
	HEAT PUMPS WITH ELECTRIC HEAT		
Above 7000	38	17	19
6001-7000	38	17	19
5001-6000	30	17	19
4501-5000	30	12	11
3501-4500	30	12	11
2501-3500	22	11	0
1001-2500	19	11	0
1000 and under	19	11	0

*R-Values are a measure of insulating ability and have the units ft^2 hr °F/Btu. The higher the R-value, the greater the ability to retard heat flow. For a complete definition, see Section 2.1.

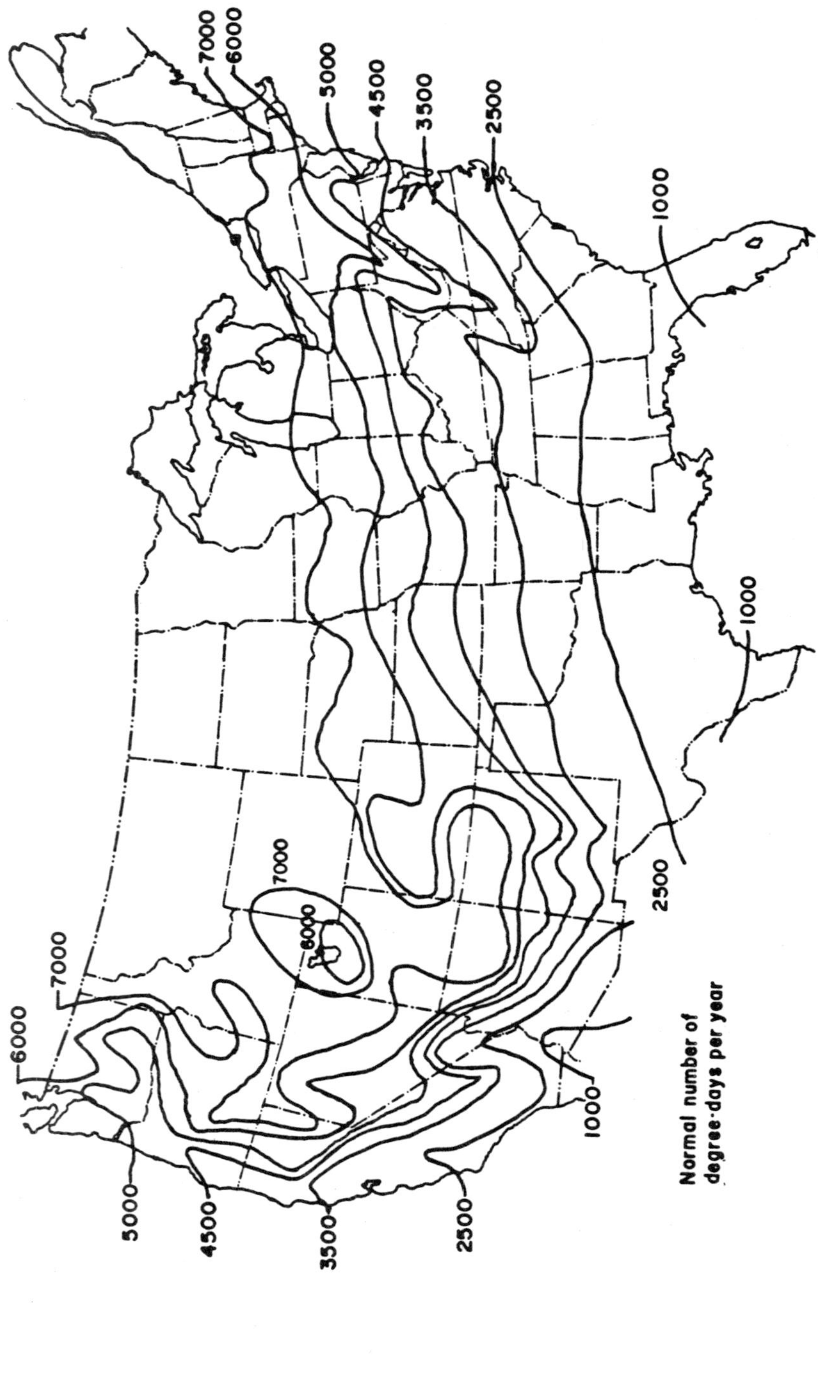

Figure 1-1 Degree-day map of the United States

REFERENCES FOR SECTION 1.0

[1] "Standards and Tentatives Relating to Thermal and Cryogenic Insulating Materials, Building Seals and Sealants, Fire Standards, Building Constructions, Environmental Acoustics," Part 18, 1980 Annual Book of ASTM Standards, American Society for Testing and Materials, Philadelphia, PA. 1980.

[2] Hilado, Carlos J. and Cumming, Heather J. "Standards for Thermal Insulation." Journal of Thermal Insulation. Vol. 1, pp. 129-148. Oct. 1977.

[3] Clyde, Gordon F. "Model Codes and Thermal Insulation Requirements." ASHRAE Journal. p. 55. Oct. 1980.

[4] Hildebrand, Floyd C. "Development of New Federal Criteria for Underground Heat Distribution Systems." ASHRAE Journal. pp. 49-53. Oct. 1980.

[5] Mineral Insulation Manufacturers Association, Inc. "Federal Trade Commission's Trade Regulation Rule - Labeling and Advertising of Home Insulation." Insulation Facts #6. Summit, NJ. Nov. 1980.

[6] Hillier, Ray. "California Insulation Quality Standards." Journal of Thermal Insulation. Vol. 3, pp. 67-70. Oct. 1979.

[7] Achenback, P.R., et. al. The National Program Plan for Building Thermal Envelope Systems and Insulating Materials: Technology and Implementation for Energy Conservation. DOE/CS-0059, Department of Energy. Jan. 1979.

[8] ASHRAE Journal. "The National Program Plan for Building Thermal Envelope Systems and Insulating Materials." pp. 39-43. March 1979.

[9] Powell, Frank J. "Aspects of a National Program Insulation for Mechanical." ASHRAE Journal. pp. 58-59. Oct. 1980.

[10] Achenback, P.R., and Freeman, E.C., Jr. "The National Program for the Thermal Performance of Building Envelope Systems and Materials." Presented at the DOE-ORNL/ASTM Conference on Thermal Insulation, Materials, and Systems for Energy Conservation in the '80s. Clearwater Beach, FL. Dec. 1981.

[11] Tye, R. P. "Thermal Insulation Evaluation: Present Status and Future Requirements." Journal of Thermal Insulation. Technomic Publishing Co., Inc. Vol. 2, pp. 109-130. Jan. 1979.

[12] ASHRAE Handbook, 1981 Fundamentals. American Society of Heating, Refrigerating, and Air-Conditioning Engineers, Inc. Atlanta, GA 1981.

REFERENCES FOR SECTION 1.0 (cont.)

[13] NAHB Research Foundation, Inc. Insulation Manual-Homes, Apartments. Rockville, MD 1979.

[14] Brick Institute of America. "Heat Gain." Tech. Note 4A Rev. Apr/May 1974; "Energy Conservation-Estimating Energy Use - Part 1." Tech. Note 4C Rev. May/Jun 1979; "-Part 2." Tech. Note 4D Rev. Nov/Dec 1979; "-Part 3." Tech. Note 4E Rev. Jan. 1980; "-Part 4." Tech. Note 4F. July 1980; "-Part 5." Tech. Note 4G. Aug. 1980; Technical Notes on Brick Construction. McLean, VA.

[15] Portland Cement Association. Simplified Thermal Design of Building Envelopes for Use with ANSI/ASHRAE/IES Standard 90A-80. 1981.

[16] National Concrete Masonry Association. "Concrete Masonry Passive Solar Design - Rules of Thumb." NCMA-TEK 116. Herndon, VA. 1980.

[17] National Concrete Masonry Association. "Estimating Temperature Swings in Direct Gain Passive Solar Buildings." NCMA-TEK 118. Herndon, VA. 1981.

[18] Department of Defense. "Construction Criteria." DOD 4270.1M. Advance Edition, June 1, 1978.

[19] ASHRAE. "Energy Conservation in New Building Design." Standards 90A-1980, 90B-1975, and 90C-1977. 1980.

2.0

Heat Transfer Fundamentals

To understand how thermal insulation can help conserve energy, one must first understand how energy is transported. There are two basic forms of energy flows, namely work and heat. Work generally involves a physical force acting across a distance. For example, if a compressed spring is allowed to move a mass attached to it as it expands, the spring is said to have done work on the mass. Compressing the spring initially, perhaps by motion of an attached piston that is subject to a compressed gas, is also said to require work. If "physical force" in the definition of work is replaced by "temperature difference," the energy flow is then called heat, or thermal energy, and the study of the processes by which such energy flows can occur is called heat transfer.

There are only three basic ways that heat transfer can take place. Conduction is a term applied to thermal energy transferred within a single body or between two bodies in direct contact with each other. Convection involves transferring energy by physically transporting masses of fluid or gas that contain energy from one point to another. Thermal radiation is a portion of the electromagnetic radiation spectrum (which includes such items as visible light, radio waves, and X-rays) and refers to energy emitted from the surface of a thermally excited object in the form of electromagnetic waves.

Whereas conduction and convection require an actual substance to carry the energy, radiation does not; in fact, radiation exchange between two objects is hindered by any intermediate material. The most obvious example of radiation is the fact that the earth is warmed by the sun through 93 million miles of vacuum.

2.1 Conduction

Conduction is actually a process which takes place on an atomic-particle level. In metals, thermal conduction results from the motion of free electrons (similar to electrical conduction). In liquids and poorly-conducting solids, oscillation of the molecular lattice is thought to be the cause. In gases, conduction occurs through collisions between molecules.[1] Fortunately, the overall effects of conduction heat transfer can be described on a much larger scale. Experiments have shown that the rate of heat flow is proportional to the temperature difference across an object and the area available for heat to flow through (perpendicular to the direction of heat flow), but inversely proportional to the thickness of the object. Thus, for the flat object shown in Figure 2.1 on the next page,

$$Q = kA \frac{T_1 - T_2}{x}$$

where k, the proportionality constant, is commonly known as the thermal conductivity of the material. Examining the equation above reveals that k

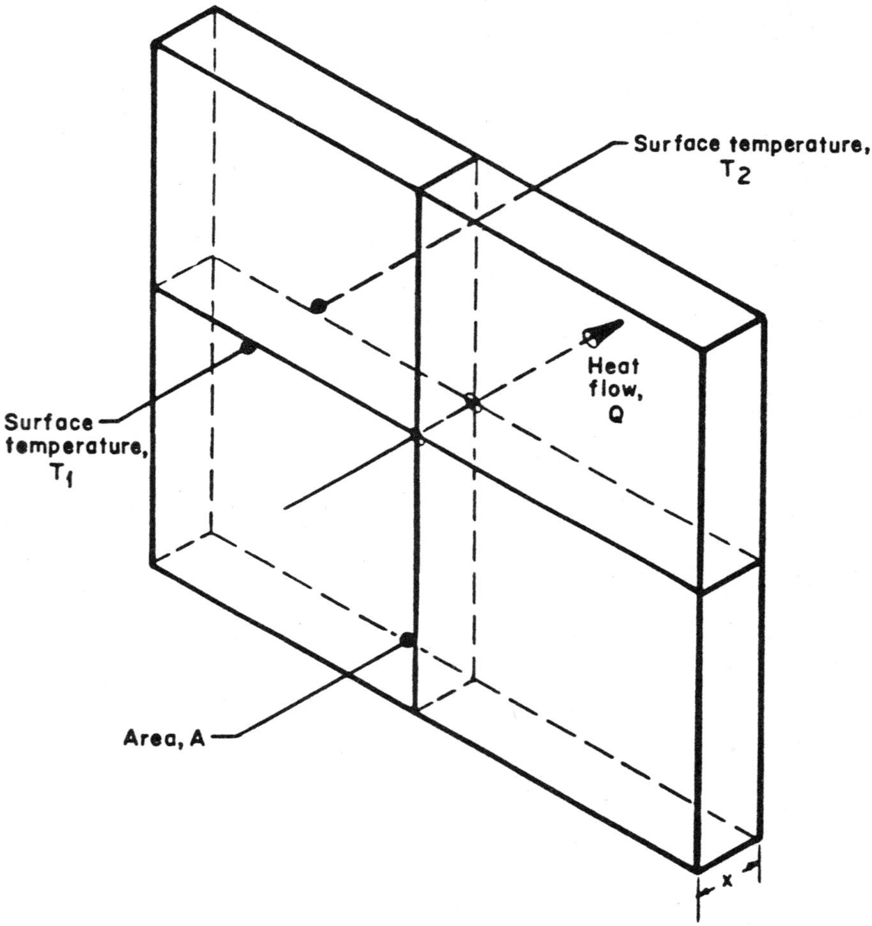

Figure 2-1 Heat conduction through a plane wall

is a measure of how rapidly heat may be conducted through a unit area and thickness of a material when driven by a 1° temperature difference. If area is given in square feet and thickness in inches, then k has the units of Btu-in/ft^2-hr-°F. Figure 2.2 on the following page shows some ranges of thermal conductivity for various types of materials. The inverse of thermal conductivity is called thermal resistivity, (r), therefore,

$$r = 1/k$$

and r has the units ft^2-hr-°F/Btu-in.

Thermal conductivity is strictly a material property. Thermal conductance (C), however, is sometimes used to describe a particular size and thickness of a material. In this case,

$$Q = C\ (T_1 - T_2)$$

where C = thermal conductance (Btu/hr-°F), and thus

$$C = kA/x$$

Note that for a unit area of materials, the higher the value of C, the more rapidly the material will conduct heat across its thickness. In the insulation field, it is more common to think in terms of how well a material resists heat flow. The thermal resistance (R) of a specific shape and material is simply the inverse of its conductance, or

$$R = 1/C = x/kA$$

where R has the units hr °F/Btu. Thermal resistance is analogous to electrical resistance if heat flow (Q) is considered similar to current flow (I), and temperature difference (T_1-T_2), is related to voltage difference (V_1-V_2). In this case,

$$Q = \frac{T_1 - T_2}{R}$$

is the appropriate equation.

Even more common when discussing insulation is the concept of R-value. Sometimes called the unit thermal resistance, the R-value is the thermal resistance per unit area of a material and is therefore dependent only on the thickness:

$$\text{R-value} = RA = x/k$$

R-values in the United States are almost always given in the units used here, ft^2 hr °F/Btu.* Note that values of R-value per inch are actually

*The reader is urged to use caution when deriving R-values from thermal conductivity values, as k is often given in Btu/ft^2 hr °F; such values will be a factor of 12 smaller than values in Btu-in/ft^2-hr-°F. The SI units for k are W/m-K.

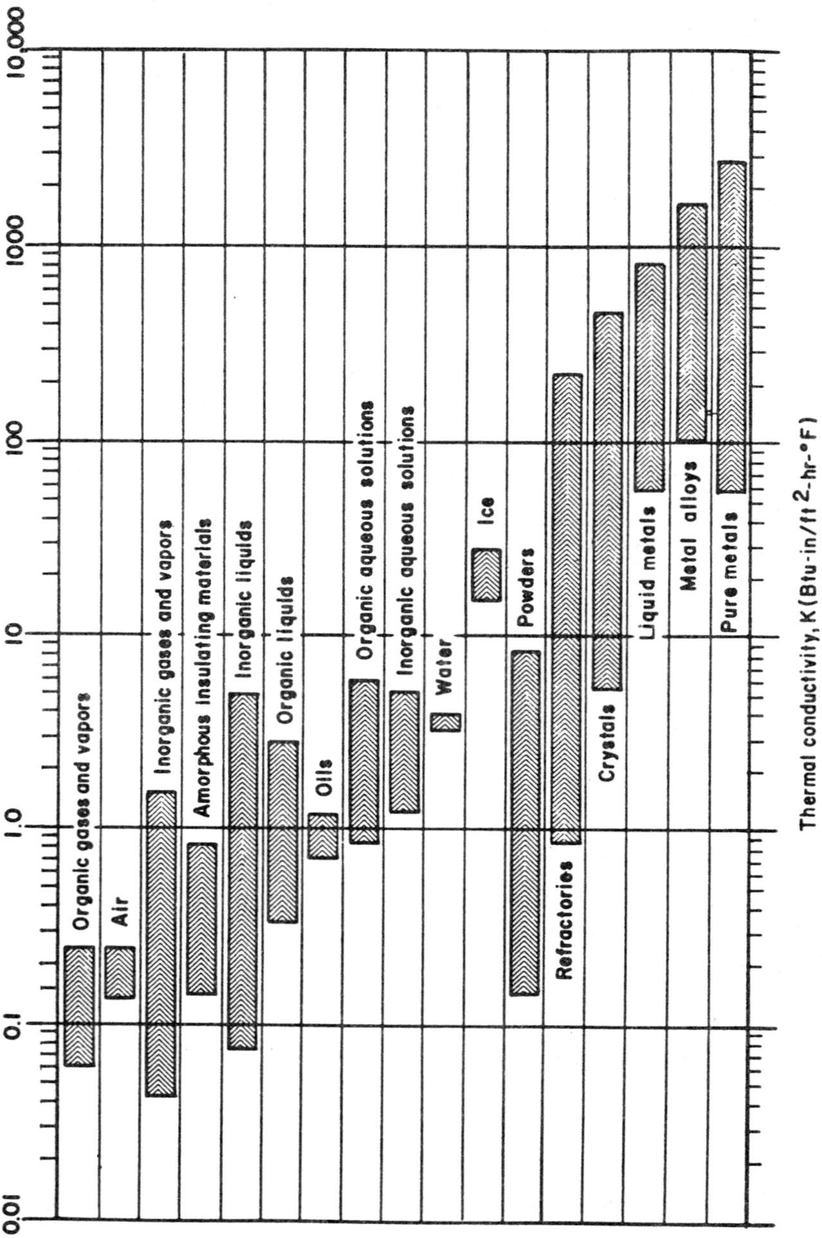

Figure 2-2 Approximate order of magnitude of thermal conductivities

values of thermal resistivity. The practice of giving R-values in SI units has only now begun; the notation in this case is RSI and the units are m^2-K/W.

Within a material having a constant thermal conductivity, the temperature will change at a constant linear rate, illustrated in Figure 2-3 on page 12. If it is necessary to determine the heat flow through a plane wall composed of two different materials (as shown in Figure 2-4 also on page 12) the following procedure may be used; again, the temperature gradients are linear, but the temperature at the interface of the two material will depend on the relative values of the two conductivities and thicknesses. The conduction equation will, of course, be valid for each of the two layers since Q is the same for both in steady state. Thus,

$$Q = \frac{T_1 - T_2}{x_1/k_1 A} \quad \text{and} \quad Q = \frac{T_2 - T_3}{x_2/k_2 A}$$

Solving each equation for the temperature difference gives

$$T_1 - T_2 = Q \frac{x_1}{k_1 A}$$

and

$$T_2 - T_3 = Q \frac{x_2}{k_2 A}$$

If the two equations are added, the unknown temperature T_2 is eliminated; thus,

$$T_1 - T_3 = Q \left(\frac{x_1}{k_1 A} + \frac{x_2}{k_2 A} \right)$$

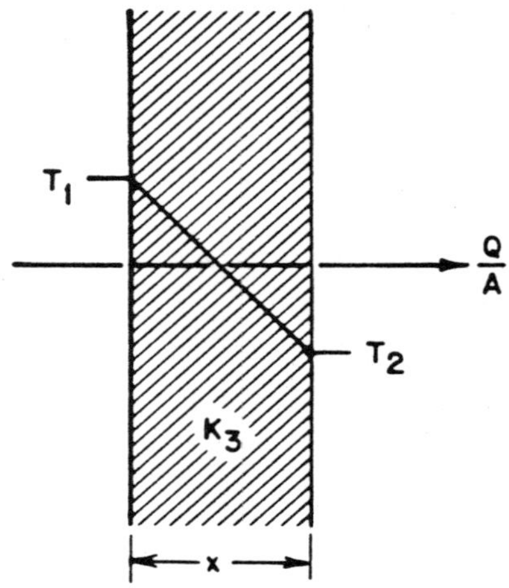

Figure 2-3 Temperature profile through a single slab

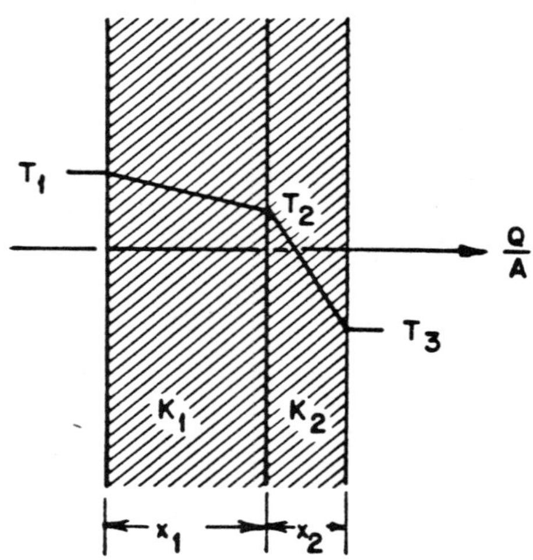

Figure 2-4 Temperature profile through a two layer slab (different materials)

or

$$T_1 - T_3 = Q(R_1 + R_2)$$

Therefore, the total heat flow becomes

$$Q = \frac{T_1 - T_3}{(R_1 + R_2)}$$

Another facet of the thermal-electrical analogy is now evident: thermal resistances in series, like electrical resistances, simply add together. This fact is true no matter how many layers a structure may have.

Similarly, the inverses of thermal resistances of parallel heat flow paths, illustrated in Figure 2-5 on page 14, add together to form the net inverse resistance (i.e., parallel conductances add directly to form the net conductance):

$$\frac{1}{R_{net}} = \frac{1}{R_1} + \frac{1}{R_2}$$

Thus, for the example shown in Figure 2-5,

$$Q = \frac{T_1 - T_4}{R_1 + \dfrac{1}{1/R_2 + 1/R_3} + R_4}$$

or

$$Q = \frac{T_1 - T_4}{\dfrac{x_1}{k_1A_1} + \dfrac{1}{k_2A_2/x_2 + k_3A_3/x_3} + \dfrac{x_4}{k_4A_4}}$$

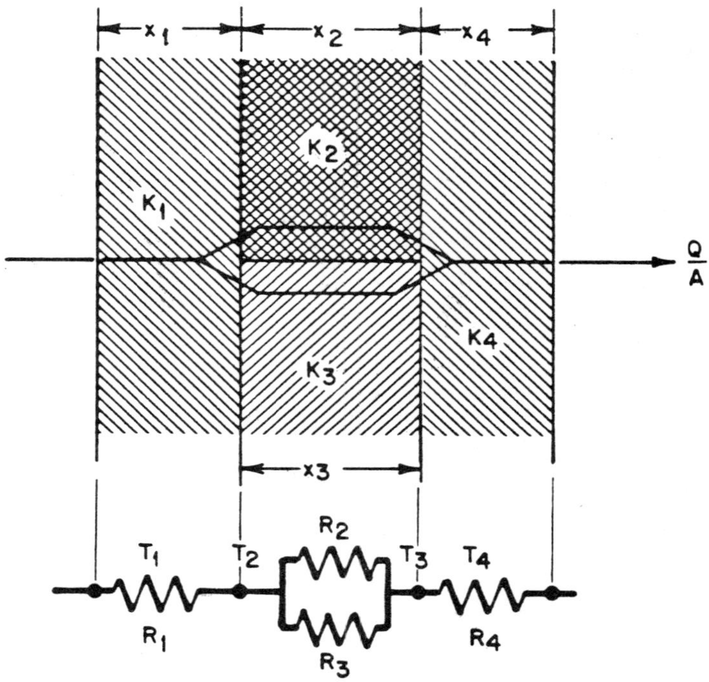

Figure 2-5 Example of parallel path heat flow

This very important concept applies to such instances as the treatment of wood studs within a frame wall or to a window within a wall.

In radial heat flow, such as from a pipe carrying a hot fluid, the area normal to the heat flow changes with distance from the center. For the pipe section shown in Figure 2-6 on page 16, the equation for heat flow becomes

$$Q = \frac{2\pi Lk}{\ln(r_2/r_1)}(T_1 - T_2)$$

Thus, for cylindrical geometry, the thermal resistance is

$$R = \frac{\ln(r_2/r_1)}{2\pi Lk}$$

Again, resistances in series add, so for the composite cylinder shown in Figure 2-7 on page 16, the equation for heat flow is

$$Q = \frac{T_1 - T_3}{R_1 + R_2}$$

or

$$Q = \frac{T_1 - T_3}{\frac{\ln(r_2/r_1)}{2\pi Lk_1} + \frac{\ln(r_3/r_2)}{2\pi Lk_2}}$$

This last equation is usually simplified even further:

$$\frac{Q}{L} = \frac{2\pi(T_1 - T_3)}{\frac{\ln(r_2/r_1)}{k_1} + \frac{\ln(r_3/r_2)}{k_2}}$$

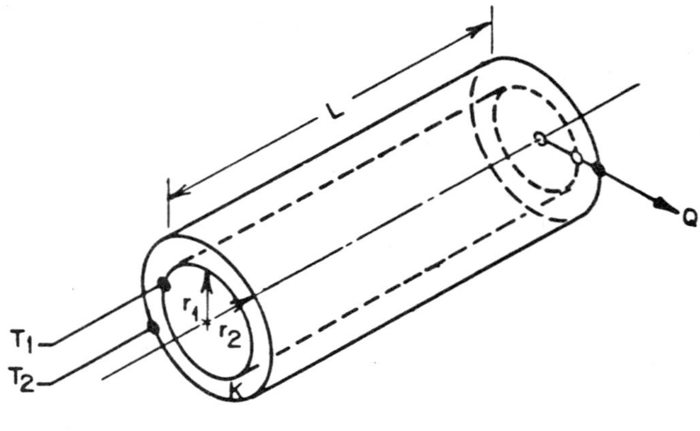

Figure 2-6 Radial heat flow

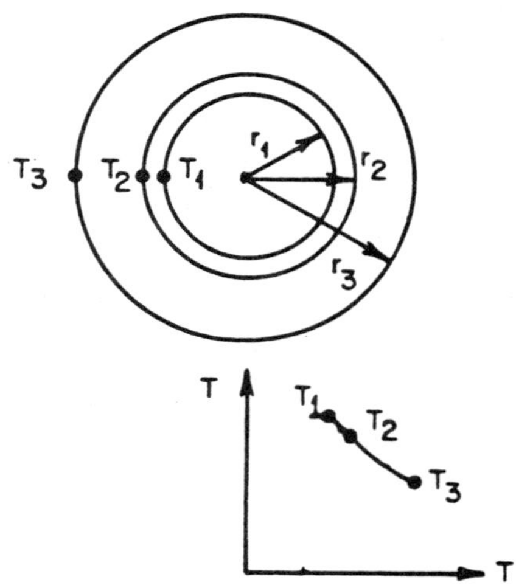

Figure 2-7 Temperature profile through a two layer hollow cylinder

Note that in Figure 2-7 on the previous page, the temperature profile is not linear as it is in the plane wall case, but instead drops off logarithmically.

The discussion of conduction so far has considered only steady state conduction; i.e., the temperatures (such as T_1 and T_2 in Figure 2.1) and the heat flow rate have constant values. In transient situations, where conditions are changing with time, another parameter is also important. This parameter is the thermal diffusivity, (α):

$$\alpha = k/\rho c_p$$

where ρ is the material density (lb/ft^3), and c_p is the specific heat of the material (Btu/lb-F) (c_p represents the amount of energy required to raise 1 pound of the material by 1 degree F). The thermal diffusivity is thus the ratio of thermal conductivity and volumetric thermal capacity, and has units of ft^2/hr. The inverse of thermal diffusivity, with units of hr/ft^2, can be thought of as the "heating time,"[2] which is the time required to heat a material to some specified temperature. This time is directly proportional to the square of the thickness of the material. A substance with a high thermal diffusivity will respond to changing conditions more rapidly than one with a low diffusivity.

Transient heat conduction problems are somewhat more complicated than steady state problems. Solutions to certain problems of special interest (for example, the immersion of a slab or cylinder at a uniform initial temperature into a fluid at a different temperature) have been put into the form of charts, from which the temperature at any point in the object at any time can be determined. These charts, and instructions for their use, are available in a number of heat transfer textbooks [1-7].

Thermal diffusivity is important in a number of cases, such as in the use of masonry materials for thermal storage. In this case, the problem of transient heat conduction has also been simplified by the development of an M-factor, a simple multiplier accounting for the time lag in temperature transmission through masonry materials [8]. The M-factor and its use is discussed further in Section 2.8.

2.2 Convection

Convection, as previously described, involves movement of fluid masses. Two major types of convection can occur. Free (or natural) convection takes place when a temperature difference results in a local density difference within a fluid, causing pockets of fluid to be more buoyant or dense than the surrounding fluid. Forced convection is produced by mechanical means such as pumps and fans. Convection resulting from wind speeds greater than a few miles per hour is considered forced. Both types of convection can be either laminar, where the heat flow into the fluid is dominated by thermal conduction from a solid boundary, or turbulent, where significant eddy mixing occurs. If forced convection is weak enough, free convection may occur along with it; such a situation is called mixed convection.

The definition of a parameter which can be used to describe the magnitude of convection arises from the fact that the temperature gradient in convection occurs over a thin boundary layer of the fluid, as shown in Figures 2-8a and 2-8b on the following page. The boundary layer represents a transition region between conditions at a solid surface (temperature = T_s and fluid as required by consideration of friction) and conditions in the free stream (temperature = T_f and fluid velocity = V_f), a region that is beyond the influence of the surface. Note that V_f is zero in free convection (Figure 2-8b) and nonzero in forced convection (Figure 2-8a). In general, the velocity boundary layer thickness (δ_v) is not equal to the thermal boundary layer thickness(δ_T), although they are related.

If the temperature gradient through the thermal boundary layer were uniform (i.e., linear in the plane wall case), the heat transfer from the surface could be defined using the conduction equation, with k being the conductivity of the fluid and x equalling δ_T. The gradient is not generally linear, however. In addition, the boundary layer thickness is not sharply defined, and is known to increase with distance along the surface. For these reasons, convection is characterized using a parameter "h" in place of k/x:

$$Q = hA(T_s - T_f)$$

Names for h, which has the units of Btu/ft^2 hr °F, include the heat transfer coefficient, convection coefficient, film coefficient, unit film conductance, and unit surface conductance. As opposed to k, h is not entirely a material property, but depends on the geometry of the surface and the nature of the fluid flow as well. Methods of calculating h are not given in this handbook; the reader is again referred to References [1-7] for such correlations. The ranges of h for a number of conditions are given in Figure 2-9 on page 20.

As with conduction, a thermal resistance may be defined:

$$R = 1/hA$$

and thus

$$Q = \frac{T_s - T_f}{R}$$

2.3 Radiation

The relationship between thermal radiation and the rest of the electromagnetic spectrum is shown in Figure 2-10 on page 21. The micron, equal to one millionth of a meter, is the most common unit of wavelength for thermal radiation and is given the symbol μ. As seen in the figure, thermal radiation spans wavelengths of 0.1 to 100 μ.

All bodies having a temperature greater than absolute zero possess internal energy. At the surface of a body, some of this energy is released as thermal radiation. This radiation will travel through space in a straight line until intercepted by another body. Unless a body emitting radiation

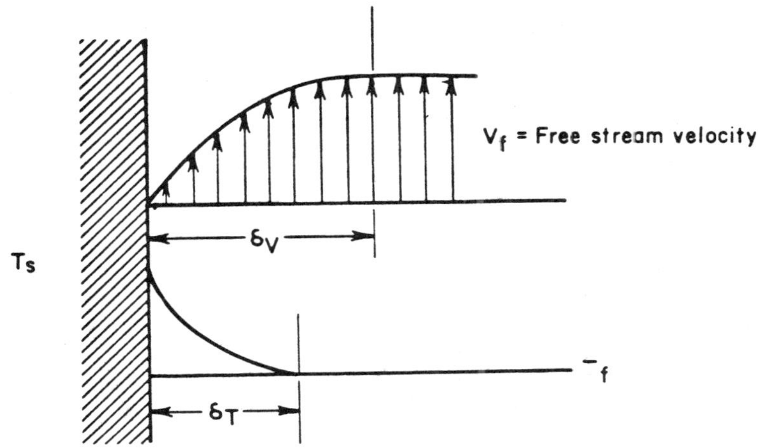

a. Forced convection

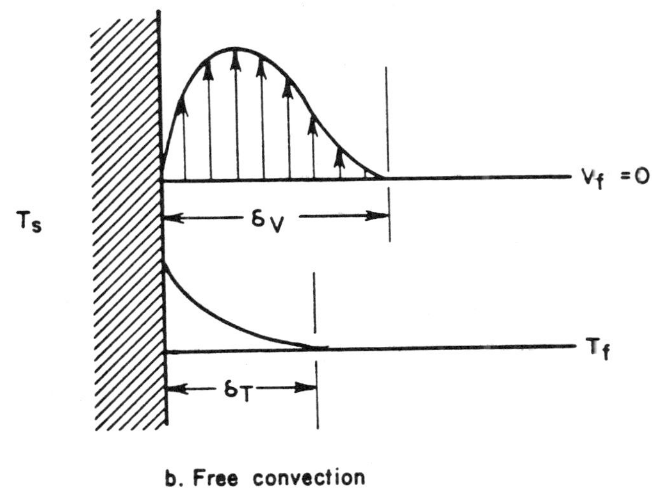

b. Free convection

Figure 2-8 Convection boundary layers

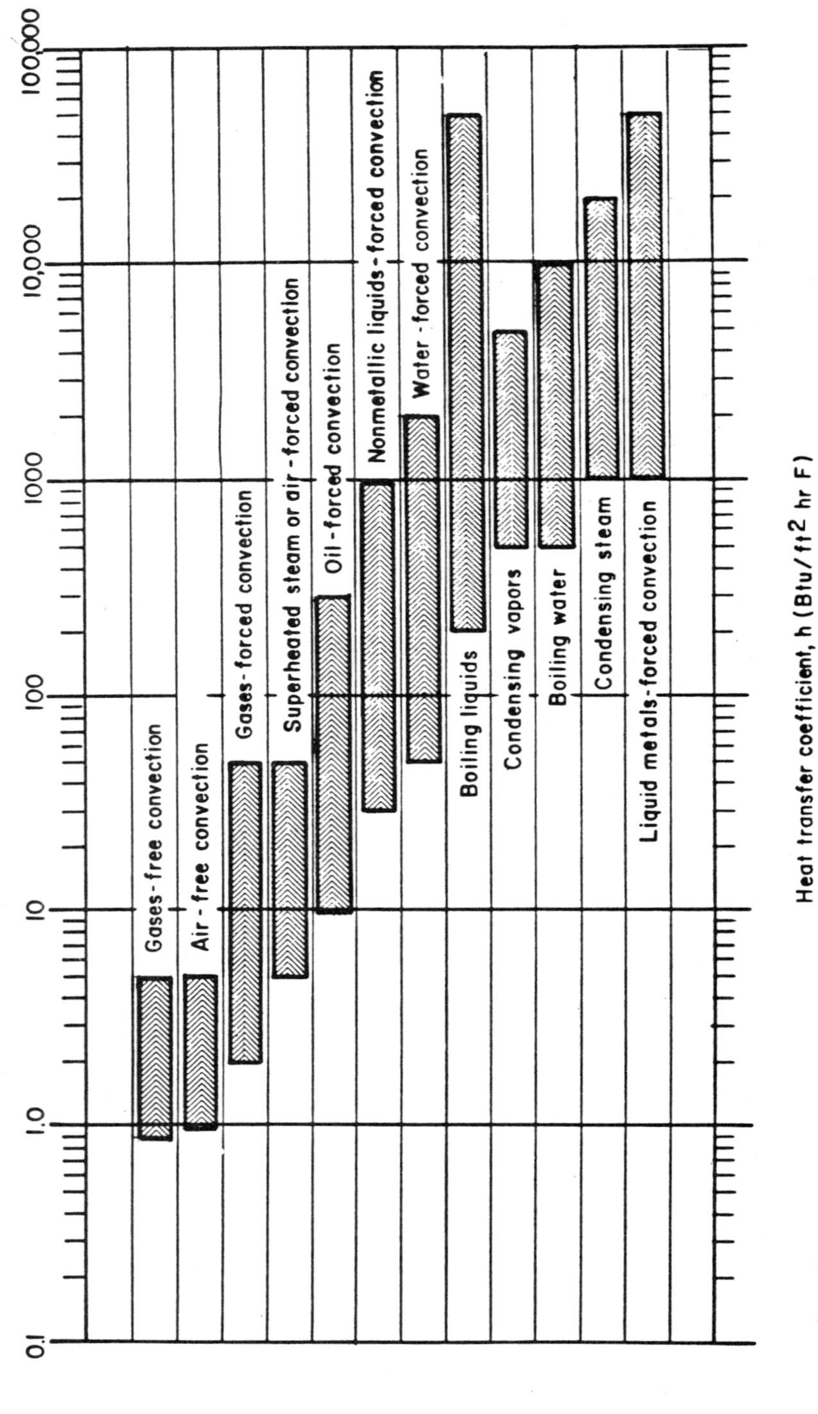

Figure 2-9 Approximate order of magnitude of convection heat transfer coefficients

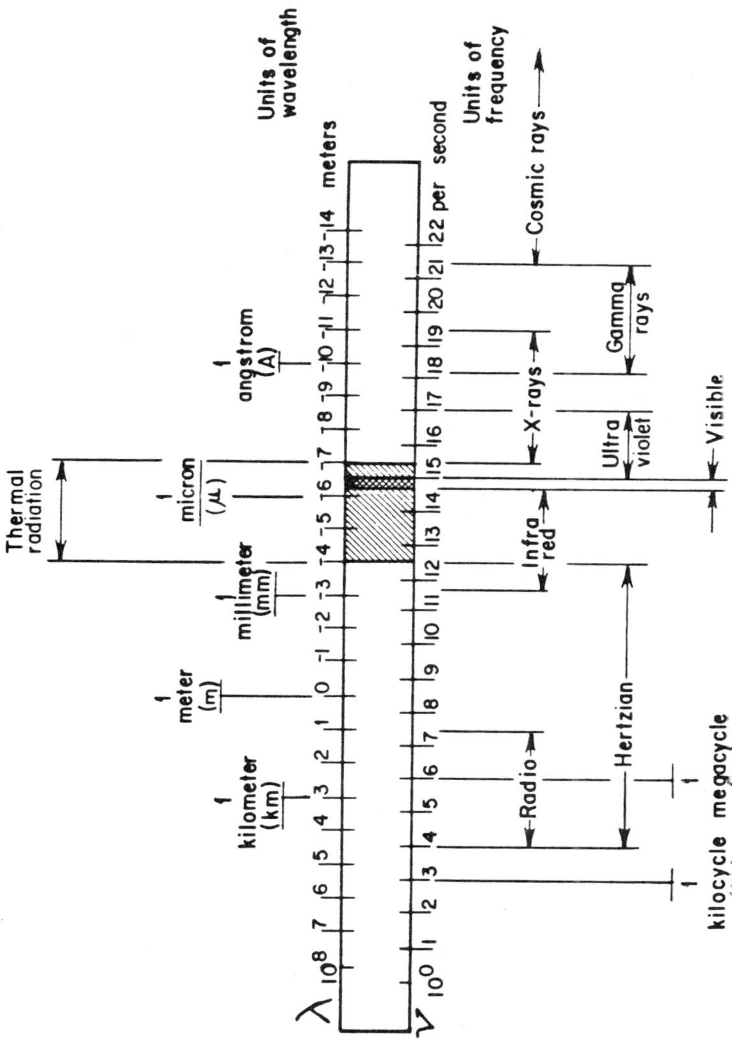

Figure 2-10 Electromagnetic spectrum

receives radiation from another source, its temperature will decrease. Likewise, a body that receives more radiation than it can give off will increase in temperature. Radiation reaching an object can be partially absorbed by, partially transmitted through, and partially reflected away from the object, as illustrated in Figure 2-11 on page 23. The relative portions of absorbed, transmitted, and reflected energy depend entirely on the object itself and its surface characteristics. For an opaque object, the transmitted portion is zero. Radiation reflected from a polished surface is called <u>specular</u>, and from a rough surface, <u>diffuse</u> (see Figure 2-12 on page 23).

The rate of energy emission per unit area of a surface is called the <u>emissive power</u> (W) and is proportional to the fourth power of the absolute temperature of the surface:

$$W = \sigma\varepsilon\ T^4$$

where ε is the <u>emissivity</u> of the surface (a fraction between 0 and 1), and σ is the Stefan-Boltzmann constant,

$$\sigma = 0.1714\text{x}10^{-8}\ \text{Btu/hr-ft}^2\text{-R}^4$$

It must be stressed that absolute temperatures must be used in radiation calculations.* A convenient form of the equation for emissive power results when the value of σ is substituted into the equation:

$$W = 0.1714\ \varepsilon\ \left(\frac{T}{100}\right)^4$$

A <u>black body</u> is defined as a body which absorbs all incident radiation (its <u>absorptivity</u> α = 1); such a body would appear black to the eye (hence the name) and is considered to be a perfect absorber. A black body is also a perfect emitter, however, so ε = 1 also. The intensity of radiation emitted from a black body has been shown to vary smoothly with wavelength, as illustrated in Figure 2-13 on page 24.

Very few real surfaces behave as black bodies, however. A <u>gray body</u> is one with $\varepsilon < 1$; because ε is constant with wavelength, intensity versus wavelength is again a smooth curve. Most real surfaces are <u>nongray</u> (though they are usually considered gray for computational purposes), and the emitted radiation may vary widely with wavelength. Radiation intensity distributions for both gray and nongray surfaces are also shown in Figure 2-13. For most materials, it is sufficient to define a total hemispherical emissivity, which is an integrated value covering all wavelengths as well as all directions from the surface. This value may still vary with surface temperature. Values of ε for several common surfaces at temperatures near 75°F are presented in Table 2-1 on page 25.

*To convert degrees Fahrenheit to the absolute scale, degrees Rankine, add 460 to the Fahrenheit temperature: °R = °F + 460.

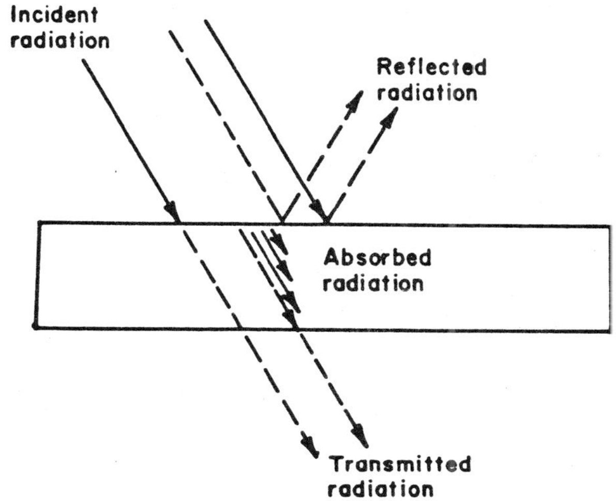

Figure 2-11 Reflection, absorption and transmission of radiation

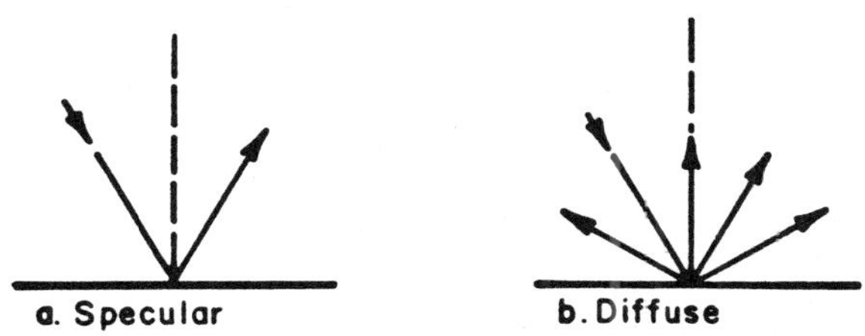

Figure 2-12 Reflecting surfaces

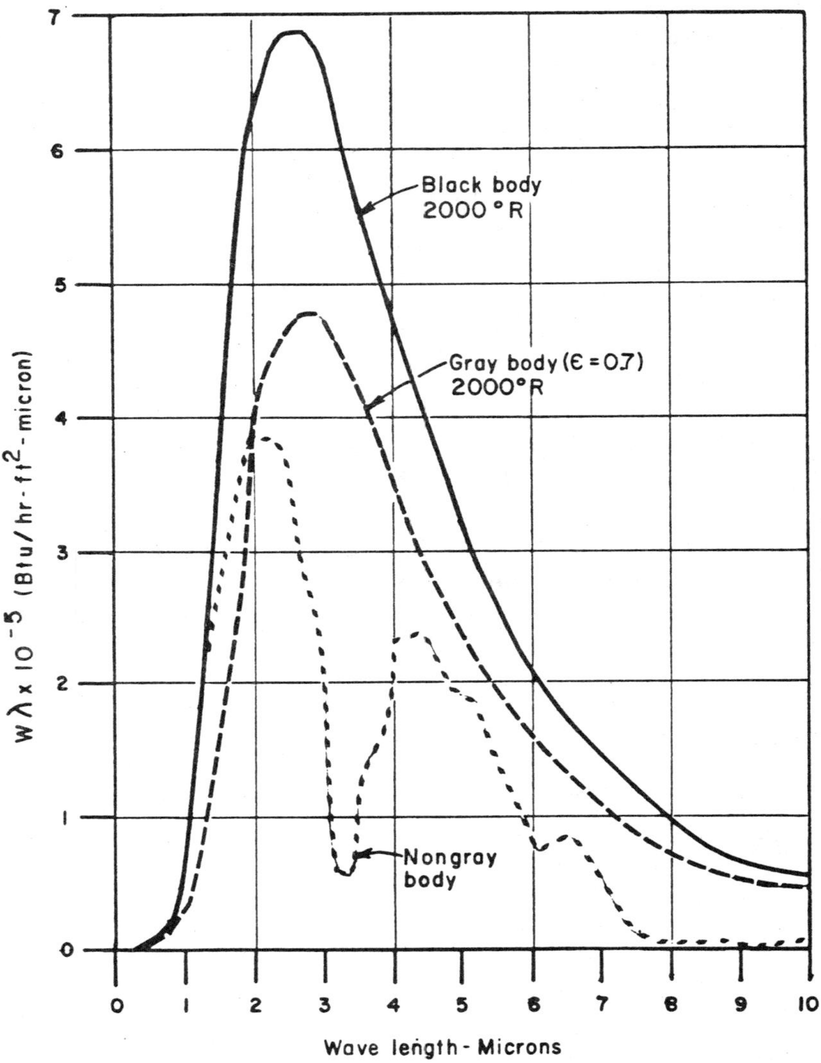

Figure 2-13 Spectral distribution of emissive power for three surfaces

Table 2-1

SURFACE EMISSIVITIES
(at approximately 75°F)

Material	Emissivity
METALS	
Aluminum foil, bright	0.05
Aluminum sheet	0.12
Brass, rolled plate	0.06
Brass, polished	0.10
Brass, oxidized	0.61
Wrought iron, polished	0.28
Wrought iron, oxidized	0.94
Iron plate, completely rusted	0.67
Steel, galvanized bright	0.25
Steel, stainless type 301	0.14
Steel, rough plate	0.94
PAINTS	
Aluminum paint	0.50
Flat black lacquer	0.95
International orange on aluminum	0.74
White enamel	0.90
BUILDING MATERIALS	
Asbestos board	0.96
Red brick, rough	0.93
Glass	0.84-0.94
Plaster	0.91
Roofing paper	0.91
Most building materials (wood, paper, masonry)	0.90
Water, ice (32°F)	0.96

The rate of energy exchange between two radiating black bodies is governed by the equation

$$Q = \sigma A_1 F_{12} (T_1^4 - T_2^4)$$

Here, A_1, is the surface area of the first body that is able to "see" the second body, and F_{12} may be interpreted as the fraction of all the radiation leaving A_1 in all directions that is intercepted by A_2. F_{12} is called the shape factor, or configuration factor, and can be determined from the geometry of any particular problem. Shape factors have been worked out for many cases; the reader is referred to heat transfer textbooks for examples. It can be shown that $A_1F_{12} = A_2F_{21}$.

For gray bodies,

$$Q = \sigma A_1 \mathcal{F}_{12} (T_1^4 - T_2^4)$$

Where $\mathcal{F}_{12}$ is an exchange factor for gray bodies and is similar to F_{12} for black bodies. In fact $\mathcal{F}_{12}$ depends on F_{12}, but also depends on the factors ε_1, and ε_2.

In the simple case of an enclosure formed by only two surfaces, if A_1 cannot see itself, $\mathcal{F}_{12}$ can be found from

$$\mathcal{F}_{12} = \left[\frac{1}{\varepsilon_1} + \frac{A_1}{A_2}\left(\frac{1}{\varepsilon_2} - 1\right)\right]^{-1}$$

Furthermore, if the two surfaces are large parallel planes, A_1 is essentially equal to A_2 and

$$\mathcal{F}_{12} = \left(\frac{1}{\varepsilon_1} + \frac{1}{\varepsilon_2} - 1\right)^{-1}$$

This equation is often assumed to hold for radiation through air spaces within plane walls.* If A_1 is enclosed by a very large surface (as would be the case for an object radiating to the sky) A_1/A_2 is essentially zero, and

$$\mathcal{F}_{12} = \varepsilon_1$$

By noting that

$$(T_1^4 - T_2^4) = (T_1^2 + T_2^2)(T_1 + T_2)(T_1 - T_2)$$

one can determine a thermal resistance for radiation problems, just as was done for conduction and convection:

$$Q = \frac{T_1 - T_2}{R}$$

*The reader familiar with the ASHRAE Handbook of Fundamentals will recognize $\mathcal{F}_{12}$ here as the same as ASHRAE's "effective emittance of an air space" (E), used in determining the thermal resistance of plane air spaces.

where

$$R = \frac{1}{\sigma A_1 \mathcal{F}_{12}(T_1{}^2 + T_2{}^2)\ (T_1 + T_2)}$$

A radiative heat transfer coefficient may be defined as $h_r = 1/RA$. Note that the above radiation resistance (R) depends explicitly on temperature. This fact has implications for the solution of problems involving combined modes of heat transfer, the subject of the next section.

2.4 Combined Heat Transfer

In actual heat transfer problems, conduction, convection, and radiation almost never occur individually. Most actual systems involve some arrangement of series and parallel heat flow paths and combinations of the three modes of heat transfer. If one follows the concept of thermal resistance, however, such problems can be analyzed in a straightforward manner by remembering that resistances in series add, and that conductances in parallel add.

As an example, consider the problem shown in Figure 2-14 on page 28. Here, a wall is composed of three materials having conductivities of k_1, k_2, and k_3 and thicknesses x_1, x_2, and x_3. The wall is bounded by convective air layers on the inside and outside, for which the convection coefficients are h_i and h_o respectively. The outside surface also radiates to T_o and has an emissvity of ε . Below the diagram of the wall are shown the thermal resistances involved. Following the rules for combining resistances,

$$Q = \frac{T_i - T_o}{R_{total}}$$

which becomes:

$$Q = \frac{T_i - T_o}{R_1 + R_2 + R_3 + R_4 + \left(\dfrac{1}{1/R_5 + 1/R_6}\right)}$$

or

$$Q = \frac{T_i - T_o}{\dfrac{1}{h_i A} + \dfrac{x_1}{k_1 A} + \dfrac{x_2}{k_2 A} + \dfrac{x_3}{k_3 A} + \dfrac{1}{(h_o + h_r)A}}$$

Finding the overall heat flow rate is now a matter of inserting the proper value of each variable into the equation. Recall that h_r will depend on T_4, the surface temperature of the outside material. The ASHRAE Handbook of Fundamentals, however, gives a table of values of (h_o+h_r) for selected situations. A more rigorous approach would be to make an initial

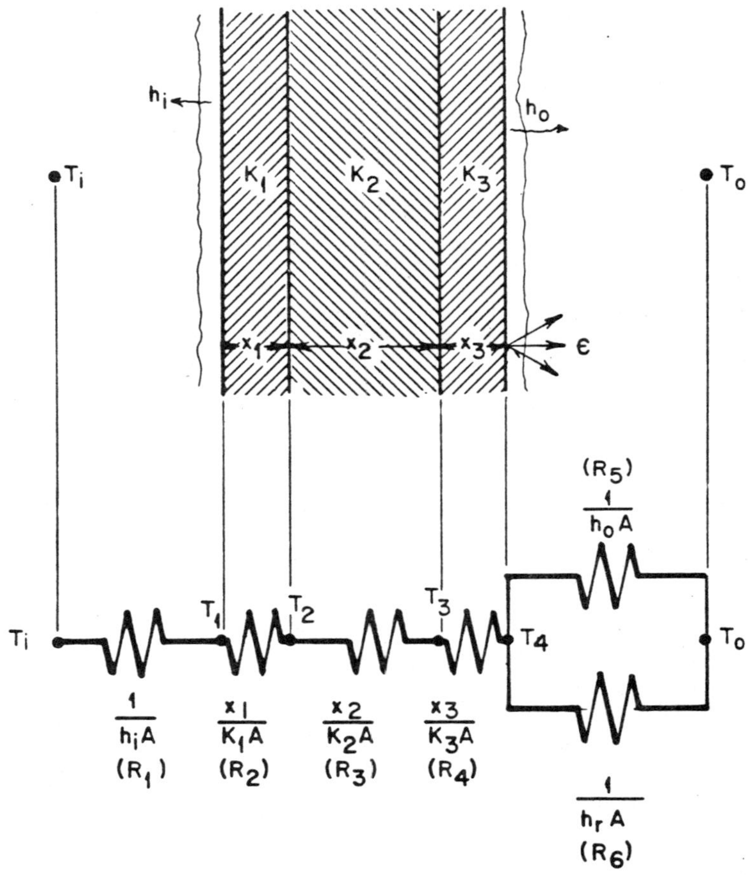

Figure 2-14 Plane example of overall heat transfer

estimate of h_r ($h_r = h_o$ is often a good starting point), and then to determine the resulting value of Q/A. Then, since Q/A must be the same (under steady state conditions) for the outside air layer as it is for the entire panel,

$$\frac{Q}{A} = \frac{T_4 - T_o}{\left(\dfrac{1}{h_o + h_r}\right)}$$

which, when solved for T_4, gives

$$T_4 = T_o + \frac{Q}{A}\left(\frac{1}{h_o + h_r}\right)$$

This value for T_4 can be used to calculate a new value for h_r. If the old h_r is not the same as the new h_r, return to the original equation and use the new h_r to find a new value for Q/A. Keep finding updated values for T_4, h_r, and Q/A until the change is not significant. Usually, three or four iterations will be sufficient.

The general equation for Q is often written as:

$$Q = UA\,\Delta T$$

where U is called the overall heat transfer coefficient. Thus:

$$U = \frac{1}{R_{total}\,A}$$

Combined heat transfer problems in cylindrical cases are very similar, except that A changes as one goes further away from the center of the pipe, since $A = 2\,\pi rL$. Thus, for the problem shown in Figure 2-15 on the following page,

$$Q = \frac{T_i - T_o}{\dfrac{1}{h_i A_i} + \dfrac{\ln(r_2/r_1)}{2\,\pi L k_1} + \dfrac{\ln(r_3/r_2)}{2\,\pi L k_2} + \dfrac{1}{(h_c + h_r)\,A_o}}$$

With $A_i = 2\,\pi L r_1$ and $A_o = 2\,\pi L r_3$,

$$Q = \frac{2\,\pi\,L(T_i - T_o)}{\dfrac{1}{h_i r_1} + \dfrac{\ln(r_2/r_1)}{k_1} + \dfrac{\ln(r_3/r_2)}{k_2} + \dfrac{1}{(h_o + h_r)r_3}}$$

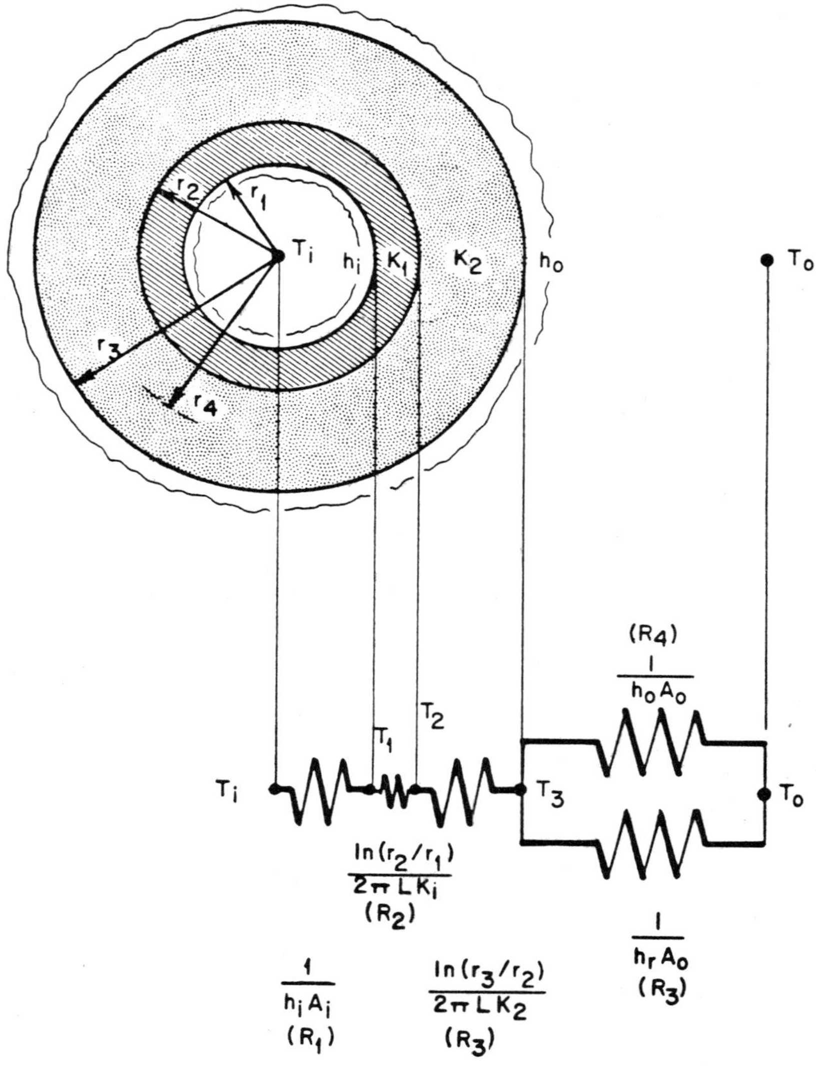

Figure 2-15 Cylindrical example of overall heat transfer

The value of the overall heat transfer coefficient, $U = 1/R_{total}A$, will depend on which A one chooses to evaluate, but the product UA will always be the same. If U is evaluated at the outermost surface, then

$$Q = \frac{(2\pi L r_3)(T_i - T_o)}{\dfrac{r_3}{h_i r_1} + \dfrac{r_3 \ln(r_2/r_1)}{k_1} + \dfrac{r_3 \ln(r_3/r_2)}{k_2} + \dfrac{1}{(h_o + h_r)}} = UA_o(T_i - T_o)$$

and therefore

$$U = \frac{1}{\dfrac{r_3}{h_i r_1} + \dfrac{r_3 \ln(r_2/r_1)}{k_1} + \dfrac{r_3 \ln(r_3/r_2)}{k_2} + \dfrac{1}{(h_o + h_r)}}$$

Again, as in the case of the plane, if a combined value of (h_o+h_r) is not readily available, an iterative procedure can be used to solve the problem.

Defining U for a structure composed of two or more parallel heat paths requires knowing the areas of each path and finding an individual U for each type of construction. An example would be a wall containing a window and a door, for which

$$A_{total} = A_{wall} + A_{window} + A_{door},$$

$$Q_{total} = Q_{wall} + Q_{window} + Q_{door},$$

and

$$UA_{total}(T_i - T_o) = U_{wall}A_{wall}(T_i - T_o) + U_{window}A_{window}(T_i - T_o) + U_{door}A_{door}(T_i - T_o)$$

Thus:

$$U = \frac{U_{wall}A_{wall} + U_{window}A_{window} + U_{door}A_{door}}{A_{total}}$$

In a general sense,

$$U = \frac{\sum_{i=1}^{n} U_i \ A_i}{\sum_{i=1}^{n} A_i}$$

where n is the number of parallel heat flow paths.

For a structure containing a highly conductive element, such as a metal beam extending into an insulating material, U should be determined using the ASHRAE Zone Method [1]. The Zone Method allows the assignment of an area A_i to the metal component that is greater than its actual area. This method helps account for two-dimensional heat transfer effects near the metal, where the temperature profile will actually be different than it is away from the metal. A recent study [9] has verified the accuracy of the Zone Method for construction such as this.

2.5 Solving For Temperatures

The method of finding the temperature at the interface of two materials has been alluded to in the discussion of how to find Q when a radiative resistance is present. In steady state, Q/A (or Q/L for cylindrical geometry) is the same for any portion of the structure as it is for the overall structure. For example, if one wants to know the temperature at the interface between the two solid materials in Figure 2-15 on page 30 (i.e., at r_2), either of the two following equations can be solved for T_2:

$$UA_o(T_i - T_o) = \frac{T_i - T_2}{R_1 + R_2}$$

or

$$UA_o(T_i - T_o) = \frac{T_2 - T_o}{R_3 + \dfrac{1}{(1/R_4 + 1/R_5)}}$$

This method actually works for finding the temperature at any point within the structure, not just at an interface, if the total resistance up to the point of interest is used. If, for example, one wants to find the temperature T_r at a distance r that is part way into the outer solid layer in Figure 2-15 on page 30, solve this equation for T_r:

$$UA_o(T_i - T_o) = \frac{T_i - T_r}{R_1 + R_2 + \frac{\ln(r/r_2)}{2\pi Lk_2}}$$

2.6 Insulation Thickness To Prevent Condensation

Insulation can be used to prevent condensation of water vapor in the air onto pipes, ducts, and equipment operating at temperatures below ambient. Condensation will occur if the temperature of the exposed surface is below the dew point (or wet-bulb temperature) of the air. Dew point temperatures are given in Table 2-2 on page 34 for various combinations of dry bulb temperature and relative humidity.

An example of an insulated chilled water pipe is shown in Figure 2-16 on page 35. If the outside surface of the pipe is assumed to be at the same temperature as the chilled water (a good approximation), T_i, then the following equation will give the insulation thickness required ($r_o - r_i$) when solved for r_o:

$$r_o \ln(r_o/r_i) = \frac{k}{h_c + h_r}\left(\frac{T_s - T_i}{T_o - T_s}\right)$$

In this equation, T_s should be set to the dew point temperature corresponding to the design relative humidity at T_o. As the relative humidity increases, T_s is closer to T_o (and farther from T_i) and the required r_o increases. Similarly, using an insulation with a low k will decrease r_o from that required with a high-k insulation.

For flat surfaces (and cylindrical surfaces greater than 3 feet in diameter), the left-hand side of the above equation is replaced with the thickness of flat insulation required (x):

$$x = \frac{k}{h_c + h_r}\left(\frac{T_s - T_i}{T_o - T_s}\right)$$

Table 2-2

DEW-POINT TEMPERATURE

Dry-Bulb temp. °F	Relative Humidity (%) 10	15	20	25	30	35	40	45	50	55	60	65	70	75	80	85	90	95	100
5	-35	-30	-25	-21	-17	-14	-12	-10	-8	-6	-5	-4	-2	-1	1	2	3	4	5
10	-31	-25	-20	-16	-13	-10	-7	-5	-3	-2	0	2	3	4	5	7	8	9	10
15	-28	-21	-16	-12	-8	-5	-3	-1	1	3	5	6	8	9	10	12	13	14	15
20	-24	-16	-11	-8	-4	-2	2	4	6	8	10	11	13	14	15	16	18	19	20
25	-20	-15	-8	-4	0	3	6	8	10	12	15	16	18	19	20	21	23	24	25
30	-15	-9	-3	2	5	8	11	13	15	17	20	22	23	24	25	27	28	29	30
35	-12	-5	1	5	9	12	15	18	20	22	24	26	27	28	30	32	33	34	35
40	-7	0	5	9	14	16	19	22	24	26	28	29	31	33	35	36	38	39	50
45	-4	3	9	13	17	20	23	25	28	30	32	34	36	38	39	41	43	44	45
50	-1	7	13	17	21	24	27	30	32	34	37	39	41	42	44	45	47	49	50
55	3	11	16	21	25	28	32	34	37	39	41	43	45	47	49	50	52	53	55
60	6	14	20	25	29	32	35	39	42	44	46	48	50	52	54	55	57	59	60
65	10	18	24	28	33	38	40	43	46	49	51	53	55	57	59	60	62	63	65
70	13	21	28	33	37	41	45	48	50	53	55	57	60	62	64	65	67	68	70
75	17	25	32	37	42	46	49	52	55	57	60	62	64	66	79	70	72	74	75
80	20	29	35	41	46	50	54	57	60	62	65	67	69	72	74	75	77	78	80
85	23	32	40	45	50	54	58	61	64	67	69	72	74	76	78	80	82	83	85
90	27	36	44	49	54	58	62	66	69	72	74	77	79	81	83	85	87	89	90
95	30	40	48	54	59	63	67	70	73	76	79	82	84	86	88	90	91	93	95
100	34	44	52	58	63	68	71	75	78	81	84	86	88	91	92	94	96	98	100
105	38	48	56	62	67	72	76	79	82	85	88	90	93	95	97	99	101	103	105
110	41	52	60	66	71	77	80	84	87	90	92	95	98	100	102	104	106	108	110
115	45	56	64	70	75	80	84	88	91	94	97	100	102	105	107	109	111	113	115
120	48	60	68	74	79	85	88	92	96	99	102	105	107	109	112	114	116	118	120
125	52	63	72	78	84	89	93	97	1001	104	107	109	111	114	117	119	121	123	125

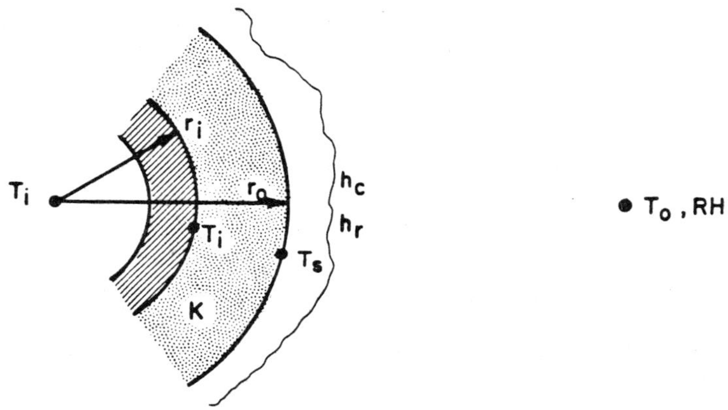

Figure 2-16 Insulated pipe carrying a chilled fluid

2.7 Heat Transfer Inside Insulation

Now that the basics of heat transfer have been explained, it is appropriate to discuss how thermal insulation inhibits heat transfer.

Since insulation is usually considered "solid," one would expect it to transfer heat by conduction and thus be characterized by a thermal conductivity. However, because heat transfer in insulation does not occur only by conduction - convection and radiation also contribute - the term "apparent thermal conductivity" is often used.

A warm surface openly exposed to cooler surroundings loses energy by all three modes of heat transfer. By placing a barrier of some sort near the warm surface, one can reduce the radiative heat loss. It is also known that if a volume of air is physically small enough, convection will be suppressed. If one therefore places a number of barriers near the warm surfaces so that the barriers form small pockets, both radiation and convection will be inhibited. With the proper distribution of barriers, the heat lost from the warm surface will be reduced to the point at which it results almost entirely from pure conduction through pockets of still air.

This point is illustrated in Figure 2-17 on the next page, which shows an example of heat transfer through a 1-inch thickness of a fibrous insulating material as a function of insulation density [10-12]. With no insulation (i.e., at a bulk density of zero), the total apparent conductivity on the vertical scale would be 1.67. With a very low density material, say 0.5 lb/ft^3, the figure indicates that convection has been suppressed almost entirely. The surface-to-surface radiation has been significantly reduced, though partially replaced with radiation heat transfer between individual fibers within the insulation itself. This fiber-to-fiber radiation occurs because each fiber is slightly warmer than the fiber next to it in the direction of the heat flow. As the density increases, fiber-to-fiber radiation peaks; it then diminishes as the fibers get closer together and the temperature difference between adjacent fibers gets smaller.

As the density keeps increasing, fiber-to-fiber radiation and surface-to-surface radiation continue to decrease. The total apparent thermal conductivity would therefore continue to decrease also, except for one important factor: as the density increases, the fibers are packed closer together, and contact between fibers increases. This leads to an increased opportunity for conductive heat transfer between fibers. Thus, the total effect of increasing the density from zero is that, at some point where convection has been eliminated and radiation between the insulation surfaces and between individual fibers has been drastically reduced, but before conduction between fibers becomes significant, the total apparent conductivity has a minimum value. Most insulating materials experience this kind of behavior.

As the mean temperature of an insulation increases, its apparent thermal conductivity also increases. This is due primarily to the rising thermal conductivity of air, although the other components may increase as well. Another effect of higher temperatures is that higher densities are required

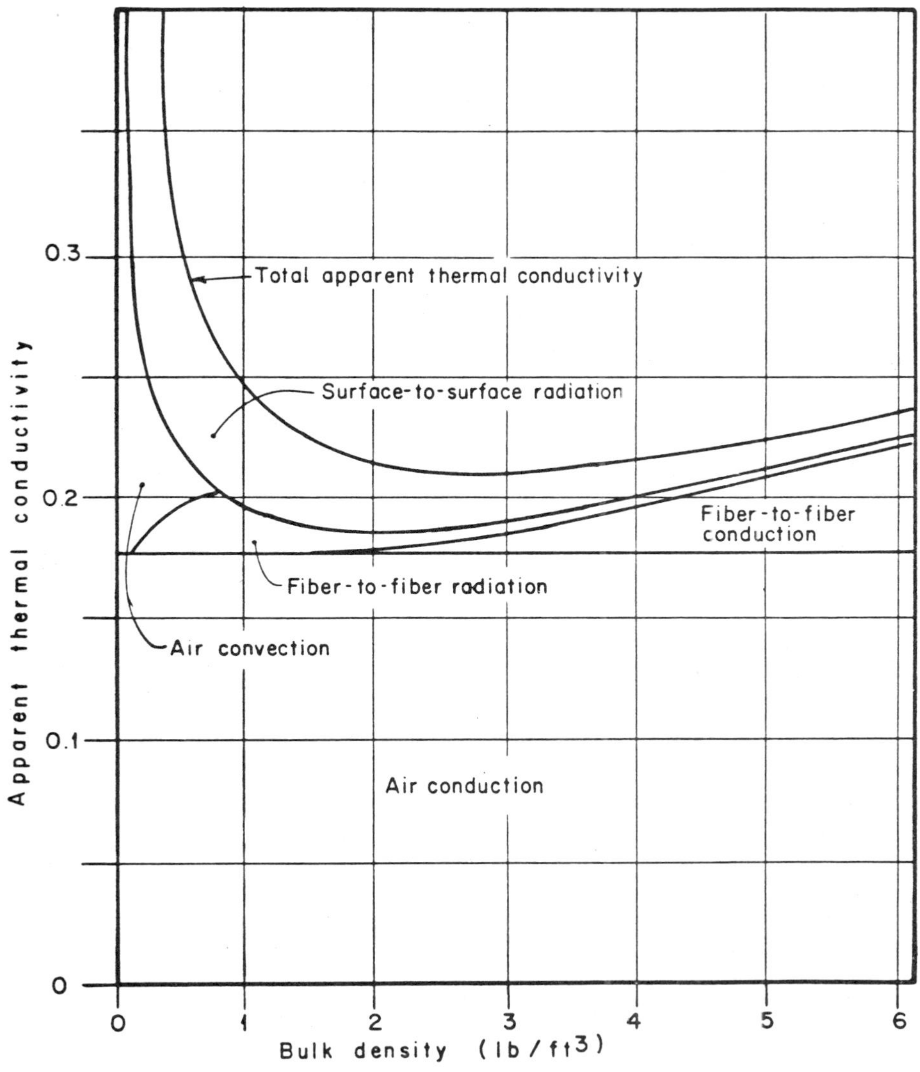

Figure 2-17 Effect of density on apparent thermal conductivity of a lightweight insulation (100°F hot face, 50°F cold face, 1-inch thick glass fiber)

to reach the minimum apparent thermal conductivity for that temperature since more material is needed to reduce the radiative heat transfer to the point where a minimum apparent conductivity can occur.

A "thickness effect" also occurs in insulation and is most important at lower densities. This effect is analogous to the fact that radiation through a translucent material can be reduced by increasing the thickness, even though other properties remain the same. Similarly, a low-density insulation lets more radiation through a thin section than a thick one. Test methods now recommend that apparent thermal conductivity be determined using insulation thicknesses which will occur in actual practice [13].

Reducing apparent thermal conductivity through the use of a low-conductivity gas instead of air is the principal behind several insulating foams. A closed-cell structure is required to ensure that the alternate gas is retained within the insulation. Forced convection effects on open-cell insulations have been found to increase apparent conductivities [14].

Analytical equations for heat transfer through insulation have been proposed [15-18], although such equations are likely to be more valuable to persons involved in developing new insulations rather than those interested in applying insulation to a specific use.

2.8 Thermal Mass

The effects of mass on heat transfer were briefly discussed in Section 2.1, where the concept of thermal diffusivity was introduced. The thermal behavior of masonry buildings - warmer in winter and cooler in summer than comparable lightweight buildings - has long been recognized. The cause of this behavior is the thermal capacity of masonry, i.e., its ability to store heat.

Steady state U-values for masonry construction are given in References [19-23]. These values alone, however, will not predict the time delay of heat transfer through the masonry resulting from the thermal storage ability of the masonry. As noted in Section 2.1, heat conduction through a material is proportional to the temperature difference across it when no storage is present; therefore, the peak heat transfer rate coincides with the maximum temperature difference. In a material with mass, however, steady state heat transfer does not often occur; the material may absorb heat when the temperature on one side is high, and release the heat to the other side at a later time. Thus, the time of the maximum heat transfer rate from a massive wall to a conditioned space, for example, is delayed relative to the time of the maximum heat transfer rate into the wall from the outside. Reference [23] lists the time lag, in hours, which can be expected with a variety of masonry wall constructions.

A number of references discuss the above concept of masonry wall behavior [19,23-32]*. Several of these works attempt to show that using steady state U-values for masonry buildings will result in over-estimating the actual building heat loss rate at design temperature differences, resulting in needlessly oversized heating equipment. A concept known as the "M factor" has been developed to take some credit for the inherent insulating ability of masonry (sometimes described as "mass insulation") [26, 33]. M factors have values between 0 and 1. By multiplying design U-values by an M factor in a heat loss calculation, the design U-value is essentially lowered. The M-factor approaches 1 for lighter-weight masonry and greater numbers of degree-days (i.e., colder climates benefit less from thermal mass than warmer climates).

Considerable controversy exists as to the validity of the M factor approach. Recent articles [34-37] have criticized the M factor as being arbitrarily based on the relative heat fluxes through heavy and lightweight walls at 8:00 a.m. on an "average" January day, and have presented results of studies indicating significant deviations in recommended and calculated M factors. In defense of M factors, counterarguments (see "Comments", Reference [35]) state other hours and other months were investigated in the original studies and the results were rather insensitive to hour or month.

M factor critics and proponents apparently agree on one point: M factors should not affect daily or yearly energy use calculations. Although a massive wall will delay heat loss at night, for example, it will also delay heat gain during the morning; the net result is the same actual energy use over the day for any wall weight. Wall weight therefore affects only the schedule of the energy use, not the amount.

Criticisms of the M factor should not be construed as attempts to underrate the importance of thermal mass in building design [37-39]. Critics do not deny that the use of thermal mass reduces peak loads; they contend only that the M factor is not the proper method of quantifying the reduction. Accurately determining building energy requirements and equipment sizes is best done with detailed dynamic analysis. Such analysis should include, in addition to envelope design, the effects of building internal mass, building occupancy schedules, internal gains from lighting and equipment, building and equipment operating modes such as night or weekend setback, local weather conditions, and solar gains.

Allowing the internal temperature to float is an important idea in maximizing the benefit from thermal mass. As excess heat gains occur within a building, the interior temperature rises above the mass temperature, and heat flows into the mass. Recovery of the energy thus stored can occur only if the internal temperature later drops below the mass temperature, allowing the energy flow to reverse directions. This

---

*Caution is urged in considering one of the conclusions of Reference [29], which says insulation on a masonry wall will never be cost effective. This reference considered the case of adding R-4 insulation and adding the cost of the job to an 8-1/2%, 30 year mortgage, and did not consider other financing methods, rising cost savings with rising fuel prices, nor reinvestment of cost savings at some appropriate interest rate.

oscillating process can reduce instantaneous heating and cooling requirements within a building, compared to a lightweight building in which either heating, cooling, or both might be required.

During the heating season, thermal mass will store excess gains, including solar gains, for release at night when a conventional heating system might otherwise be used. Heating equipment, if used in the daytime, will operate more efficiently since the temperature of infiltrating outside air is higher. Heat pumps in the heating mode will also operate more efficiently for the same reason. In the cooling season, mass absorbs excess gains and reduces the need for cooling. Advantage can then be taken of cooler night temperatures to dissipate the heat, thus using mechanical cooling more efficiently, or even possibly eliminating mechanical cooling with night ventilation [40].

Reducing peak heating demands also means heating equipment can be lower in capacity, and thus lower in size and cost. Such equipment is likely to operate near its design point more often than equipment sized to meet a higher peak demand, and will experience less cycling because of the slower response of a massive structure. Both of these effects will result in higher-efficiency operation. Thus, even though the total heating load may not change with massive construction, the amount of energy required to meet the load could be reduced.

Insulation can be applied to masonry walls to reduce energy consumption. Whether the insulation should be inside or outside the wall depends partially on how the building temperature is controlled [38-39]. For buildings predominantly maintained at a constant temperature, the insulation should generally be outside the mass in most cases if possible. This internal mass then helps stabilize loads by storing heat or releasing heat within the conditioned space, as is the case, for example, in a passively heated and cooled building. However, buildings conditioned during normal daytime hours only should generally have insulation placed near the inside wall surfaces with the wall mass outside the insulation if possible. Energy savings resulting from night setback of space temperature are generally greater if the space air temperature is more responsive to the thermostat setting. When night setback begins, the temperature of responsive buildings falls rapidly to the setpoint and in the morning when the thermostat is set back up, it rises quickly to the daytime value. Thus, energy savings for the responsive building result because heat transfers through the building walls across a smaller temperature differential for a longer period of time.

The foregoing are generalized statements which there are exceptions to. Thus, the placement of insulation in the wall of a building should be evaluated on an individual building-by-building basis and should include the effects of climate, hours of operation, type of mechanical system, building orientation, and building function. The optimal placement of insulation may require the use of a sophisticated hour-by-hour building computer simulation program for evaluation purposes.

REFERENCES FOR SECTION 2.0

[1] ASHRAE Handbook, 1981 Fundamentals. American Society of Heating, Refrigerating, and Air-Conditioning Engineers, Inc. Atlanta, GA. 1981.

[2] Schneider, P.J. Conduction Heat Transfer. Addison-Wesley Publishing Company, Reading, MA. 1955.

[3] Chapman, Alan J. Heat Transfer. The Macmillan Company, New York, NY. 1967.

[4] Holman, J.P. Heat Transfer. 4th edition. McGraw-Hill Book Company, New York, NY. 1976.

[5] Kreith, Frank. "Principles of Heat Transfer." 3rd edition. Harper & Row, Publishers. New York, NY. 1973.

[6] Rohsenhow, Warren M. and Choi, Harry. Heat, Mass, and Momentum Transfer. Prentice-Hall, Inc. Englewood Cliffs, NJ. 1961.

[7] Rohsenhow, Warren M. and Hartnett, James P. (eds.) Handbook of Heat Transfer. McGraw-Hill Book company, New York, NY. 1973.

[8] Brick Institute of America. "Thermal Transmission Corrections for Dynamic Conditions - M Factor." Technical Notes on Brick Construction. Technical Note 4B. McLean, VA. Mar/Apr 1977.

[9] Miller, R.G., and Sherman, M. "Thermal Performance of Insulated Metal Building Roof Deck Constructions." Presented at the DOE-ORNL/ASTM Conference on Thermal Insulation, Materials, and Systems for Energy Conservation in the '80s. Clearwater Beach, FL. Dec. 1981.

[10] Pelanne, Charles M. "Thermal Insulation. What Is It? - How Does It Work?" Johns Manville Corp., Research and Development Center, Denver, CO. Sept. 1976.

[11] Pelanne, Charles M. "Heat Flow Principles in Thermal Insulations." Journal of Thermal Insulation. Vol. 1, pp. 49-80. July 1977.

[12] Pelanne, Charles M. "Thermal Insulation: What It Is and How It Works." Journal of Thermal Insulation. Vol. 1, pp. 223-235. April 1978.

[13] Pelanne, Charles M. "Thermal Insulation Heat Flow Measurements: Requirements for Implementation." ASHRAE Journal. pp. 51-58. March 1979.

[14] Berlad, A. L., Jaung, R., Joshi, N., and Westerinen, W. J. "Free and Forced Convective Effects on Air-Permeable Insulation Systems." Paper No. 81-HT-47, presented at the 20th Joint ASME/AIChE National Heat Transfer Conference, Milwaukee, WI. Aug. 1981.

REFERENCES FOR SECTION 2.0 (cont.)

[15] Rennex, Brian G. "Thermal Parameters as a Function of Thickness for Combined Radiation and Conduction Heat Transfer in Low-Density Insulation." Journal of Thermal Insulation. Technomic Publishing Co., Inc. Vol. 1, pp. 37-61. July 1979.

[16] Bejan, Adrian. "A General Variational Principle for Thermal Insulation System Design." International Journal Heat Mass Transfer. Pergamon Press Ltd. Great Britain. Vol. 22, pp. 219-228. 1979.

[17] Marciano, Jorge H., Rojas, Alfredo J., and Williams, Roberto J.J. "A Theoretical Model for the Thermal Conductivity of Plastic Foams." European Journal of Cellular Plastics. Technomic Publishing Co., Inc. Vol. 3, No. 3. 1981.

[18] Tong, T. W., and Tien, C. L. "Radiative Heat Transfer in Fibrous Insulations - Part I: Analytical Study." Paper No. 81-HT-42, and "-Part II: Experimental Study." Paper No. 81-HT-43, presented at the 20th Joint ASME/AIChE National Heat Transfer Conference, Milwaukee, WI. Aug. 1981.

[19] National Concrete Masonry Association. "Thermal Insulation of Concrete Masonry Walls." NCMA TEK 38-A. Herndon, VA. 1980.

[20] National Concrete Masonry Association. "Estimating U-Factors for Concrete Masonry Construction." NCMA TEK 12. Herndon, VA 1969.

[21] National Concrete Masonry Association. "Tables of U-Values for Concrete Masonry Walls." NCMA TEK 67. Herndon, VA. 1975.

[22] National Concrete Masonry Association. "'U' Values for Reinforced Concrete Masonry Walls." NCMA TEK 101. Herndon, VA. 1978.

[23] National Concrete Masonry Association. "Design of Solar Energy Walls with Concrete Masonry." NCMA TEK 97. Herndon, VA. 1978.

[24] National Concrete Masonry Association. "Energy Conservation with Concrete Masonry." NCMA TEK 58. Herndon, VA. 1974.

[25] National Concrete Masonry Association. "New Findings on Energy Conservation with Concrete Masonry." NCMA TEK 68. Herndon, VA. 1975.

[26] National Concrete Masonry Association. "Energy Conscious Design for Buildings." NCMA TEK 82. Herndon, VA. 1976.

[27] National Concrete Masonry Association. "Second Generation Passive Solar Systems Using Concrete Masonry." NCMA TEK 120. Herndon, VA. 1981.

[28] Aho, Arnold, J. "The Role of Brick Masonry in Energy Conservation Design." Brick Institute of America. McLean, VA. 1974.

[29] National Concrete Masonry Association. "A Comparison of Very Lightweight Walls of Wood, Metal and Glass Versus Concrete Masonry in Energy Conservation." CM-230. McLean, VA. 1976.

[30] Berkeley Solar Group. "Passive Design Saves Energy and Money." Concrete Masonry Association of California and Nevada. Sacramento, CA. 1979.

[31] Fiorato, A.E. and Cruz, C.R. "Thermal Performance of Masonry Walls." Portland Cement Association. Research and Development Bulletin RD071.01M. Skokie, Il. 1980.

[32] Peavy, Bradley, A., Powell, Frank J., and Burch, Douglas M. "Dynamic Thermal Performance of an Experimental Masonry Building." NBS Building Science Series 45. National Bureau of Standards. Washington, D.C. July 1973.

[33] Brick Institute of America. "Thermal Transmission Corrections for Dynamic Conditions - M Factor." Technical Note 4B. Technical Notes on Brick Construction. McLean, VA. Mar/Apr. 1977.

[34] Godfrey, R.D., Wilkes, K.E. and Lavine, A.G. "A Technical Review of the "M" Factor Concept." ASHRAE Journal. pp. 47-50. March 1980.

[35] Godfrey, R.D., Wilkes, Kenneth E., and Lavine, Adrienne G. "A Technical Review of the 'M' Factor Concept." ASHRAE SP28. Proceedings of the ASHRAE/DOE-ORNL Conference, Thermal Performance of the External Envelopes of Buildings. Kissimmee, FL. Dec. 3-5, 1979.

[36] Wilkes, K.E., Godfrey, R.D., and Lavine, A.G. "Wall Massiveness - What is Known Today About Annual Energy Requirements?" ASHRAE Journal, Vol. 24, No. 2, pp. 21-23. Feb. 1982.

[37] Childs, K.W. "An Appraisal of the M Factor and the Role of Building Thermal Mass in Energy Conservation." ORNL/CON-46. Oak Ridge National Laboratory. Oak Ridge, TN. July 1980.

[38] Dexter, Michael E. "Energy Conservation Design Guidelines for Including Mass and Insulation in Buildings Walls." ASHRAE SP28. Proceedings of the ASHRAE/DOE-ORNL Conference. Thermal Performance of the Exterior Envelopes Of Buildings. Kissimmee, FL. Dec. 3-5, 1979.

[39] Dexter, Michael E. "Energy Conservation Design Guidelines: Including Mass and Insulation in Building Walls." ASHRAE Journal. pp. 35-38. March 1980.

[40] Gujral, Parambir S., Clark, Raymond J. and Burch, Douglas M. "Transient Thermal Behavior of Externally Insulated Massive Building." ASHRAE Transactions. DV-80-4. No. 2. 1980.

3.0

Components

3.1 Generic Insulation: Types and Material Properties

This section presents typical data for the thermophysical properties of commonly used generic thermal insulating materials. There are six basic types of thermal insulation. They are:

- o Air film or air layers.
- o Closed cellular material.
- o Fibrous materials.
- o Flake materials.
- o Granular materials.
- o Reflective foils.

Certain insulations may be a combination of the above. Air films or air layers may be composed simply of a single surface, or possibly of multiple surfaces between which only air exists.

Resistance to heat flow is exhibited by surfaces utilizing surface film resistance, as well as by the layers of air as a result of conduction or convection across the layers.

Cellular insulation is composed of parent material containing many small voids of air or gas. Usually, this material is of the closed cell variety, where each cell is separated from the others by cell windows or thin membranes. Cellular insulation is produced from glass, plastics, and rubber. Common generic thermal insulations of this type are cellular glass, expanded elastomeric foam, polystyrene foam, polyisocyanurate and polyurethane foams, and urea-formaldehyde foam.

Fibrous insulation is composed of many small-diameter fibers. These fibers may be made from organic materials such as hair, wood, and cane, or may be made from synthetic materials such as glass, rock wool, slag wool, aluminun silicate, asbestos, and carbon.

Flake insulation is composed of small particles. These particles, or flakes, may be poured into an air space or bonded together to provide a rigid form of the insulation. Rigid form flake insulation can be used for pipe insulation or for other applications in block or board form. The two types of flake insulation commonly used are perlite and vermiculite.

Granular insulation is composed of small particles which contain voids or hollow spaces. These hollow spaces can transfer air between the individual voids. The parent material can be magnesia, calcium silicate, diatomaceous earth, or vegetable cork. The first three are commonly used as industrial piping insulation, while the cork is used in low-temperature refrigeration applications.

Reflective foil insulation is composed of parallel thin sheets of foil with either high thermal reflectance or low emittance. These thin sheets are spaced to reflect radiant heat back to the source. Each separate sheet provides two heat transfer film coefficients; the air space between two sheets causes a reduction in conduction and convection.

Most applications of foil insulation system are of an industrial nature; reflective foils are commonly used in specially designed environmental chambers and in high-temperature applications where radiative heat transfer is the predominate mode of heat transfer.

Because foil insulation is often misapplied and misunderstood, it has met with very limited success in building insulation systems. To be effective, uniform spaces must be used in conjunction with reflective foils.

Other uses for reflective foils are found in cryogenic (extremely low-temperature) environments and in spacecraft applications with an evacuated environment. In this usage, it is called "multilayer insulation" or "superinsulation."

3.1.1 Calcium Silicate Insulation

Calcium silicate insulation, which is granular in nature, is composed of hydrous calcium silicate. During the manufacture of this insulation, live steam converts lime and silica into hydrous calcium silicate. This substance is noted for its tough, hard composition, as well as for its ability to withstand repeated wetting.

The apparent thermal conductivity of calcium silicate insulation at a material density of 13 lb/ft^3 and at room temperature is 0.35 Btu in/hr ft^2 °F. This yields a thermal resistance of 2.86 hr ft^2 °F/Btu per inch of thickness.

Calcium silicate has a maximum service temperature of 1200°F, is nontoxic, and does not exhibit reduced thermal performance as a result of aging. Like other materials, it does exhibit some thermal degradation as a result of moisture absorption; after it dries, however, the thermal performance is unchanged. Calcium silicate has excellent fire resistance properties; it is noncombustible and has a flame spread of 0.

Typically, this material is used for industrial piping insulation in high-temperature processes. In certain applications where a high compressive strength is needed, calcium silicate is used to insulate industrial tanks.

The properties of calcium silicate are summarized in Table 3-1 on the following page.

Table 3-1

CALCIUM SILICATE

Property	Calcium Silicate
Apparent thermal conductivity (k) Btu in/hr ft^2 °F	0.35
Apparent thermal resistance (R) hr ft^2 °F/Btu in.	2.86
Material density (ρ) lb/ft^3	13
Specific heat (Cp) Btu/lb °F	0.28
Thermal diffusivity (α) ft^2/hr	0.008
Maximum service temperature °F	1200
Coefficient of thermal expansion in/in °F	0
Closed cell content percent	N/A
Water vapor permeability perm-in	very high
Water absorption percent by volume	90
Corrosiveness	none
Fire resistance	
flame spread	0
fuel contributed	0
smoke developed	0
Degradation as a result of:	
temperature cycling	none
vermin	none
moisture	transient
fungal/bacterial	none
weathering	none
wind	none
Human factors	
toxicity	none
odor	none
sound absorption	N/A
Effect of aging	
dimensional stability	will shrink 1.1% after 24 hr soak at 1200°F
thermal performance	none
fire resistance	none

3.1.2 Cellular Glass Insulation

Cellular glass insulation is a rigid foam material formed by blowing glass with H_2S to produce a foam with a very small cell size. This glass foam has a closed cell content of 100 percent; because it is made from an inorganic material, it is completely impervious to moisture and is noncombustible.

The apparent thermal conductivity of cellular glass at a density of 8.5 lb/ft^3 is about 0.38 Btu in/hr ft^2 °F. This yields a thermal resistance of 2.63 hr ft^2 °F/Btu per inch of thickness.

Cellular glass has a maximum service temperature of 800°F, is nontoxic, and exhibits no reduced thermal performance or dimensional change as a result of aging; however, repeated freeze-thaw cycling in the presence of water can induce fracturing of the insulation.

Typical applications of cellular glass insulation include building overdeck assemblies and load-bearing floors and wall siding material. With its high compressive strength of 75-100 psi and its impermeability to moisture, it can be used as a plaza or parking deck insulation. It also has applications as a pipe covering insulation.

The properties of cellular glass are summarized in Table 3-2 on page 49.

3.1.3 Cellulosic Fiber Insulation

Cellulosic fiber insulation is manufactured by shredding and pulling recycled paper or wood pulp into a fluffy, low density material. Boric acid, borax 5 mol, or a mixture of these two materials is added to provide resistance to fire, moisture absorption, and fungal growth.

During the last few years, a tremendous surge in sales of this type of insulation has taken place. This popularity results from the good thermal properties, its relative ease of installation, low cost, and simplicity of production. Problems have arisen because of the influx of a large number of unsophisticated newcomers in both the manufacturing and installation of this material, including inadequate installed densities and insufficient fire protection. As a result, all cellulosic fiber insulation sold within the United States is required by the Consumer Product Safety Commission to meet the Federal Specification, General Services Administration Standard HH-I-515D, dated June 15, 1978.

For the most part, cellulosic fiber is used as a loose fill material to insulate residential attics and wall cavities. It is available in batts and blankets, or in a spray-in-place form with a binder; in this form, it is used to insulate roof underdecks or large underground tanks.

As a loose fill material, cellulosic fiber has an apparent thermal conductivity ranging between 0.27 and 0.31 Btu in/hr ft^2 °F. This value corresponds to a thermal resistance of 3.2 to 3.7 ft^2 hr°F/Btu per inch of thickness at a density of 2.6 to 3.0 lb/ft^3.

If cellulosic fiber insulation is applied at densities significantly less than the nominal density (resulting from the introduction of too much air in the blowing machine), it will gradually settle up to 20 percent because of temperature changes, vibration and moisture effects. Such settling causes both a reduction in installed thickness and a decrease in the material thermal resistance.

These effects are direct results of a change in application technique; when the material is blown in strict accordance with the manufacture's instructions, settling should not be a problem, either in horizontal application or in blown-in wall applications using the two-hole technique.

Without a fire retardant, cellulosic fiber is a naturally combustible material. Tests were performed on two cellulosic fiber samples, one without a fire retardant and one with 25 percent by weight boric acid. Both produced severe surface corrosion when placed in contact with low carbon steel; however, when applied in the correct proportions, a mixture of cellulosic fiber, boric acid, and borax 5 mol consistently passes the corrosion test specified by GSA HH-I-515D.

When tested according to ASTM C739-73, cellulosic fiber insulation should have a weight gain from water absorption not exceeding 15 percent by volume. Loose fill cellulosic fiber insulation has a high degree of water vapor permeability and will absorb up to 98 percent by weight; consequently, this could cause problems in high-humidity environments.

The properties of cellulosic fiber are summarized in Table 3-2 on the following page.

Table 3-2

CELLULAR GLASS AND CELLULOSIC FIBER

Property	Cellular Glass	Cellulosic Fiber
Apparent thermal conductivity (k) Btu in/hr ft^2 °F	0.38	0.27-0.31
Apparent thermal resistance (R) hr ft^2 °F/Btu in	2.63	3.7-3.2
Material density (ρ) lb/ft^3	8.5	2.6-3.0
Specific heat (Cp) Btu/lb F	0.18	0.33
Thermal diffusivity (α) ft^2/hr	0.021	0.023
Maximum service temperature °F	800	180
Coefficient of thermal expansion in/in °F	4.6×10^{-6}	N/A
Closed cell content, percent	100	N/A
Water vapor permeability perm-in	0.00	High
Water absorption percent by volume	5 (Surface only)	98 (by weight)
Corrosiveness	none	May corrode steel, Al, Cu
Fire resistance		
flame spread	5	15-40
fuel contributed	0	0-40
smoke developed	0	0-45
Degradation as a result of		
temperature cycling	Possible freeze damage in presence of water	may cause settling
vermin	none	depends on treatment
moisture	none	reduces thermal performance
fungal/bacterial	none	may support growth
weathering	none	do not expose
wind (convection effect)	none	slight
Human Factors		
toxicity	none	none
odor	slight H_2S odor if cells rupture	none
sound absorption	fair	good
Effect of aging		
dimensional stability	none	will settle
thermal performance	none	reduced with settling
fire resistance	none	depends on initial treatment

3.1.4 Elastomeric Foam

Elastomeric foam, also known as expanded rubber foam, is a flexible pipe insulation that is blown and expanded in a mold. The apparent thermal conductivity of this material is between 0.26 and 0.27 Btu in/hr ft^2 °F, which corresponds to a thermal resistance between 3.85 and 3.70 hr ft^2 °F/Btu per inch of thickness at a density of about 4.0 lb/ft^3.

Elastomeric foam has a closed cell content of greater than 90 percent; its maximum service temperature is 220°F and its minimum service temperature is -40°F. The water vapor permeability of elastomeric foam is very low (between 0.1 and 0.14 perm-in) and its water absorption percentage by weight is between 3 and 6 percent. When subjected to 200°F for 7 days, elastomeric foam exhibits a shrinkage of approximately 5 percent. Aging has no apparent effect on its properties; however, when used in an outside environment, a protective coating or material is recommended.

Two types of elastomeric foam are available; the types differ in their flammability ratings. For "standard" or "type I" material, the flame spread rating is 50 to 75, and smoke developed rating is 330 to 450. "Type II' material has a flame spread rating of 25 and smoke developed rating of 100 to 150.

Elastomeric foam is available in pipe insulation form or as preformed sheet. In piping, this material is used to retard heat flow and control condensation on refrigerant lines and chilled water lines; it is commonly used on both hot and cold water plumbing lines. The sheet form of this type of insulation is adaptable for insulating duct work, large pipes, tanks and vessels.

The properties of elastomeric foam are summarized in Table 3-3 on the following page.

Table 3-3

ELASTOMERIC (EXPANDED RUBBER) FOAM

Property	Elastomeric Foam	
Apparent thermal conductivity (k) Btu in/hr ft^2 °F	0.26	
Apparent thermal resistance (R) hr ft^2 °F/Btu in	3.85	
Material density (ρ) lb/ft^3	4.0	
Specific heat (Cp) Btu/lb F	0.20	
Thermal diffusivity (α) ft^2/hr	0.027	
Maximum service temperature °F	220	
Coefficient of thermal expansion in/in °F	N/A	
Closed cell content percent	>90	
Water vapor permeability (perm-in)	0.1-0.14	
Water absorption percent by volume	3-6	
Corrosiveness	none	
Fire resistance	Type I	Type II
flame spread	50-75	25
fuel contributed	combustible	combustible
smoke developed	330-450	100-150
Degradation as a result of		
temperature cycling	5% shrinkage when exposed to 200°F for 7 days.	
vermin	none	
moisture	none	
fungal/bacterial	none	
weathering	needs protective finish UV sensitive	
wind	none	
Human factors		
toxicity	none	
odor	slight rubber odor	
sound absorption	very good	
Effect of aging		
dimensional stability	none	
thermal performance	none	
fire resistance	none	

3.1.5 Glass Fiber Insulation

Glass fiber insulation is manufactured by spinning thin strands of glass from preformed glass "marbles." This insulation is available as loose fill, boards, and batts or blankets.

Glass fiber batts or blankets usually have a density between 0.6 and 1.0 lb/ft^3. Because of the relatively long fibers, this form of the material tends to recover to the design thickness after packaging. When used in this form, glass fiber insulation yields an R-value of about 3.2 per inch of thickness.

Loose fill glass fiber is made by hammer-milling glass fiber batts and usually provides an R-value of about 2.8 per inch of thickness. Both loose fill and batt or blanket forms of glass fiber insulation are permeable to water vapor to the extent of over 100 perm-inches. Water absorption is typically no more than 1 percent when tested by weight, by ASTM C553-70.

Glass fiber itself is an inorganic, noncombustible material; however, flammable organic binders are used in the production of batts and blowing wool. For the material with binder, ASTM E-84 assigns the following approximate ratings: flame spread: 15-20; fuel contributed: 5-15; smoke developed: 0-20. Facings on fiberglass building insulation usually consist of an asphalt-coated kraft or foil-kraft paper laminate. Because this facing is a flammable surface, it must not be exposed to open flames or temperatures exceeding 180°F. Any burning of facings or organic binders could produce hazardous fumes.

Glass fiber batt insulation does not appear to settle or shrink with age. However, loose fill may settle if applied at densities below the manufacturer's specifications. Other properties of the material, such as thermal peformance and resistance to fire, are reportedly unaffected by age and temperature cycling at normal installed temperatures. Glass fiber does not promote bacterial or fungal growth, and provides no sustenance to vermin. Insulation products made from glass fiber are noncorrosive (Federal Specification HH-I-558D) and have no objectionable odor (ASTM C-553 - Section 16).

Glass fiber board is manufactured by several companies. The properties of the board are dependent on the material of the substrate and the percentage of glass fiber present. A typical R-value is of the order of 4 hr ft^2 °F/Btu at a density of 4.0 lb/ft^3. Common applications for glass fiber insulation include building roofs, walls, floors, basements, and piping systems. This material is also used in industrial applications to insulate industrial tanks, mechanical systems, and air ducts.

The properties of glass fiber insulation are summarized in Table 3-4 on the following page.

Table 3-4

GLASS FIBER

Property	Glass Fiber (Batt)	Glass Fiber (Loose Fill)	Glass Fiber (Bonded)
Apparent thermal conductivity (k), Btu in/hr ft^2 °F	0.31	0.35	0.25
Apparent thermal resistance (R), hr ft^2 °F/Btu in	3.2	2.85	4.0
Material density (ρ) lb/ft^3	1.0	1.0	4.9
Specific heat (Cp) Btu/lb °F	0.20	0.20	0.30
Thermal diffusivity (α) ft^2/hr	0.129	0.1458	0.017
Maximum service temperature °F	370	1000	400
Coefficient of thermal expansion in/in °F	N/A	N/A	N/A
Closed cell content percent	N/A	N/A	N/A
Water vapor permeability perm-in	>100	>100	>100
Water absorption percent by weight	1	1	1
Corrosiveness	None	None	None
Fire resistance			
flame spread	15-20	15-20	15-20
fuel contributed	5-15	5-15	5-15
smoke developed	0-20	0-20	0-20
Degradation as a result of			
temperature cycling	none	none	none
vermin	none	none	none
moisture	transient	transient	transient
fungal/bacterial	none	none	none
weathering	none	none	none
wind (convection)	medium	moderate	slight
Human factors			
toxicity	nontoxic	nontoxic	nontoxic
odor	none	none	none
sound absorption	good	good	good
Effect of aging			
dimensional stability	none	will settle	none
thermal performance	none	settling causes some degradation	none
fire resistance	none	none	none

3.1.6 Mineral Fiber

Mineral fiber insulation (also called mineral rock, or slag wool) is produced similarly to glass fiber insulation. In the United States, the material most commonly used to manufacture mineral fiber is slag, from the production of steel, copper, or lead.

Mineral fiber and glass fiber are similar forms of insulation. They are often used for the same applications in residential, commercial, and industrial buildings.

Mineral fiber batts and blowing wool are produced with densities in the range of 1.5 to 2.5 lb/ft^3. For batts, reported unit thermal resistances (R-values) are 3.2 to 3.7 hr ft^2 °F/Btu in at 75°F (k factor 0.31 to 0.27 Btu in/hr ft^2 °F). For blowing wool, the R-value is 2.9 at 75°F (k factor 0.34). Water vapor permeability is reported to be >100 perm-in, and water absorption up to 2 percent by weight.

Mineral fiber is made from rock or slag; therefore, the base material is non-combustible and its melting point is above 1200°C. However, binders added to the wool may be flammable. The flame spread rating of this material is reported to be less than 25 when tested according to ASTM-84. Asphalt coated or foil-laminated kraft paper may be used as a vapor retardant facing on batts. Because these facings are flammable, they should be protected from open flames or high temperatures. Burning of facings or organic binders could produce toxic vapors.

Properties such as dimensional stability, thermal performance, and fire resistance are reportedly unaffected by age, temperature cycling, or weathering. Because mineral fiber does not have the resiliency of glass, it may not recover to design thickness after packaging; thus, lower than design R-values may result. Thermal conductivity is affected by moisture content, but the change is transient and the material returns to its original properties upon drying.

Mineral fiber does not support the growth of fungus, bacteria, or vermin, exudes no odor and is noncorrosive. The thermal properties of the material are affected by "shot" content, i.e., pieces of slag that spin off as particles rather than fibers. High shot content results in a higher apparent thermal conductivity with a corresponding density increase.

Mineral fiber is also produced in boards; in this form, it has a thermal resistance of 3.57 hr ft^2 °F/Btu per inch of thickness at a density ranging from 9 to 11 lb/ft^3. Mineral fiber boards are used primarily in roofing insulation, as a building sheath material, and other industrial applications.

The properties of mineral fiber insulation are summarized in Table 3-5 on the following page.

Table 3-5

MINERAL FIBER

Property	Mineral Fiber (Batt)	Mineral Fiber (Loose fill)	Mineral Fiber (bonded)
Apparent thermal conductivity (K), Btu in/hr ft^2 °F	0.31	0.32	0.28
Apparent thermal resistance (R), hr ft^2-°F/Btu in	3.23	3.10	3.57
Material density (ρ) lb/ft^3	2.0	1.7	9-11
Specific heat (Cp) Btu/lb °f	0.20	0.20	0.20
Thermal diffusivity (α) ft^2/hr	0.065	0.078	0.030
Maximum service temperature °F	400	1200	500
Coefficient of thermal expansion in/in °F	N/A	N/A	N/A
Closed cell content percent	N/A	N/A	N/A
Water vapor permeability perm-in	high	high	high
Water absorption, percent by weight	90	90	90
Corrosiveness	none	none	none
Fire resistance			
flame spread	15	0	15
fuel contributed	0	0	0
smoke developed	0	0	0
Degradation as a result of			
temperature cycling	none	none	none
vermin	none	none	none
moisture	transient	transient	transient
fungal/bacterial	none	none	none
weathering	none	none	none
wind	medium	moderate	slight
Human factors			
toxicity	non-toxic	non-toxic	non-toxic
odor	none	none	none
sound absorptivity	good	medium	good
Effect of aging			
dimensional stability	none	none	none
thermal performance	none	settling causes some degradation	none
fire resistance	none	none	none

3.1.7 Perlite

Perlite loose fill insulation is made from siliceous volcanic glass pellets, expanded to between 4 and 20 times their original volume. These pellets contain glass-enclosed dead air spaces. Expanded perlite can be produced with densities between 2 and 11 lb/ft^3.

The thermal conductivity of loose fill perlite insulation is dependent on the applied density. The conductivity ranges from 0.27 Btu in/ft^2 hr °F at a density of 2 lb/ft^3 to 0.40 Btu in/ft^2 at 11 lb/ft^3. Perlite boardstock is available with a conductivity of 0.36 to 0.38 Btu in/ft^2 hr °F at a density of about 10 lb/ft^3.

Usually, a nonflammable silicone treatment is used to increase the resistance of perlite to water penetration; consequently, the material is claimed to be water repellent and impervious to moisture. Because it is inorganic, perlite is noncombustible and resists bacterial and fungal growth and vermin. Loose fill perlite has a maximum service temperature of 1800°F, while the perlite roof board has a maximum service temperature of 200°F.

Expanded perlite is mixed with portland cement to form a lightweight insulating concrete. Density is varied by controlling the perlite to cement ratio, and a range of 20 to 40 lb/ft^3 is typical. Perlite concrete, which may be precast in a number of shapes or cast-in-place, possesses sufficient mechanical strength to be load-bearing at high densities. It has a k-factor of 0.51 to 2.00 Btu in/hr ft^2 °F; k increases with increasing density.

Perlite is used primarily in industrial and commercial buildings as a roof insulation board material. The next most frequent use is in lightweight insulating concrete. Perlite insulating concrete, both preformed and cast-in-place, is used primarily for roof decks, floor slabs, and wall systems. Low density expanded perlite is used as a loose fill insulation.

The properties of perlite are summarized in Table 3-6 on the following page.

Table 3-6

PERLITE

Property	Perlite (loose fill)	Perlite (roof board)
Apparent thermal conductivity (k), Btu in/hr ft^2 °F	0.276	0.38
Apparent thermal resistance (R), hr ft^2 °F/Btu in	3.62	2.63
Material density (ρ) lb/ft^3	5	10
Specific heat (Cp) Btu/lb-f	0.30	0.25
Thermal diffusivity (α) ft^2/hr	0.015	0.013
Maximum service temperature °F	1800	200
Coefficient of thermal expansion in/in °F		
Closed cell content percent	N/A	N/A
Water vapor permeability perm-in	high	25
Water absorption, percent by weight	high	1.5
Corrosiveness	none	none
Fire resistance		
flame spread	0	0
fuel contributed	0	0
smoke developed	0	0
Degradation as a result of		
temperature cycling	none	none
vermin	none	none
moisture	transient	transient
fungal/bacterial	none	none
weathering	none	none
wind	none	none
Human factors		
toxicity	nontoxic	nontoxic
odor	none	none
sound absorptivity	medium	good
Effect of aging		
dimensional stability	none	none
thermal performance	none	none
fire resistance	none	none

3.1.8 Phenolic Foam

Phenolic foam insulation is a molded rigid pipe insulation made from chemically neutral phenolic material. The manufacturing process consists of placing the cold liquid resin mix in a closed mold, then applying a vacuum . The vacuum produces a low boiling azeotropic mixture used as a blowing agent to cause a foaming action. Heat is then applied to the mold to drive the exothermic reaction to completion.

The apparent thermal conductivity of phenolic foam at a nominal density of 2 lb/ft^3 is about 0.23 Btu in/hr ft^2 °F. This yields a thermal resistance of 4.35 hr ft^2 °F/Btu per inch of thickness.

Phenolic foam, which has a maximum service temperature of 275°F, is very low in water vapor permeability and water absorptance. The material does not exhibit changes in thermal performance with aging and temperature cycling. It has a neutral pH of 6.5 to 7.5. According to manufacturers' literature, phenolic foam has a flame spread rating of 25 or less and a smoke developed rating of 50 or less.

Primary applications for phenolic foam are to control condensation drip on chilled water piping and cold water plumbing lines, and to reduce heat flow for hot water plumbing and low pressure steam lines at temperatures less than 275°F.

The properties of phenolic foam are summarized in Table 3-7 on the following page.

Table 3-7

PHENOLIC FOAM

Property	Phenolic Foam
Apparent thermal conductivity (k), Btu in/hr ft^2 °F	0.23
Apparent thermal resistance (R), hr ft^2 °F/Btu in	4.35
Material density (ρ) lb/ft^3	2-3
Specific heat (Cp) Btu/lb-f	unknown
Thermal diffusivity (α) ft^2/hr	unknown
Maximum service temperature °F	275
Coefficient of thermal expansion in/in °F	50×10^{-6}
Closed cell content percent	>75
Water vapor permeability perm-in	high
Water absorption, percent by weight	0.4-0.8
Corrosiveness	none
Fire resistance	
flame spread	0-25
fuel contributed	combustible
smoke developed	0-50
Degradation as a result of	
temperature cycling	shrinks 1.3% when exposed to 104°F for 7 days
vermin	none
moisture	none
fungal/bacterial	none
weathering	none
wind	none
Human factors	
toxicity	none
odor	none
sound absorptivity	N/A
Effect of aging	
dimensional stability	none
thermal performance	none
fire resistance	none

3.1.9 Polystyrene Foam

Polystyrene foam insulation is manufactured in two forms: (a) extruded and (b) molded expanded bead. Foam produced by the extrusion process has a more consistent density, a more uniform appearance, and greater compressive and tensile strength than that produced by the molding process. Extruded density is usually in the range of 1.8 to 2.6 lb/ft^3. The reported k factor is 0.12 Btu in/hr ft^2 °F as manufactured; however, as the air diffuses in, the k factor rises to 0.20 Btu in/hr ft^2 °F. This value (with an equivalent R-value of 5 hr ft^2 °F/Btu per inch thickness) is normally accepted for this material in use.

Extruded polystyrene foam shows a permeability to water vapor of 0.6 to 0.9 perm-in when tested by ASTM-C355-64 and a volumetric water absorptance of 0.5 to 0.7 percent when tested by ASTM-C2842-69.

Molded polystyrene foam has densities in the range of 1.0 to 1.5 lb/ft^3. Because of the molding process, variations of about 10 percent from the average density can be found in a piece of molded polystyrene. The thermal conductivity of this material is directly proportional to density, and is usually in the range of 0.23 to 0.26 Btu in/hr ft^2 °F. This value does not change with age.

The R-value for molded polystyrene foam is lower than the R-value for extruded polystyrene foam because the former has air in the cells while the latter has a mixture of air and fluorocarbon gas. Water vapor permeability for the molded material is reported to be 1.2 to 3.0 perm-in when tested by ASTM-C355, and water absorption less than 2 percent by weight when tested by ASTM-C272.

Other properties of polystyrene insulation are independent of the manufacturing process. Because polystyrene is combustible, it must be covered with a flame resistant covering such as gypsum board. It must also be protected from direct exposure to ultraviolet light, which causes yellowing. Insulating properties, however, are not affected by short-term exposure to UV light. The maximum service temperature of polystyrene is 165°F; exposure to higher temperatures will cause the plastic to soften. According to survey responses, there are no effects of cycling or weathering on the insulation in the service temperature range. Polystyrene does not promote the growth of fungus or bacteria, and contains no sustenance for vermin. This insulation has no odor, and is noncorrosive.

Polystyrene foam insulating boards and sheathing are used in residential, commercial and industrial applications. When used as an external sheathing material, the entire area of the building envelope may be insulated, thus reducing the heat loss through the more conductive structural members. Plastic foam sheathings provide a barrier to air infiltration superior to conventional sheathings; however, they are nonstructural materials with low nail-holding capabilities. Other uses for polystyrene insulation include building foundation, perimeter slab insulation, roofing insulation, cold storage facilities and other industrial applications.

The properties of polystyrene foam are summarized in Table 3-8 on the following page.

Table 3-8

POLYSTYRENE FOAM

Property	Polystyrene Foam (extruded)	Polystyrene Foam (bead, board)
Apparent thermal conductivity (k), Btu in/hr ft^2 °F	0.20	0.25
Apparent thermal resistance (R) hr ft^2 F/Btu in	5.0	4.0
Material density (ρ) lb/ft^3	1.8-2.6	1.5
Specific heat (Cp) Btu/lb-f	0.27	0.27
Thermal diffusivity (α) ft^2/hr	0.034	0.052
Maximum service temperature °F	180	180
Coefficient of thermal expansion in/in °F	35×10^{-6}	35×10^{-6}
Closed cell content percent	85	85
Water vapor permeability perm-in	0.6-0.9	1.2-3.0
Water absorption percent by weight	0.5-0.7	3.8-4.0
Corrosiveness	none	none
Fire resistance		
flame spread	5-25	5-25
fuel contributed	5-80	5-80
smoke developed	10-400	10-400
Degradation as a result of		
temperature cycling	none	none
vermin	none	none
moisture	none	slight
fungal/bacterial	none	none
weathering	UV degrades	UV degrades
wind (convection)	none	none
Human factors		
toxicity	nontoxic	nontoxic
odor	none	none
sound absorptivity	medium	medium
Effect of aging		
dimensional stability	none	none
thermal performance	none	none
fire resistance	combustible	combustible

3.1.10 Polyurethane/Polyisocyanurate Foam

Polyurethane and polyisocyanurate foams are fluorcarbon-blown materials which possess a rigid structure upon curing. These foams are available as precast boardstock, with or without felt surfacing, and as either foamed-in-place or sprayed-in-place insulation.

These materials have a thermal conductivity (k factor) of 0.11 to 0.15 Btu in/ft hr°F when new, and a density of 2.0 lb/ft^3. The sprayed-in-place foam is readily available in densities of 2.0 to 3.0 lb/ft^3. The closed cell content of these rigid foams is approximately 90 percent. The fluorocarbon gas (usually Freon-11) within the cells has a significantly lower thermal conductivity than air, which explains the low k factor of the material.

It is known that "as manufactured" foam will have values of 0.11 to 0.12 Btu in/hr ft^2 °F; however, the thermal conductivity will increase as the foam ages and air diffuses into the cells. This process is reduced or eliminated when a relatively airtight facing is used on the foam. The ASTM Standard Specification C-591-69, for rigid preformed cellular urethane thermal insulation shows initial values of 0.11 to 0.12 Btu in/hr ft^2 °F, and values of 0.16 to 0.17 for material aged over 300 days.

Polyurethane and polyisocyanurate foams show dimensional change upon curing and aging. The degree of expansion or shrinkage is related to conditions of temperature and humidity and the duration of exposure to extreme conditions. For polyurethane, results of ASTM-D-2126 Procedure F (160°F and 100% RH) indicate a change in volume of up to 12 percent after 14 days. For polyisocyanurate, results with this same test indicate a 3 percent change in volume after 14 days. Because of the high closed cell content, water absorption and premeability are very low; permeability is typically 2 to 3 perm-in. Polyurethane and polyisocyanurate foams are resistant to fungal and bacterial growth, and are nontoxic except during fires.

Polyurethane and polyisocyanurate foams are flammable and must be covered with a fire retardant material when used for thermal insulation in most applications. Certain polyisocyanurate foams have been approved for exposed use in some industrial or commercial buildings. Typical burning characteristics for polyurethane are a flame spread rating of 25 to 75, fuel contributed value of 10 to 25, and smoke developed rating of 155 to over 500. For polyisocyanurate, the flame spread rate is less than 25, fuel contributed value is less than 5, and smoke developed rating is 55 to 200. Most compositions of these foams begin to decompose above 250°F.

Polyurethane or polyisocyanurate board stock can be used as frame sheathing in building construction to provide insulation over the whole building frame, thus minimizing the effect of the more conductive structural members. To allow escape of water vapor which has penetrated the inner face vapor barrier, one major manufacturer of polyisocyanurate foam sheathing specifies vent strips. This may lessen the benefits of reduced air infiltration.

In commercial or industrial buildings, rigid polyurethane and polyisocyanurate foams are used primarily as roof insulation, floor and foundation insulation, cavity wall insulation, and interior and exterior wall insulation. These foams are also used in residential construction, principally as sheathing. Usually, they are used in combination with a reflective surface on the outside of the insulation sheathing. Composites are available for building insulation retrofits; these are polyurethane foam board stock and gypsum board laminated together and attached to the inside wall.

The properties of polyurethane and polyisocyanurate foams are summarized in Table 3-9 on the following page.

Table 3-9

POLYURETHANE/POLYISOCYANURATE FOAM

Property	Polyurethane foams	Polyisocyanurate foam
Apparent thermal cond-tivity (k), Btu in/hr ft^2 °F	0.16	0.16
Apparent thermal resis-tance (R), hr ft^2 °F/Btu in	6.25	6.25
Material density (ρ) P_1 lb/ft^3	2.00	2.00
Specific heat (Cp) Btu/lb °F	0.38	0.38
Thermal diffusivity (α) ft^2/hr	0.0175	0.0175
Maximum service temp-erature °F	250	250
Coefficient of thermal expansion in/in °F	50×10^{-6}	50×10^{-6}
Closed cell content percent	90	90
Water vapor permeability perm-in	2-3	2-3
Water absorption, percent by weight	1.3	1.3
Corrosiveness	none	none
Fire resistance		
flame spread	25-75	0-25
fuel contributed	10-25	0-5
smoke developed	155-500	55-200
Degradation as a result of		
temperature cycling	none	none
vermin	none	none
moisture	negligible	negligible
fungal/bacterial	none	none
weathering	none	none
wind	none	none
Human factors		
toxicity	produces toxic gases when burned	produces toxic gases when burned
odor	none	none
sound absorptivity	medium	medium
Effect of aging		
dimensional stability	none	none
thermal performance	0.11 new 0.16 aged 300 days	0.11 new 0.16 aged 300 days
fire resistance	none	none

3.1.11 Urea-Formaldehyde Foam

Urea-formaldehyde is one of the oldest of the cellular plastics. Discovered in 1933, it has been commercially available in the United States since the 1950s. However, the use of urea-formaldehyde foam insulation (UFFI) has not been very extensive and it appears to be one of the least understood insulating materials, especially in regard to its installation. Since the advent of the energy crisis and the emphasis on conserving energy, its use has been rapidly increasing. The primary uses of urea-formaldehyde have been in retrofitting existing walls in residences, and within the cavities of new masonry walls in both residential and commercial buildings.

The recommended apparent thermal conductivity for urea-formaldehyde foam is 0.24 Btu in/ft^2 hr °F, which yields a thermal resistance of 4.17 ft^2 hr °F/Btu per inch of thickness at a nominal density of 0.60 to 0.90 lb/ft^3.

UFFI has a maximum service temperature of 392°F, after which decomposition takes place. Closed cell content of the material is between 60 and 80 percent; its water vapor permeability is not clearly defined, but is between 4.5 and 100 perm-in. The material is combustible; however, its flame spread classification should not exceed 25.

UFFI has shown a linear shrinkage of between 0.8 and 4 percent during curing. Recent National Bureau of Standards tests have shown extremely high shrinkage rates and a high degree of cell reticulation when the material is subjected to an environment with high temperature and high humidity. HUD specifications recommend a 28 percent degradation in thermal performance to account for shrinkage effects.

The quality of installation of UFFI is critical to its acceptability as an insulation material. Installation techniques have been developed by the major resin producers, who have also initiated training programs for their applicators. Conformance to these installation guidelines should reduce the risk of faulty applications.

Excessive shrinkage and the possibility of lingering formaldehyde vapors within the insulated structure are among the effects of improper installation. If a major odor problem does occur, it could be costly to fix, especially if the UFFI has to be removed.

Because of the formaldehyde problem, many municipalities and state governments have banned the use of urea-formaldehyde foam insulation in public buildings. On February 22, 1982, the Federal Consumer Product Safety Commission voted to ban further installation of UFFI in buildings on the grounds that UFFI "presents an unreasonable risk of injury to the consumer because of the acute and chronic toxicity of the formaldehyde gas it releases [1]." Formaldehyde fumes reportedly can create skin irritation and breathing problems, and formaldehyde has caused cancer in animals. Evidence of unsafe formaldehyde levels occurring even in homes in which the foam was properly installed was presented to the Commission. The ban is expected to be challenged in court; however, other federal agencies (EPA, OSHA, FDA) have found no human health reason to regulate formaldehyde, and the American Cancer Society has no convincing evidence indicating formaldehyde is a carcinogen in humans.

At present, no standards or specifications exist in the United States for UFFI. No means are available to assure a minimum level of quality; however, the ASTM (American Society for Testing Materials) has appointed a task group to initiate development of a standard for this foam.

The properties of urea-formaldehyde foam are summarized in Table 3-10 on the following page.

Table 3-10

UREA-FORMALDEHYDE FOAM

Property	Urea-Formaldehyde Foam
Apparent thermal conductivity (k), Btu in/hr ft^2 °F	0.24
Apparent thermal resistance (R), hr ft^2 F/Btu in	4.17
Material Density (ρ) lb/ft^3	0.6 - 0.9
Specific heat (C_p) Btu/lb f	0.30
Thermal diffusivity (α) ft^2/hr	0.11
Maximum service temperature °F	decomposes at 392°F
Coefficient of thermal expansion in/in °F	50×10^{-6}
Closed cell content percent	60-80
Water vapor permeability perm-in	4.5 - 100
Water absorption, percent by weight	32%
Corrosiveness	none
Fire resistance	
flame spread	0-25
fuel contributed	0-30
smoke developed	0-10
Degradation as a result of	
temperature cycling	not established
vermin	none
moisture	not established
fungal/bacterial	none
weathering	susceptible to photodegradation
wind	none
Human factors	
toxicity	no more toxic upon ignition than burning wood
odor	only danger from possibility of inhaled gases
sound absorptivity	medium
Effect of aging	
dimensional stability	1-7% shrinkage during curing
thermal performance	shrinkage may cause degradation
fire resistance	none

3.1.12 Vermiculite

Vermiculite insulation is made from mica-like hydrated silicate particles. In the production process, the particles are heated quickly to temperatures between 700 and 1000°C; this causes the occluded water to vaporize and exfoliate the micaceous layers.

By controlling the degree of exfoliation, a density range of 4 to 10 lb/ft^3 is typically produced in the expanded material. The lower density material has an average particle size of 6.5 millimeters.

The normal thermal conductivity of exfoliated vermiculite is 0.33 to 0.46 Btu in/hr ft^2 °F at ambient temperatures, which translates to R-values of 3.0 to 2.4 per inch of thickness.

Vermiculite is treated to ensure water repellency. It is noncombustible and has a maximum service temperature of 760°F. Because it is an inorganic material, it is resistant to bacterial and fungal growth, and vermin. The material is not affected by age, temperature, or humidity. Vermiculite is chemically inert, and therefore noncorrosive, and exudes no odors.

Common applications of vermiculite include its use as a general purpose pouring type insulation to fill block and cavity walls. Higher density material is used as plaster aggregate and as a high-temperature loose fill insulation.

Vermiculite is also mixed with portland cement and sometimes sand to produce vermiculite concrete. Densities of this material usually range from 19 to 35 lb/ft^3; higher densities result in higher thermal conductivities, which range from 0.64 to 0.79 Btu in/hr ft^2 °F (R-value of 1.6 to 1.3).

The properties of vermiculite are summarized in Table 3-11 on the following page.

Table 3-11

VERMICULITE

Property	Vermiculite
Apparent thermal conductivity (k), Btu in/hr ft^2 °F	0.46
Apparent thermal resistance (R), hr ft^2 °F/Btu in	2.08
Material density (ρ) lb/ft^3	4-10
Specific heat (C_p) Btu/lb °F	0.32
Thermal diffusivity (α) ft^2/hr	0.039
Maximum service temperature °F	760
Coefficient of thermal expansion in/in °F	N/A
Closed cell content percent	N/A
Water vapor permeability perm-in	N/A
Water absorption, percent by weight	300
Corrosiveness	none
Fire resistance	
flame spread	0
fuel contributed	0
smoke developed	0
Degradation as a result of	
temperature cycling	none
vermin	none
moisture	none
fungal/bacterial	none
weathering	none
wind	none
Human factors	
toxicity	none
odor	none
sound absorptivity	medium
Effect of aging	
dimensional stability	none
thermal performance	none
fire resistance	none

3.2 Comparison of Selected Generic Building Insulation Materials

Table 3-12 on the following page summarizes the R-values per inch, conductivity, and density of the thermal insulations presented in Section 3.1, and also lists the numbers of applicable standards of the American Society for Testing and Materials (ASTM: see also Appendix A) and federal specifications.

Table 3-13 on page 73 qualitatively summarizes the advantages, disadvantages, and limitations of the various forms of insulation included in this document.

The cost of purchasing and installing insulation is often the criterion used to decide between insulations which are otherwise acceptable for a particular job. However, because costs change frequently, installed cost per R-value is listed in Table 3-12 only as low, moderate, or high for building, industrial, and piping applications. For 1981, these costs cover the ranges shown below:

1981 Installed Cost Ranges*

	Application		
	Buildings	Industrial	Piping**
Low	1.5-4.5	20-35	30-320
Moderate	4.5-10	35-150	40-660
High	10-25	150-500	120-1250

*1981¢/ft^2-R for building and industrial categories, 1981¢/linear foot-R for piping.

**Piping ranges overlap because cost per R-value depends on pipe size and insulation thickness: thick insulation on a small pipe has a low cost; thin insulation on a large pipe has a high cost. When comparing various pipe insulation materials, compare the cost of the same thicknesses on the same size pipe; the low moderate-high relationships are then valid.

Cost is also listed as an advantage or disadvantage in Table 3-13.

Table 3-12

GENERIC INSULATION PROPERTIES
AND SPECIFICATION SUMMARY*

Generic Insulation	R per Inch	k_{App}	ρ	Federal Specifications and/or ASTM Standards	Relative Cost Application ** Bldg	Ind	Piping
Batts and Blankets							
Glass Fiber	3.2	0.31	0.6-1.0	HH-I-521E C 553, C 592 C 665, C 892	Low		
Mineral Fiber	3.2	0.31	1.5-2.5	HH-I-521E C 553, C 592 C 665, C 892	Low		
Loose Fill							
Cellulose	3.2-3.7	0.27-0.31	2.2-3.2	HH-I-515D C 739	Low-Mod		
Glass Fiber	2.9	0.35	0.6-1.0	HH-I-103A C 764	Low		
Mineral Fiber	3.1	0.32	1.5-2.5	HH-I-1030A C 764	Low		
Perlite	2.5-3.7	0.27-0.40	2-11	HH-I-574B C 549	Mod		
Vermiculite	2.1-2.3	0.44-0.47	4-12	HH-I-585 C 516	Mod		
Foamed In Place							
Polyurethane and Polyisocyanurate Foam	5.9-6.2	0.16-0.17	2.0		Mod		
Urea-formaldehyde Foam	4.2	0.24	0.6-0.9		Low		

*The units of thermal resistance per inch (R per inch) are hr ft^2 °F/Btu in.
The units of apparent thermal conductivity (k_{App}) are Btu in/hr ft^2 °F.
The units of density (ρ) are lb/ft^3.

**Application categories are Buildings (Bldg), Industrial (Ind), and Piping (Pipe).

Table 3-12 (Cont'd)

GENERIC INSULATION PROPERTIES
AND SPECIFICATION SUMMARY*

Generic Insulation	R per Inch	k_{App}	ρ	Federal Specifications and/or ASTM Standards	Relative Cost Application ** Bldg	Ind	Piping
Blocks, Boards, and Pipe Insulation							
Calcium Silicate	2.6	0.38	13	C 656		High	High
Cellular Glass	2.6	0.38	8.5	HH-I-551E C 552	High		
Elastomeric Foam	3.8	0.26	4.5	C 534			Low
Glass Fiber	4.0	0.25	4.9	MIL-I-742 C 720 C 726 C 892	High	Mod-High	Low
Mineral Fiber	3.6	0.28	9-11	HH-I-55B C 612 C 726		Mod-High	
Perlite	2.6	0.38	10	C 610 C 728	Mod-High		
Phenolic Foam	4.3	0.23	2.0-3.0				Low
Polystyrene Foam	extruded: 5.0 molded: 3.9-4.4	extruded: 0.20 molded: 0.23-0.26	extruded: 1.8-2.6 molded: 1.5	HH-I-524B MIL-P-40619 MIL-P-43110 C 578	Mod-High	Mod	Mod
Polyurethane and Polyisocyanurate Foam	unfaced: 5.8-6.2 imper-meable skin faced: 7.1-7.7	unfaced: 0.16-0.17 imper-meable skin faced: 0.13-0.14	2.0	HH-I-530A C 591	Mod-High	Mod	Mod

*The units of thermal resistance per inch (R per inch) are hr ft^2 °F/Btu in.
The units of apparent thermal conductivity (k_{App}) are Btu in/hr ft^2 °F.
The units of density (ρ) are lb/ft^3.

**Application categories are Buildings (Bldg), Industrial (Ind), and Piping (Pipe).

Table 3-13

INSULATION COMPARISONS

GENERIC DESCRIPTION	ADVANTAGES	DISADVANTAGES	LIMITATIONS
Calcium Silicate	Noncombustible; high maximum service temperature. Will dry out without physical or thermal degradation; stable when dry.	Moderate thermal conductivity; high water vapor permeability and water absorption. High cost; high density.	Not for use in the presence of water such as in direct burial; will not take physical abuse when wet.
Cellular Glass	Noncombustible; impermeable to moisture; stable; high compressive strength; high service temperature nontoxic.	Moderate thermal conductivity; high cost.	May fracture during repeated freeze-thaw cycling in presence of water.
Cellulosic Fiber	Low thermal conductivity; low to moderate cost; nontoxic.	Combustible, but combustibility reduced somewhat with fire retardants; high water vapor permeability and water absorption; may settle with aging.	Fire retardant used may accelerate corrosion of steel, Al, and Cu; certain states and municipalities have limited its use in public buildings.
Elastomeric (Rubber) Foam	Low thermal conductivity; low cost; easy to install and seal; low water vapor permeability; stable; nontoxic.	Combustible; develops high smoke content on burning; UV sensitive.	Maximum service temperature limited to 220°F; code requirements may prevent its use inside buildings as a result of combustibility.
Glass Fiber	Low thermal conductivity; low cost in batt, blanket, and loose fill forms and for pipe insulation; noncombustible without facings; stable; nontoxic; low water absorption.	Batt facings may be combustible and binders may burn out; moving air may degrade thermal performance unless used with air infiltration barrier; moderate to high cost in board form.	Requires vapor barrier, as permeability is high when used with binder or facing; has moderate temperature limitation.

Table 3-13 (Con't)

INSULATION COMPARISONS

GENERIC DESCRIPTION	ADVANTAGES	DISADVANTAGES	LIMITATIONS
Mineral Fiber	Low thermal conductivity; low cost in batt, blanket, and loose fill forms; noncombustible without facing; stable; nontoxic.	Batt facings may be combustible and binders may burn out; moving air may degrade thermal performance unless used with air infiltration barrier; high water absorption; moderate to high cost in block and board forms.	Requires vapor barrier because of high permeability; when used with binder or facing, has moderate temperature limitations.
Perlite	Low thermal conductivity; noncombustible; stable; nontoxic.	High moisture permeability and water absorption; moderate cost.	Requires a vapor barrier and physical containment device in loose fill.
Phenolic Foam	Low thermal conductivity; noncombustible; low water absorption; stable; low cost.	UV sensitive; high water vapor permeability.	Maximum service temperature limited to 275°F.
Polystyrene Foam	Low thermal conductivity; may provide infiltration and moisture seal; low water vapor permeability and low water absorption; stable; nontoxic.	Moderate cost; combustible; low service temperature.	Maximum service temperature limited to 180°F.
Polyurethane/ Polyisocyanurate Foam	Lowest thermal conductivity; provides infiltration and moisture seal; low water vapor permeability and low water absorption.	Moderate to high cost; combustible; requires fire barrier when installed to meet building codes.	Will produce toxic fumes while burning.

Table 3-13 (Con't)

INSULATION COMPARISONS

GENERIC DESCRIPTION	ADVANTAGES	DISADVANTAGES	LIMITATIONS
Urea-Formaldehyde Foam	Low thermal conductivity; easy to install; noncombustible; low cost.	May liberate formaldehyde gas until cured; may shrink during curing, especially under high humidity environment; high water vapor permeability and water absorption.	Many municipalities, state governments, and the CPSC have banned its use in public buildings; requires trained and experienced installer; removal of the foam may be required if formaldehyde fume problems develop.
Vermiculite	High maximum service temperature; noncombustible; non-toxic; stable.	High thermal conductivity; high water absorption; moderate cost.	Requires a physical containment device.

3.3 Thermophysical Properties of Building and Industrial Insulation and Materials

The thermophysical properties of building insulation and materials are presented in Tables 3-14, 3-15, and 3-16 beginning on pages 77, 82, and 87 respectively. Properties of industrial insulation are presented in Table 3-17 on page 92. The values shown are represented as nominal values; when available, these were selected from References [2 - 6]. If not available in the references, manufacturers' data were used on a very limited basis as the source for thermal properties.

It is anticipated that the thermal properties from these tables will be used to compute overall heat transfer coefficients for building assemblies. With this in mind, it must be noted that the values in these tables are representative of properties obtained in idealized laboratory conditions; their installed thermal performance may deviate from the results as calculated. In actual situations, significant changes in installed heat transmission coefficients may be attributed to such factors as material shrinkage, moisture effects, framing members, settling, and improper or faulty installation.

The building thermal properties are presented in three different ways in order that comparisons can be easily made. Table 3-14 shows building insulation and materials properties sorted alphabetically within major building category.

Thermal insulation building categories shown are:

- Air films.
- Air layers (non-reflective and reflective).
- Building boards.
- Building insulation.
- Flooring materials.
- Masonry materials.
- Masonry units.
- Plastering materials.
- Roofing materials.
- Siding materials.
- Woods.

Table 3-15 shows the same properties as Table 3-14, but is sorted within each category in order of increasing thermal conductivity. Table 3-16 shows the properties sorted in order of increasing density. Table 3-17 shows representative values of the density, maximum service temperature, and thermal conductivity as a function of temperature for various industrial insulation materials. The categories shown are:

- Blankets and felts.
- Blocks, board, and pipe insulation.
- Loose fill insulation.

Table 3-14
THERMOPHYSICAL PROPERTIES OF TYPICAL
INSULATION AND BUILDING MATERIALS
(IN ALPHABETICAL ORDER BY CATEGORY)

DESCRIPTION	THERMAL COND(K) (BTU/ HR-FT-F)	THERMAL COND(K) (BTU-IN/ HR-SQFT-F)	DENSITY (LB/CUFT)	SPECIFIC HEAT (BTU/LB-F)	RESISTANCE (SQFT-HR-F/ BTU)	RESISTANCE PER INCH (SQFT-HR-F/ BTU-IN)
AIR FILMS						
MOVING AIR,15MPH WIND,WINTER					0.17	
MOVING AIR,7.5MPH WIND,SUMMER					0.25	
STILL AIR,HORIZ SURFACE,HEAT FLOW DOWN					0.92	
STILL AIR,HORIZ SURFACE,HEAT FLOW UP					0.61	
STILL AIR,VERTICAL SURFACE					0.68	
AIR LAYERS						
HORIZ REFLECT SURFACE,SUMMER,LYR=.5IN					1.89	
HORIZ REFLECT SURFACE,SUMMER,LYR=.75IN					2.41	
HORIZ REFLECT SURFACE,SUMMER,LYR=1.5IN					3.27	
HORIZ REFLECT SURFACE,SUMMER,LYR=3.5IN					4.09	
HORIZ REFLECT SURFACE,WINTER,LYR=.5IN					1.60	
HORIZ REFLECT SURFACE,WINTER,LYR=.75IN					1.70	
HORIZ REFLECT SURFACE,WINTER,LYR=1.5IN					1.81	
HORIZ REFLECT SURFACE,WINTER,LYR=3.5IN					1.95	
HORIZONTAL SURFACE,SUMMER,LAYER=.5IN					0.92	
HORIZONTAL SURFACE,SUMMER,LAYER=.75IN					1.02	
HORIZONTAL SURFACE,SUMMER,LAYER=1.5IN					1.15	
HORIZONTAL SURFACE,SUMMER,LAYER=3.5IN					1.24	
HORIZONTAL SURFACE,WINTER,LAYER,1.5IN					0.89	
HORIZONTAL SURFACE,WINTER,LAYER=.5IN					0.84	
HORIZONTAL SURFACE,WINTER,LAYER=.75IN					0.87	
HORIZONTAL SURFACE,WINTER,LAYER=3.5IN					0.93	
VERTICAL REFLECTIVE SURF,LAYER=.5IN					2.35	
VERTICAL REFLECTIVE SURF,LAYER=1.5IN					2.39	
VERTICAL REFLECTIVE SURF,LAYER=3.5IN					2.32	
VERTICAL REFLECTIVE SURFACE,LAYER=.5IN					1.88	
VERTICAL SURFACE,LAYER=.5IN					0.91	
VERTICAL SURFACE,LAYER=.75IN					1.01	
VERTICAL SURFACE,LAYER=1.5IN					1.02	
VERTICAL SURFACE,LAYER=3.5IN					1.01	
BUILDING BOARD						
ASBESTOS-CEMENT BOARD	0.345	4.140	120.0	0.24		0.242
GYPSUM BOARD	0.093	1.111	50.0	0.26		0.900
HARDBOARD,HI-DENS,SERV TEMP UNDERLAY	0.068	0.820	55.0	0.32		1.220
HARDBOARD,HIGH DENSITY,STD TEMP	0.083	1.000	63.0	0.32		1.000
HARDBOARD,MEDIUM DENSITY	0.054	0.653	50.0	0.31		1.532

Table 3-14 (cont.)

THERMOPHYSICAL PROPERTIES OF TYPICAL INSULATION AND BUILDING MATERIALS (IN ALPHABETICAL ORDER BY CATEGORY)

DESCRIPTION	THERMAL COND(K) (BTU/ HR-FT-F)	THERMAL COND(K) (BTU-IN/ HR-SQFT-F)	DENSITY (LB/CUFT)	SPECIFIC HEAT (BTU/LB-F)	RESISTANCE (SQFT-HR-F/ BTU)	RESISTANCE PER INCH (SQFT-HR-F/ BTU-IN)
BUILDING BOARD (Cont.)						
NAIL BASE SHEATHING	0.036	0.438	25.0	0.31		2.283
PARTICLEBOARD, LOW DENSITY	0.045	0.540	37.0	0.31		1.852
PARTICLEBOARD,HIGH DENSITY	0.098	1.180	62.5	0.31		0.848
PARTICLEBOARD,MEDIUM DENSITY	0.078	0.940	50.0	0.31		1.064
PARTICLEBOARD,UNDERLAYMENT	0.180	2.155	40.0	0.29		0.464
PLYWOOD (DOUGLAS FIR)	0.080	0.960	34.0	0.29		1.042
SHEATHING,INTERMEDIATE DENSITY	0.034	0.410	22.0	0.31		2.437
SHEATHING,REGULAR DENSITY	0.032	0.379	18.0	0.31		2.637
SHINGLE BACKER	0.033	0.398	18.0	0.31		2.510
BUILDING INSULATION						
CELLULAR GLASS	0.032	0.380	8.5	0.18		2.629
CELLULOSE,LOOSE FIBER	0.024	0.288	3.2	0.33		3.472
GLASS FIBER,BATT	0.026	0.318	1.0	0.20		3.145
GLASS FIBER,LOOSE FILL	0.029	0.348	1.0	0.20		2.874
GLASS FIBER,ORGANICALLY BONDED	0.021	0.250	5.0	0.23		4.006
MINERAL FIBER,BATTS	0.026	0.318	2.0	0.20		3.145
MINERAL FIBER,LOOSE FILL	0.027	0.322	1.7	0.17		3.109
MINERAL FIBER,RESIN BINDER	0.024	0.290	3.5	0.20		3.444
MINERAL FIBERBD,WET FELTED,ACOUS TILE	0.029	0.350	18.0	0.19		2.854
MINERAL FIBERBD,WET FELTED,ROOF INSUL	0.028	0.340	17.0	0.19		2.945
MINERAL FIBERBD,WET MOLDED,ACOUS TILE	0.035	0.420	23.0	0.14		2.381
MINERAL FIBERBOARD,PREFORMED	0.024	0.290	15.0	0.17		3.444
PERLITE,EXPANDED LOOSE FILL	0.023	0.276	5.0	0.30		3.623
PERLITE,EXPANDED,ORG BOND ROOF INSUL	0.030	0.360	10.0	0.25		2.778
POLYISOCYANURATE,R-11 EXPANDED	0.013	0.160	2.0	0.38		6.266
POLYSTYRENE EXTRUDED,SMOOTH SKIN	0.017	0.200	1.8	0.29		4.990
POLYSTYRENE,MOLDED BEADS	0.021	0.250	1.5	0.29		4.006
POLYURETHANE,R-11 EXP,FOIL FACED	0.012	0.140	2.0	0.38		7.123
POLYURETHANE,R-11 EXPANDED	0.013	0.160	2.0	0.38		6.266
UREA FORMALDEHYDE	0.020	0.240	0.6	0.30		4.167
VERMICULITE,EXPANDED LOOSE FILL	0.040	0.480	9.0	0.32		2.083
WOOD FIBERBOARD ACOUSTIC TILE	0.033	0.400	23.0	0.31		2.503
FLOORING MATERIALS						
CARPET AND FIBER PAD					2.08	
CARPET AND RUBBER PAD					1.23	
CORK TILE					0.28	
TILE,TERRAZO,AHPHLT,VINYL,LINEOLEUM,ETC					0.05	

Table 3-14 (cont.)

THERMOPHYSICAL PROPERTIES OF TYPICAL
INSULATION AND BUILDING MATERIALS
(IN ALPHABETICAL ORDER BY CATEGORY)

DESCRIPTION	THERMAL COND (K) (BTU/ HR-FT-F)	THERMAL COND (K) (BTU-IN/ HR-SQFT-F)	DENSITY (LB/CUFT)	SPECIFIC HEAT (BTU/LB-F)	RESISTANCE (SQFT-HR-F/ BTU)	RESISTANCE PER INCH (SQFT-HR-F/ BTU-IN)
MASONRY MATERIALS						
CONCRETE,HEAVYWEIGHT,DRIED AGGREGATE	0.750	9.000	140.0	0.22		0.111
CONCRETE,HEAVYWEIGHT,UNDRIED AGGREGATE	1.000	12.000	140.0	0.22		0.083
CONCRETE,LIGHTWEIGHT,INSULATING	0.075	0.901	30.0	0.21		1.110
CONCRETE,LIGHTWEIGHT,STRUCTURAL	0.439	5.263	110.0	0.21		0.190
GRANITE	1.000	12.000	165.0	0.20		0.083
LIMESTONE	0.720	8.640	155.0	0.22		0.116
MARBLE	1.600	19.200	160.0	0.19		0.052
SANDSTONE	0.940	11.280	140.0	0.17		0.089
STUCCO	0.417	5.000	116.0	0.21		0.200
MASONRY UNITS						
BRICK,12IN COMMON	0.400	4.800	120.0	0.20		0.208
BRICK,3IN FACE	0.750	9.000	130.0	0.22		0.111
BRICK,4IN COMMON	0.400	4.800	120.0	0.20		0.208
BRICK,4IN FACE	0.750	9.000	130.0	0.22		0.111
BRICK,8IN COMMON	0.400	4.800	120.0	0.20		0.208
CLAY TILE,10IN,2CELLS	0.375	4.499	70.0	0.21		0.222
CLAY TILE,12IN,3CELLS	0.400	4.800	70.0	0.21	2.00	0.208
CLAY TILE,3IN,1CELL	0.312	3.750	70.0	0.21		0.267
CLAY TILE,4IN,1CELL	0.300	3.599	70.0	0.21		0.278
CLAY TILE,6IN,2CELLS	0.330	3.960	70.0	0.21		0.253
CLAY TILE,8IN,2CELLS	0.360	4.320	70.0	0.21		0.231
CLAY TILE,PAVER	1.042	12.499	120.0	0.21		0.080
CMU,4IN,HW,CONCRETE FILLED	0.757	9.090	140.0	0.21	0.44	0.110
CMU,4IN,HW,CONCRETE & LOOSE FILL INS(2)	0.477	5.726	115.0	0.21	0.70	0.175
CMU,4IN,HW,HOLLOW	0.469	5.633	101.0	0.21	0.71	0.178
CMU,4IN,HW,LOOSE FILL INSULATION	0.300	3.601	103.0	0.21	1.11	0.278
CMU,4IN,HW,PARTIALLY FILLED CONCRETE(1)	0.584	7.013	114.0	0.21	0.57	0.143
CMU,4IN,LW,CONCRETE FILLED	0.369	4.434	104.0	0.20	0.90	0.226
CMU,4IN,LW,CONCRETE & LOOSE FILL INS(2)	0.208	2.495	79.0	0.20	1.60	0.401
CMU,4IN,LW,HOLLOW	0.222	2.666	65.0	0.20	1.50	0.375
CMU,4IN,LW,LOOSE FILL INSULATION	0.127	1.525	67.0	0.20	2.62	0.656
CMU,4IN,LW,PARTIALLY FILLED CONCRETE(1)	0.281	3.370	78.0	0.20	1.19	0.297
CMU,6IN,HW,CONCRETE FILLED	0.757	9.090	140.0	0.21	0.66	0.110
CMU,6IN,HW,CONCRETE & LOOSE FILL INS(2)	0.424	5.086	104.0	0.21	1.18	0.197
CMU,6IN,HW,HOLLOW	0.555	6.666	85.0	0.21	0.90	0.150
CMU,6IN,HW,INSULATION FILLED	0.222	2.666	88.0	0.21	2.25	0.375
CMU,6IN,HW,PARTIALLY FILLED CONCRETE(1)	0.612	7.343	104.0	0.21	0.82	0.136
CMU,6IN,LW,CONCRETE & LOOSE FILL INS(2)	0.193	2.315	74.0	0.21	2.59	0.432
CMU,6IN,LW,CONCRETE FILLED	0.382	4.583	110.0	0.21	1.31	0.218
CMU,6IN,LW,HOLLOW	0.278	3.332	55.0	0.21	1.80	0.300

Table 3-14 (cont.)

THERMOPHYSICAL PROPERTIES OF TYPICAL INSULATION AND BUILDING MATERIALS (IN ALPHABETICAL ORDER BY CATEGORY)

DESCRIPTION	THERMAL COND (K) (BTU/ HR-FT-F)	THERMAL COND (K) (BTU-IN/ HR-SQFT-F)	DENSITY (LB/CUFT)	SPECIFIC HEAT (BTU/LB-F)	RESISTANCE (SQFT-HR-F/ BTU)	RESISTANCE PER INCH (SQFT-HR-F/ BTU-IN)
MASONRY UNITS (Cont.)						
CMU,6IN,LW,LOOSE FILL INSULATION	0.098	1.182	57.0	0.21	5.08	0.846
CMU,6IN,LW,PARTIALLY FILLED CONCRETE(1)	0.319	3.827	73.0	0.21	1.57	0.261
CMU,6IN,MW,CONCETE FILLED	0.444	5.332	119.0	0.21	1.13	0.188
CMU,6IN,MW,CONCRETE & LOOSE FILL INS(2)	0.226	2.711	84.0	0.21	2.21	0.369
CMU,6IN,MW,HOLLOW	0.357	4.285	65.0	0.21	1.40	0.233
CMU,6IN,MW,LOOSE FILL INSULATION	0.117	1.399	67.0	0.21	4.29	0.715
CMU,6IN,MW,PARTIALLY FILLED CONCRETE(1)	0.369	4.423	83.0	0.21	1.36	0.226
CMU,8IN,HW,CONCRETE FILLED	0.757	9.090	140.0	0.21	0.88	0.110
CMU,8IN,HW,CONCRETE & LOOSE FILL INS(2)	0.416	4.992	93.0	0.21	1.60	0.200
CMU,8IN,HW,HOLLOW	0.606	7.272	69.0	0.21	1.10	0.138
CMU,8IN,HW,LOOSE FILL INSULATION	0.227	2.726	70.0	0.21	2.93	0.367
CMU,8IN,HW,PARTIALLY FILLED CONCRETE(1)	0.675	8.095	93.0	0.21	0.99	0.124
CMU,8IN,LW,LOOSE FILL INSULATION	0.114	1.369	56.0	0.21	5.84	0.730
CMU,8IN,LW,CONCRETE FILLED	0.436	5.231	115.0	0.21	1.53	0.191
CMU,8IN,LW,CONCRETE & LOOSE FILL INS(2)	0.209	2.514	69.0	0.21	3.18	0.398
CMU,8IN,LW,HOLLOW	0.333	4.000	45.0	0.21	2.00	0.250
CMU,8IN,LW,LOOSE FILL INSULATION	0.096	1.156	48.0	0.21	6.92	0.865
CMU,8IN,LW,PARTIALLY FILLED CONCRETE(1)	0.385	4.615	68.0	0.21	1.73	0.217
CMU,8IN,MW,CONCRETE FILLED	0.496	5.948	123.0	0.21	1.34	0.168
CMU,8IN,MW,CONCRETE & LOOSE FILL INS(2)	0.241	2.896	77.0	0.21	2.76	0.345
CMU,8IN,MW,HOLLOW	0.388	4.651	53.0	0.21	1.72	0.215
CMU,8IN,MW,PARTIALLY FILLED CONCRETE(1)	0.435	5.218	76.0	0.21	1.53	0.192
CMU,12IN MW CONCRETE FILLED	0.481	5.777	121.0	0.21	2.08	0.173
CMU,12IN,HW,CONCRETE FILLED	0.757	9.090	140.0	0.21	1.32	0.110
CMU,12IN,HW,HOLLOW	0.781	9.376	76.0	0.21	1.28	0.107
CMU,12IN,HW,PARTIALLY FILLED CONCRETE(1)	0.777	9.328	98.0	0.21	1.29	0.107
CMU,12IN,LW,CONCRETE FILLED	0.419	5.033	113.0	0.21	2.38	0.199
CMU,12IN,LW,HOLLOW	0.440	5.286	49.0	0.21	2.27	0.189
CMU,12IN,LW,PARTIALLY FILLED CONCRETE(1)	0.427	5.129	70.0	0.21	2.34	0.195
CMU,12IN,MW,HOLLOW	0.496	5.951	58.0	0.21	2.02	0.168
CMU,12IN,MW,PARTIALLY FILLED CONCRETE(1)	0.492	5.903	79.0	0.21	2.03	0.169

(1) ONE CONCRETE FILLED AND REINFORCED CORE EVERY 2 FEET OF WALL LENGTH.
(2) ONE CONCRETE FILLED AND REINFORCED CORE EVERY 2 FEET OF WALL LENGTH WITH REMAINING CORES FILLED WITH LOOSE FILL INSULATION.

DESCRIPTION	THERMAL COND (K) (BTU/ HR-FT-F)	THERMAL COND (K) (BTU-IN/ HR-SQFT-F)	DENSITY (LB/CUFT)	SPECIFIC HEAT (BTU/LB-F)	RESISTANCE (SQFT-HR-F/ BTU)	RESISTANCE PER INCH (SQFT-HR-F/ BTU-IN)
PLASTERING MATERIALS						
CEMENT PLASTER,SAND AGGREGATE	0.417	5.000	116.0	0.20		0.200
GYPSUM PLASTER,LIGHTWEIGHT AGGREGATE	0.133	1.596	45.0	0.20		0.627
GYPSUM PLASTER,PERLITE AGGREGATE	0.125	1.500	45.0	0.32		0.667
GYPSUM PLASTER,SAND AGGREGATE	0.467	5.600	105.0	0.20		0.179
GYPSUM PLASTER,VERMICULITE AGGREGATE	0.142	1.700	45.0	0.22		0.588

Table 3-14 (cont.)

THERMOPHYSICAL PROPERTIES OF TYPICAL
INSULATION AND BUILDING MATERIALS
(IN ALPHABETICAL ORDER BY CATEGORY)

DESCRIPTION	THERMAL COND (K) (BTU/ HR-FT-F)	THERMAL COND (K) (BTU-IN/ HR-SQFT-F)	DENSITY (LB/CUFT)	SPECIFIC HEAT (BTU/LB-F)	RESISTANCE (SQFT-HR-F/ BTU)	RESISTANCE PER INCH (SQFT-HR-F/ BTU-IN)
ROOFING MATERIALS						
ASBESTOS-CEMENT SHINGLES			120.0	0.24	0.21	
ASPHALT ROLL ROOFING			70.0	0.36	0.15	
ASPHALT SHINGLES			70.0	0.30	0.44	
BUILT-UP ROOFING	0.094	1.126	70.0	0.35		0.888
INSULATION,PREFORMED	0.030	0.360	10.0	0.25		2.778
ROOF GRAVEL OR SLAG	0.834	10.008	55.0	0.40		0.100
SLATE			120.0	0.30	0.05	
WOOD SHNGLS,PLAIN AND PLASTIC FILM FCD			32.0	0.31	0.94	
SIDING MATERIALS						
ARCHITECTURAL GLASS	0.440	5.280	169.0	0.20		0.189
ASBESTOS CEMENT SHINGLES			12.0		0.21	
ASBESTOS CEMENT SIDING,0.25IN,LAPPED					0.24	
ASPHALT INSULATION SIDING (0.5 IN BED)					1.46	
ASPHALT ROLL SIDING					0.15	
HARDBOARD SIDING	0.124	1.488	40.0	0.28		0.672
METAL CLAD SHEATHING,HOLLOW-BACKED					0.61	
METAL CLAD SHTH,INS-BD BACKED,0.375IN					1.82	
MTL CLAD SHTH,INS-BD BCKD,0.375IN,FOIL					2.96	
PLYWOOD,0.375IN,LAPPED					0.63	
WOOD SHINGLES,16IN, 7.5IN EXPOSURE					0.87	
WOOD SHINGLES,DOUBLE,16IN, 12IN EXPOSURE					1.19	
WOOD,BEVEL,0.5IN BY 8IN,LAPPED					0.81	
WOOD,BEVEL,0.75IN BY 10IN,LAPPED					1.05	
WOOD,DROP,1IN BY 8IN					0.79	
WOOD,PLUS INS BACKER BOARD,0.3125IN					1.40	
WOODS						
WOOD,CYPRESS	0.056	0.672	29.0	0.65		1.488
WOOD,FIR	0.063	0.756	26.0	0.57		1.323
WOOD,HARD	0.085	1.020	38.0	0.57		0.980
WOOD,OAK	0.085	1.020	38.0	0.57		0.980
WOOD,SOFT	0.065	0.780	32.0	0.65		1.282
WOOD,WHITE PINE	0.065	0.780	32.0	0.67		1.282
WOOD,YELLOW PINE	0.082	0.984	40.0	0.67		1.016

Table 3-15
THERMOPHYSICAL PROPERTIES OF TYPICAL
INSULATION AND BUILDING MATERIALS
(IN ASCENDING ORDER BY CONDUCTIVITY)

DESCRIPTION	THERMAL COND (K) (BTU/ HR-FT-F)	THERMAL COND (K) (BTU-IN/ HR-SQFT-F)	DENSITY (LB/CUFT)	SPECIFIC HEAT (BTU/LB-F)	RESISTANCE (SQFT-HR-F/ BTU)	RESISTANCE PER INCH (SQFT-HR-F/ BTU-IN)
AIR FILMS						
MOVING AIR,15MPH WIND,WINTER					0.17	
MOVING AIR,7.5MPH WIND,SUMMER					0.25	
STILL AIR,HORIZ SURFACE,HEAT FLOW DOWN					0.92	
STILL AIR,HORIZ SURFACE,HEAT FLOW UP					0.61	
STILL AIR,VERTICAL SURFACE					0.68	
AIR LAYERS						
HORIZ REFLECT SURFACE,SUMMER,LYR=.5IN					1.89	
HORIZ REFLECT SURFACE,SUMMER,LYR=.75IN					2.41	
HORIZ REFLECT SURFACE,SUMMER,LYR=1.5IN					3.27	
HORIZ REFLECT SURFACE,SUMMER,LYR=3.5IN					4.09	
HORIZ REFLECT SURFACE,WINTER,LYR=.5IN					1.60	
HORIZ REFLECT SURFACE,WINTER,LYR=.75IN					1.70	
HORIZ REFLECT SURFACE,WINTER,LYR=1.5IN					1.81	
HORIZ REFLECT SURFACE,WINTER,LYR=3.5IN					1.95	
HORIZONTAL SURFACE,SUMMER,LAYER=.5IN					0.92	
HORIZONTAL SURFACE,SUMMER,LAYER=.75IN					1.02	
HORIZONTAL SURFACE,SUMMER,LAYER=1.5IN					1.15	
HORIZONTAL SURFACE,SUMMER,LAYER=3.5IN					1.24	
HORIZONTAL SURFACE,WINTER,LAYER,1.5IN					0.89	
HORIZONTAL SURFACE,WINTER,LAYER=.5IN					0.84	
HORIZONTAL SURFACE,WINTER,LAYER=.75IN					0.87	
HORIZONTAL SURFACE,WINTER,LAYER=3.5IN					0.93	
VERTICAL REFLECTIVE SURF,LAYER=.5IN					2.35	
VERTICAL REFLECTIVE SURF,LAYER=1.5IN					2.39	
VERTICAL REFLECTIVE SURF,LAYER=3.5IN					2.32	
VERTICAL REFLECTIVE SURFACE,LAYER=.5IN					1.88	
VERTICAL SURFACE,LAYER=.5IN					0.91	
VERTICAL SURFACE,LAYER=.75IN					1.01	
VERTICAL SURFACE,LAYER=1.5IN					1.02	
VERTICAL SURFACE,LAYER=3.5IN					1.01	
BUILDING BOARD						
SHEATHING,REGULAR DENSITY	0.032	0.379	18.0	0.31		2.637
SHINGLE BACKER	0.033	0.398	18.0	0.31		2.510
SHEATHING,INTERMEDIATE DENSITY	0.034	0.410	22.0	0.31		2.437
NAIL BASE SHEATHING	0.036	0.438	25.0	0.31		2.283
PARTICLEBOARD, LOW DENSITY	0.045	0.540	37.0	0.31		1.852

Table 3-15 (cont.)

THERMOPHYSICAL PROPERTIES OF TYPICAL INSULATION AND BUILDING MATERIALS (IN ASCENDING ORDER BY CONDUCTIVITY)

DESCRIPTION	THERMAL COND (K) (BTU/HR-FT-F)	THERMAL COND (K) (BTU-IN/HR-SQFT-F)	DENSITY (LB/CUFT)	SPECIFIC HEAT (BTU/LB-F)	RESISTANCE (SQFT-HR-F/BTU)	RESISTANCE PER INCH (SQFT-HR-F/BTU-IN)
BUILDING BOARD (Cont.)						
HARDBOARD,MEDIUM DENSITY	0.054	0.653	50.0	0.31		1.532
HARDBOARD,HI-DENS,SERV TEMP UNDERLAY	0.068	0.820	55.0	0.32		1.220
PARTICLEBOARD,MEDIUM DENSITY	0.078	0.940	50.0	0.31		1.064
PLYWOOD (DOUGLAS FIR)	0.080	0.960	34.0	0.29		1.042
HARDBOARD,HIGH DENSITY,STD TEMP	0.083	1.000	63.0	0.32		1.000
GYPSUM BOARD	0.093	1.111	50.0	0.26		0.900
PARTICLEBOARD,HIGH DENSITY	0.098	1.180	62.5	0.31		0.848
PARTICLEBOARD,UNDERLAYMENT	0.180	2.155	40.0	0.29		0.464
ASBESTOS-CEMENT BOARD	0.345	4.140	120.0	0.24		0.242
BUILDING INSULATION						
POLYURETHANE,R-11 EXP,FOIL FACED	0.012	0.140	2.0	0.38		7.123
POLYISOCYANURATE,R-11 EXPANDED	0.013	0.160	2.0	0.38		6.266
POLYURETHANE,R-11 EXPANDED	0.013	0.160	2.0	0.38		6.266
POLYSTYRENE EXTRUDED,SMOOTH SKIN	0.017	0.200	1.8	0.29		4.990
UREA FORMALDEHYDE	0.020	0.240	0.6	0.30		4.167
GLASS FIBER,ORGANICALLY BONDED	0.021	0.250	5.0	0.23		4.006
POLYSTYRENE,MOLDED BEADS	0.021	0.250	1.5	0.29		4.006
PERLITE,EXPANDED LOOSE FILL	0.023	0.276	5.0	0.30		3.623
CELLULOSE,LOOSE FIBER	0.024	0.288	3.2	0.33		3.472
MINERAL FIBER,RESIN BINDER	0.024	0.290	3.5	0.20		3.444
MINERAL FIBERBOARD,PREFORMED	0.024	0.290	15.0	0.17		3.444
GLASS FIBER,BATT	0.026	0.318	1.0	0.20		3.145
MINERAL FIBER,BATTS	0.026	0.318	2.0	0.20		3.145
MINERAL FIBER,LOOSE FILL	0.027	0.322	1.7	0.17		3.109
MINERAL FIBERBD,WET FELTED,ROOF INSUL	0.028	0.340	17.0	0.19		2.945
GLASS FIBER,LOOSE FILL	0.029	0.348	1.0	0.20		2.874
MINERAL FIBERBD,WET FELTED,ACOUS TILE	0.029	0.350	18.0	0.19		2.854
PERLITE,EXPANDED,ORG BOND ROOF INSUL	0.030	0.360	10.0	0.25		2.778
CELLULAR GLASS	0.032	0.380	8.5	0.18		2.629
WOOD FIBERBOARD ACOUSTIC TILE	0.033	0.400	23.0	0.31		2.503
MINERAL FIBERBD,WET MOLDED,ACOUS TILE	0.035	0.420	23.0	0.14		2.381
VERMICULITE,EXPANDED LOOSE FILL	0.040	0.480	9.0	0.32		2.083
FLOORING MATERIALS						
CARPET AND FIBER PAD					2.08	
CARPET AND RUBBER PAD					1.23	
CORK TILE					0.28	
TILE,TERRAZO,ASPHLT,VINYL,LINEOLEUM,ETC					0.05	

Table 3-15 (cont.)
THERMOPHYSICAL PROPERTIES OF TYPICAL
INSULATION AND BUILDING MATERIALS
(IN ASCENDING ORDER BY CONDUCTIVITY)

DESCRIPTION	THERMAL COND(K) (BTU/ HR-FT-F)	THERMAL COND(K) (BTU-IN/ HR-SQFT-F)	DENSITY (LB/CUFT)	SPECIFIC HEAT (BTU/LB-F)	RESISTANCE (SQFT-HR-F/ BTU)	RESISTANCE PER INCH (SQFT-HR-F/ BTU-IN)
MASONRY MATERIALS						
CONCRETE,LIGHTWEIGHT,INSULATING	0.075	0.901	30.0	0.21		1.110
STUCCO	0.417	5.000	116.0	0.21		0.200
CONCRETE,LIGHTWEIGHT,STRUCTURAL	0.439	5.263	110.0	0.21		0.190
LIMESTONE	0.720	8.640	155.0	0.22		0.116
CONCRETE,HEAVYWEIGHT,DRIED AGGREGATE	0.750	9.000	140.0	0.22		0.111
SANDSTONE	0.940	11.280	140.0	0.17		0.089
CONCRETE,HEAVYWEIGHT,UNDRIED AGGREGATE	1.000	12.000	140.0	0.22		0.083
GRANITE	1.000	12.000	165.0	0.20		0.083
MARBLE	1.600	19.200	160.0	0.19		0.052
MASONRY UNITS						
CMU,8IN,LW,LOOSE FILL INSULATION	0.096	1.156	48.0	0.21	6.92	0.865
CMU,6IN,LW,LOOSE FILL INSULATION	0.098	1.182	57.0	0.21	5.08	0.846
CMU,8IN,LW,LOOSE FILL INSULATION	0.114	1.369	56.0	0.21	5.84	0.730
CMU,6IN,MW,LOOSE FILL INSULATION	0.117	1.399	67.0	0.21	4.29	0.715
CMU,4IN,LW,LOOSE FILL INSULATION	0.127	1.525	67.0	0.20	2.62	0.656
CMU,6IN,LW,CONCRETE & LOOSE FILL INS(2)	0.193	2.315	74.0	0.21	2.59	0.432
CMU,4IN,LW,CONCRETE & LOOSE FILL INS(2)	0.208	2.495	79.0	0.20	1.60	0.401
CMU,8IN,LW,CONCRETE & LOOSE FILL INS(2)	0.209	2.514	69.0	0.21	3.18	0.398
CMU,4IN,LW,HOLLOW	0.222	2.666	65.0	0.20	1.50	0.375
CMU,6IN,HW,INSULATION FILLED	0.222	2.666	88.0	0.21	2.25	0.375
CMU,6IN,MW,CONCRETE & LOOSE FILL INS(2)	0.226	2.711	84.0	0.21	2.21	0.369
CMU,8IN,HW,LOOSE FILL INSULATION	0.227	2.726	70.0	0.21	2.93	0.367
CMU,8IN,MW,CONCRETE & LOOSE FILL INS(2)	0.241	2.896	77.0	0.21	2.76	0.345
CMU,6IN,LW,HOLLOW	0.278	3.332	55.0	0.21	1.80	0.300
CMU,4IN,LW,PARTIALLY FILLED CONCRETE(1)	0.281	3.370	78.0	0.20	1.19	0.297
CLAY TILE,4IN,1CELL	0.300	3.599	70.0	0.21		0.278
CMU,4IN,HW,LOOSE FILL INSULATION	0.300	3.601	103.0	0.21	1.11	0.278
CLAY TILE,3IN,1CELL	0.312	3.750	70.0	0.21		0.267
CMU,6IN,LW,PARTIALLY FILLED CONCRETE(1)	0.319	3.827	73.0	0.21	1.57	0.261
CLAY TILE,6IN,2CELLS	0.330	3.960	70.0	0.21		0.253
CMU,8IN,LW,HOLLOW	0.333	4.000	45.0	0.21	2.00	0.250
CMU,6IN,MW,HOLLOW	0.357	4.285	65.0	0.21	1.40	0.233
CLAY TILE,8IN,2CELLS	0.360	4.320	70.0	0.21		0.231
CMU,6IN,MW,PARTIALLY FILLED CONCRETE(1)	0.369	4.423	83.0	0.21	1.36	0.226
CMU,4IN,LW,CONCRETE FILLED	0.369	4.434	104.0	0.20	0.90	0.226
CLAY TILE,10IN,2CELLS	0.375	4.499	70.0	0.21		0.222
CMU,6IN,LW,CONCRETE FILLED	0.382	4.583	110.0	0.21	1.31	0.218
CMU,8IN,LW,PARTIALLY FILLED CONCRETE(1)	0.385	4.615	68.0	0.21	1.73	0.217
CMU,8IN,MW,HOLLOW	0.388	4.651	53.0	0.21	1.72	0.215
BRICK,12IN COMMON	0.400	4.800	120.0	0.20		0.208

Table 3-15 (cont.)

THERMOPHYSICAL PROPERTIES OF TYPICAL INSULATION AND BUILDING MATERIALS (IN ASCENDING ORDER BY CONDUCTIVITY)

DESCRIPTION	THERMAL COND(K) (BTU/ HR-FT-F)	THERMAL COND(K) (BTU-IN/ HR-SQFT-F)	DENSITY (LB/CUFT)	SPECIFIC HEAT (BTU/LB-F)	RESISTANCE (SQFT-HR-F/ BTU)	RESISTANCE PER INCH (SQFT-HR-F/ BTU-IN)
MASONRY UNITS (Cont.)						
BRICK,4IN COMMON	0.400	4.800	120.0	0.20		0.208
BRICK,8IN COMMON	0.400	4.800	120.0	0.20		0.208
CLAY TILE,12IN,3CELLS	0.400	4.800	70.0	0.21	2.00	0.208
CMU,8IN,HW,CONCRETE & LOOSE FILL INS(2)	0.416	4.992	93.0	0.21	1.60	0.200
CMU,12IN,LW,CONCRETE FILLED	0.419	5.033	113.0	0.21	2.38	0.199
CMU,6IN,HW,CONCRETE & LOOSE FILL INS(2)	0.424	5.086	104.0	0.21	1.18	0.197
CMU,12IN,LW,PARTIALLY FILLED CONCRETE(1)	0.427	5.129	70.0	0.21	2.34	0.195
CMU,8IN,MW,PARTIALLY FILLED CONCRETE(1)	0.435	5.218	76.0	0.21	1.53	0.192
CMU,8IN,LW,CONCRETE FILLED	0.436	5.231	115.0	0.21	1.53	0.191
CMU,12IN,LW,HOLLOW	0.440	5.286	49.0	0.21	2.27	0.189
CMU,6IN,MW,CONCRETE FILLED	0.444	5.332	119.0	0.21	1.13	0.188
CMU,4IN,HW,HOLLOW	0.469	5.633	101.0	0.21	0.71	0.178
CMU,4IN,HW,CONCRETE & LOOSE FILL INS(2)	0.477	5.726	115.0	0.21	0.70	0.175
CMU,12IN MW CONCRETE FILLED	0.481	5.777	121.0	0.21	2.08	0.173
CMU,12IN,MW,PARTIALLY FILLED CONCRETE(1)	0.492	5.903	79.0	0.21	2.03	0.169
CMU,8IN,MW,CONCRETE FILLED	0.496	5.948	123.0	0.21	1.34	0.168
CMU,12IN,MW,HOLLOW	0.496	5.951	58.0	0.21	2.02	0.168
CMU,6IN,HW,HOLLOW	0.555	6.666	85.0	0.21	0.90	0.150
CMU,4IN,HW,PARTIALLY FILLED CONCRETE(1)	0.584	7.013	114.0	0.21	0.57	0.143
CMU,8IN,HW,HOLLOW	0.606	7.272	69.0	0.21	1.10	0.138
CMU,6IN,HW,PARTIALLY FILLED CONCRETE(1)	0.612	7.343	104.0	0.21	0.82	0.136
CMU,8IN,HW,PARTIALLY FILLED CONCRETE(1)	0.675	8.095	93.0	0.21	0.99	0.124
BRICK,3IN FACE	0.750	9.000	130.0	0.22		0.111
BRICK,4IN FACE	0.750	9.000	130.0	0.22		0.111
CMU,12IN,HW,CONCRETE FILLED	0.757	9.090	140.0	0.21	1.32	0.110
CMU,4IN,HW,CONCRETE FILLED	0.757	9.090	140.0	0.21	0.44	0.110
CMU,6IN,HW,CONCRETE FILLED	0.757	9.090	140.0	0.21	0.66	0.110
CMU,8IN,HW,CONCRETE FILLED	0.757	9.090	140.0	0.21	0.88	0.110
CMU,12IN,HW,PARTIALLY FILLED CONCRETE(1)	0.777	9.328	98.0	0.21	1.29	0.107
CMU,12IN,HW,HOLLOW	0.781	9.376	76.0	0.21	1.28	0.107
CLAY TILE,PAVER	1.042	12.499	120.0	0.21		0.080

(1) ONE CONCRETE FILLED AND REINFORCED CORE EVERY 2 FEET OF WALL LENGTH.
(2) ONE CONCRETE FILLED AND REINFORCED CORE EVERY 2 FEET OF WALL LENGTH WITH REMAINING CORES FILLED WITH LOOSE FILL INSULATION.

DESCRIPTION	THERMAL COND(K) (BTU/ HR-FT-F)	THERMAL COND(K) (BTU-IN/ HR-SQFT-F)	DENSITY (LB/CUFT)	SPECIFIC HEAT (BTU/LB-F)	RESISTANCE (SQFT-HR-F/ BTU)	RESISTANCE PER INCH (SQFT-HR-F/ BTU-IN)
PLASTERING MATERIALS						
GYPSUM PLASTER,PERLITE AGGREGATE	0.125	1.500	45.0	0.32		0.667
GYPSUM PLASTER,LIGHTWEIGHT AGGREGATE	0.133	1.596	45.0	0.20		0.627
GYPSUM PLASTER,VERMICULITE AGGREGATE	0.142	1.700	45.0	0.22		0.588
CEMENT PLASTER,SAND AGGREGATE	0.417	5.000	116.0	0.20		0.200
GYPSUM PLASTER,SAND AGGREGATE	0.467	5.600	105.0	0.20		0.179

Table 3-15 (cont.)

THERMOPHYSICAL PROPERTIES OF TYPICAL INSULATION AND BUILDING MATERIALS (IN ASCENDING ORDER BY CONDUCTIVITY)

DESCRIPTION	THERMAL COND(K) (BTU/ HR-FT-F)	THERMAL COND(K) (BTU-IN/ HR-SQFT-F)	DENSITY (LB/CUFT)	SPECIFIC HEAT (BTU/LB-F)	RESISTANCE (SQFT-HR-F/ BTU)	RESISTANCE PER INCH (SQFT-HR-F/ BTU-IN)
ROOFING MATERIALS						
ASBESTOS-CEMENT SHINGLES			120.0	0.24	0.21	
ASPHALT ROLL ROOFING			70.0	0.36	0.15	
ASPHALT SHINGLES			70.0	0.30	0.44	
SLATE			120.0	0.30	0.05	
WOOD SHNGLS,PLAIN AND PLASTIC FILM FCD			32.0	0.31	0.94	
INSULATION,PREFORMED	0.030	0.360	10.0	0.25		2.778
BUILT-UP ROOFING	0.094	1.126	70.0	0.35		0.888
ROOF GRAVEL OR SLAG	0.834	10.008	55.0	0.40		0.100
SIDING MATERIALS						
ASBESTOS CEMENT SHINGLES			120.0		0.21	
ASBESTOS CEMENT SIDING,0.25IN,LAPPED					0.24	
ASPHALT INSULATION SIDING(0.5 IN BED)					1.46	
ASPHALT ROLL SIDING					0.15	
METAL CLAD SHEATHING,HOLLOW-BACKED					0.61	
METAL CLAD SHTH,INS-BD BACKED,0.375IN					1.82	
MTL CLAD SHTH,INS-BD BCKD,0.375IN,FOIL					2.96	
PLYWOOD,0.375IN,LAPPED					0.63	
WOOD SHINGLES,16IN, 7.5IN EXPOSURE					0.87	
WOOD SHINGLES,DOUBLE,16IN, 12IN EXPOSURE					1.19	
WOOD,BEVEL,0.5IN BY 8IN,LAPPED					0.81	
WOOD,BEVEL,0.75IN BY 10IN LAPPED					1.05	
WOOD,DROP,1IN BY 8IN					0.79	
WOOD,PLUS INS BACKER BOARD,0.3125IN					1.40	
HARDBOARD SIDING	0.124	1.488	40.0	0.28		0.672
ARCHITECTURAL GLASS	0.440	5.280	169.0	0.20		0.189
WOODS						
WOOD,CYPRESS	0.056	0.672	29.0	0.65		1.488
WOOD,FIR	0.063	0.756	26.0	0.57		1.323
WOOD,SOFT	0.065	0.780	32.0	0.65		1.282
WOOD,WHITE PINE	0.065	0.780	32.0	0.67		1.282
WOOD,YELLOW PINE	0.082	0.984	40.0	0.67		1.016
WOOD,HARD	0.085	1.020	38.0	0.57		0.980
WOOD,OAK	0.085	1.020	38.0	0.57		0.980

Table 3-16

THERMOPHYSICAL PROPERTIES OF TYPICAL
INSULATION AND BUILDING MATERIALS
(IN ASCENDING ORDER BY DENSITY)

DESCRIPTION	THERMAL COND (K) (BTU/ HR-FT-F)	THERMAL COND (K) (BTU-IN/ HR-SQFT-F)	DENSITY (LB/CUFT)	SPECIFIC HEAT (BTU/LB-F)	RESISTANCE (SQFT-HR-F/ BTU)	RESISTANCE PER INCH (SQFT-HR-F/ BTU-IN)
AIR FILMS						
MOVING AIR,15MPH WIND,WINTER					0.17	
MOVING AIR,7.5MPH WIND,SUMMER					0.25	
STILL AIR,HORIZ SURFACE,HEAT FLOW DOWN					0.92	
STILL AIR,HORIZ SURFACE,HEAT FLOW UP					0.61	
STILL AIR,VERTICAL SURFACE					0.68	
AIR LAYERS						
HORIZ REFLECT SURFACE,SUMMER,LYR=.5IN					1.89	
HORIZ REFLECT SURFACE,SUMMER,LYR=.75IN					2.41	
HORIZ REFLECT SURFACE,SUMMER,LYR=1.5IN					3.27	
HORIZ REFLECT SURFACE,SUMMER,LYR=3.5IN					4.09	
HORIZ REFLECT SURFACE,WINTER,LYR=.5IN					1.60	
HORIZ REFLECT SURFACE,WINTER,LYR=.75IN					1.70	
HORIZ REFLECT SURFACE,WINTER,LYR=1.5IN					1.81	
HORIZ REFLECT SURFACE,WINTER,LYR=3.5IN					1.95	
HORIZONTAL SURFACE,SUMMER,LAYER=.5IN					0.92	
HORIZONTAL SURFACE,SUMMER,LAYER=.75IN					1.02	
HORIZONTAL SURFACE,SUMMER,LAYER=1.5IN					1.15	
HORIZONTAL SURFACE,SUMMER,LAYER=3.5IN					1.24	
HORIZONTAL SURFACE,WINTER,LAYER,1.5IN					0.89	
HORIZONTAL SURFACE,WINTER,LAYER=.5IN					0.84	
HORIZONTAL SURFACE,WINTER,LAYER=.75IN					0.87	
HORIZONTAL SURFACE,WINTER,LAYER=3.5IN					0.93	
VERTICAL REFLECTIVE SURF,LAYER=.5IN					2.35	
VERTICAL REFLECTIVE SURF,LAYER=1.5IN					2.39	
VERTICAL REFLECTIVE SURF,LAYER=3.5IN					2.32	
VERTICAL REFLECTIVE SURFACE,LAYER=.5IN					1.88	
VERTICAL SURFACE,LAYER=.5IN					0.91	
VERTICAL SURFACE,LAYER=.75IN					1.01	
VERTICAL SURFACE,LAYER=1.5IN					1.02	
VERTICAL SURFACE,LAYER=3.5IN					1.01	
BUILDING BOARD						
SHEATHING,REGULAR DENSITY	0.032	0.379	18.0	0.31		2.637
SHINGLE BACKER	0.033	0.398	18.0	0.31		2.510
SHEATHING,INTERMEDIATE DENSITY	0.034	0.410	22.0	0.31		2.437
NAIL BASE SHEATHING	0.036	0.438	25.0	0.31		2.283
PLYWOOD (DOUGLAS FIR)	0.080	0.960	34.0	0.29		1.042

Table 3-16 (cont.)
THERMOPHYSICAL PROPERTIES OF TYPICAL
INSULATION AND BUILDING MATERIALS
(IN ASCENDING ORDER BY DENSITY)

DESCRIPTION	THERMAL COND(K) (BTU/HR-FT-F)	THERMAL COND(K) (BTU-IN/HR-SQFT-F)	DENSITY (LB/CUFT)	SPECIFIC HEAT (BTU/LB-F)	RESISTANCE (SQFT-HR-F/BTU)	RESISTANCE PER INCH (SQFT-HR-F/BTU-IN)
BUILDING BOARD (Cont.)						
PARTICLEBOARD, LOW DENSITY	0.045	0.540	37.0	0.31		1.852
PARTICLEBOARD,UNDERLAYMENT	0.180	2.155	40.0	0.29		0.464
GYPSUM BOARD	0.093	1.111	50.0	0.26		0.900
HARDBOARD,MEDIUM DENSITY	0.054	0.653	50.0	0.31		1.532
PARTICLEBOARD,MEDIUM DENSITY	0.078	0.940	50.0	0.31		1.064
HARDBOARD,HI-DENS,SERV TEMP UNDERLAY	0.068	0.820	55.0	0.32		1.220
PARTICLEBOARD,HIGH DENSITY	0.098	1.180	62.5	0.31		0.848
HARDBOARD,HIGH DENSITY,STD TEMP	0.083	1.000	63.0	0.32		1.000
ASBESTOS-CEMENT BOARD	0.345	4.140	120.0	0.24		0.242
BUILDING INSULATION						
UREA FORMALDEHYDE	0.020	0.240	0.6	0.30		4.167
GLASS FIBER,BATT	0.026	0.318	1.0	0.20		3.145
GLASS FIBER,LOOSE FILL	0.029	0.348	1.0	0.20		2.874
POLYSTYRENE,MOLDED BEADS	0.021	0.250	1.5	0.29		4.006
MINERAL FIBER,LOOSE FILL	0.027	0.322	1.7	0.17		3.109
POLYSTYRENE EXTRUDED,SMOOTH SKIN	0.017	0.200	1.8	0.29		4.990
MINERAL FIBER,BATTS	0.026	0.318	2.0	0.20		3.145
POLYISOCYANURATE,R-11 EXPANDED	0.013	0.160	2.0	0.38		6.266
POLYURETHANE,R-11 EXP,FOIL FACED	0.012	0.140	2.0	0.38		7.123
POLYURETHANE,R-11 EXPANDED	0.013	0.160	2.0	0.38		6.266
CELLULOSE,LOOSE FIBER	0.024	0.288	3.2	0.33		3.472
MINERAL FIBER,RESIN BINDER	0.024	0.290	3.5	0.20		3.444
GLASS FIBER,ORGANICALLY BONDED	0.021	0.250	5.0	0.23		4.006
PERLITE,EXPANDED LOOSE FILL	0.023	0.276	5.0	0.30		3.623
CELLULAR GLASS	0.032	0.380	8.5	0.18		2.629
VERMICULITE,EXPANDED LOOSE FILL	0.040	0.480	9.0	0.32		2.083
PERLITE,EXPANDED,ORG BOND ROOF INSUL	0.030	0.360	10.0	0.25		2.778
MINERAL FIBERBOARD,PREFORMED	0.024	0.290	15.0	0.17		3.444
MINERAL FIBERBD,WET FELTED,ROOF INSUL	0.028	0.340	17.0	0.19		2.945
MINERAL FIBERBD,WET FELTED,ACOUS TILE	0.029	0.350	18.0	0.19		2.854
MINERAL FIBERBD,WET MOLDED,ACOUS TILE	0.035	0.420	23.0	0.14		2.381
WOOD FIBERBOARD ACOUSTIC TILE	0.033	0.400	23.0	0.31		2.503
FLOORING MATERIALS						
CARPET AND FIBER PAD					2.08	
CARPET AND RUBBER PAD					1.23	
CORK TILE					0.28	
TILE,TERRAZO,ASPHLT,VINYL,LINEOLEUM,ETC					0.05	

Table 3-16 (cont.)

THERMOPHYSICAL PROPERTIES OF TYPICAL INSULATION AND BUILDING MATERIALS (IN ASCENDING ORDER BY DENSITY)

DESCRIPTION	THERMAL COND (K) (BTU/ HR-FT-F)	THERMAL COND (K) (BTU-IN/ HR-SQFT-F)	DENSITY (LB/CUFT)	SPECIFIC HEAT (BTU/LB-F)	RESISTANCE (SQFT-HR-F/ BTU)	RESISTANCE PER INCH (SQFT-HR-F/ BTU-IN)
MASONRY MATERIALS						
CONCRETE,LIGHTWEIGHT,INSULATING	0.075	0.901	30.0	0.21		1.110
CONCRETE,LIGHTWEIGHT,STRUCTURAL	0.439	5.263	110.0	0.21		0.190
STUCCO	0.417	5.000	116.0	0.21		0.200
CONCRETE,HEAVYWEIGHT,DRIED AGGREGATE	0.750	9.000	140.0	0.22		0.111
CONCRETE,HEAVYWEIGHT,UNDRIED AGGREGATE	1.000	12.000	140.0	0.22		0.083
SANDSTONE	0.940	11.280	140.0	0.17		0.089
LIMESTONE	0.720	8.640	155.0	0.22		0.116
MARBLE	1.600	19.200	160.0	0.19		0.052
GRANITE	1.000	12.000	165.0	0.20		0.083
MASONRY UNITS						
CMU,8IN,LW,HOLLOW	0.333	4.000	45.0	0.21	2.00	0.250
CMU,8IN,LW,LOOSE FILL INSULATION	0.096	1.156	48.0	0.21	6.92	0.865
CMU,12IN,LW,HOLLOW	0.440	5.286	49.0	0.21	2.27	0.189
CMU,8IN,MW,HOLLOW	0.388	4.651	53.0	0.21	1.72	0.215
CMU,6IN,LW,HOLLOW	0.278	3.332	55.0	0.21	1.80	0.300
CMU,8IN,LOOSE FILL INSULATION	0.114	1.369	56.0	0.21	5.84	0.730
CMU,6IN,LW,LOOSE FILL INSULATION	0.098	1.182	57.0	0.21	5.08	0.846
CMU,12IN,MW,HOLLOW	0.496	5.951	58.0	0.21	2.02	0.168
CMU,4IN,LW,HOLLOW	0.222	2.666	65.0	0.20	1.50	0.375
CMU,6IN,MW,HOLLOW	0.357	4.285	65.0	0.21	1.40	0.233
CMU,4IN,LW,LOOSE FILL INSULATION	0.127	1.525	67.0	0.20	2.62	0.656
CMU,6IN,MW,LOOSE FILL INSULATION	0.117	1.399	67.0	0.21	4.29	0.715
CMU,8IN,LW,PARTIALLY FILLED CONCRETE(1)	0.385	4.615	68.0	0.21	1.73	0.217
CMU,8IN,HW,HOLLOW	0.606	7.272	69.0	0.21	1.10	0.138
CMU,8IN,LW,CONCRETE & LOOSE FILL INS(2)	0.209	2.514	69.0	0.21	3.18	0.398
CLAY TILE,10IN,2CELLS	0.375	4.499	70.0	0.21		0.222
CLAY TILE,12IN,3CELLS	0.400	4.800	70.0	0.21	2.00	0.208
CLAY TILE,3IN,1CELL	0.312	3.750	70.0	0.21		0.267
CLAY TILE,4IN,1CELL	0.300	3.599	70.0	0.21		0.278
CLAY TILE,6IN,2CELLS	0.330	3.960	70.0	0.21		0.253
CLAY TILE,8IN,2CELLS	0.360	4.320	70.0	0.21		0.231
CMU,12IN,LW,PARTIALLY FILLED CONCRETE(1)	0.427	5.129	70.0	0.21	2.34	0.195
CMU,8IN,HW,LOOSE FILL INSULATION	0.227	2.726	70.0	0.21	2.93	0.367
CMU,6IN,LW,PARTIALLY FILLED CONCRETE(1)	0.319	3.827	73.0	0.21	1.57	0.261
CMU,6IN,LW,CONCRETE & LOOSE FILL INS(2)	0.193	2.315	74.0	0.21	2.59	0.432
CMU,12IN,HW,HOLLOW	0.781	9.376	76.0	0.21	1.28	0.107
CMU,8IN,MW,PARTIALLY FILLED CONCRETE(1)	0.435	5.218	76.0	0.21	1.53	0.192
CMU,8IN,MW,CONCRETE & LOOSE FILL INS(2)	0.241	2.896	77.0	0.21	2.76	0.345
CMU,4IN,LW,PARTIALLY FILLED CONCRETE(1)	0.281	3.370	78.0	0.20	1.19	0.297
CMU,12IN,MW,PARTIALLY FILLED CONCRETE(1)	0.492	5.903	79.0	0.21	2.03	0.169

Table 3-16 (cont.)

THERMOPHYSICAL PROPERTIES OF TYPICAL INSULATION AND BUILDING MATERIALS (IN ASCENDING ORDER BY DENSITY)

DESCRIPTION	THERMAL COND (K) (BTU/ HR-FT-F)	THERMAL COND (K) (BTU-IN/ HR-SQFT-F)	DENSITY (LB/CUFT)	SPECIFIC HEAT (BTU/LB-F)	RESISTANCE (SQFT-HR-F/ BTU)	RESISTANCE PER INCH (SQFT-HR-F/ BTU-IN)
MASONRY UNITS (Cont.)						
CMU,4IN,LW,CONCRETE & LOOSE FILL INS(2)	0.208	2.495	79.0	0.20	1.60	0.401
CMU,6IN,MW,PARTIALLY FILLED CONCRETE(1)	0.369	4.423	83.0	0.21	1.36	0.226
CMU,6IN,MW,CONCRETE, & LOOSE FILL INS(2)	0.226	2.711	84.0	0.21	2.21	0.369
CMU,6IN,HW,HOLLOW	0.555	6.666	85.0	0.21	0.90	0.150
CMU,6IN,HW,INSULATION FILLED	0.222	2.666	88.0	0.21	2.25	0.375
CMU,8IN,HW,CONCRETE & LOOSE FILL INS(2)	0.416	4.992	93.0	0.21	1.60	0.200
CMU,8IN,HW,PARTIALLY FILLED CONCRETE(1)	0.675	8.095	93.0	0.21	0.99	0.124
CMU,12IN,HW,PARTIALLY FILLED CONCRETE(1)	0.777	9.328	98.0	0.21	1.29	0.107
CMU,4IN,HW,HOLLOW	0.469	5.633	101.0	0.21	0.71	0.178
CMU,4IN,HW,LOOSE FILL INSULATION	0.300	3.601	103.0	0.21	1.11	0.278
CMU,4IN,LW,CONCRETE FILLED	0.369	4.434	104.0	0.20	0.90	0.226
CMU,6IN,HW,CONCRETE & LOOSE FILL INS(2)	0.424	5.086	104.0	0.21	1.18	0.197
CMU,6IN,HW,PARTIALLY FILLED CONCRETE(1)	0.612	7.343	104.0	0.21	0.82	0.136
CMU,6IN,LW,CONCRETE FILLED	0.382	4.583	110.0	0.21	1.31	0.218
CMU,12IN,LW,CONCRETE FILLED	0.419	5.033	113.0	0.21	2.38	0.199
CMU,4IN,HW,PARTIALLY FILLED CONCRETE(1)	0.584	7.013	114.0	0.21	0.57	0.143
CMU,4IN,HW,CONCRETE & LOOSE FILL INS(2)	0.477	5.726	115.0	0.21	0.70	0.175
CMU,8IN,LW,CONCRETE FILLED	0.436	5.231	115.0	0.21	1.53	0.191
CMU,6IN,MW,CONCRETE FILLED	0.444	5.332	119.0	0.21	1.13	0.188
BRICK,12IN COMMON	0.400	4.800	120.0	0.20		0.208
BRICK,4IN COMMON	0.400	4.800	120.0	0.20		0.208
BRICK,8IN COMMON	0.400	4.800	120.0	0.20		0.208
CLAY TILE,PAVER	1.042	12.499	120.0	0.21		0.080
CMU,12IN MW CONCRETE FILLED	0.481	5.777	121.0	0.21	2.08	0.173
CMU,8IN,MW,CONCRETE FILLED	0.496	5.948	123.0	0.21	1.34	0.168
BRICK,3IN FACE	0.750	9.000	130.0	0.22		0.111
BRICK,4IN FACE	0.750	9.000	130.0	0.22		0.111
CMU,12IN,HW,CONCRETE FILLED	0.757	9.090	140.0	0.21	1.32	0.110
CMU,4IN,HW,CONCRETE FILLED	0.757	9.090	140.0	0.21	0.44	0.110
CMU,6IN,HW,CONCRETE FILLED	0.757	9.090	140.0	0.21	0.66	0.110
CMU,8IN,HW,CONCRETE FILLED	0.757	9.090	140.0	0.21	0.88	0.110

(1) ONE CONCRETE FILLED AND REINFORCED CORE EVERY 2 FEET OF WALL LENGTH.
(2) ONE CONCRETE FILLED AND REINFORCED CORE EVERY 2 FEET OF WALL LENGTH WITH REMAINING CORES FILLED WITH LOOSE FILL INSULATION.

DESCRIPTION	THERMAL COND (K) (BTU/ HR-FT-F)	THERMAL COND (K) (BTU-IN/ HR-SQFT-F)	DENSITY (LB/CUFT)	SPECIFIC HEAT (BTU/LB-F)	RESISTANCE (SQFT-HR-F/ BTU)	RESISTANCE PER INCH (SQFT-HR-F/ BTU-IN)
PLASTERING MATERIALS						
GYPSUM PLASTER,LIGHTWEIGHT AGGREGATE	0.133	1.596	45.0	0.20		0.627
GYPSUM PLASTER,PERLITE AGGREGATE	0.125	1.500	45.0	0.32		0.667
GYPSUM PLASTER,VERMICULITE AGGREGATE	0.142	1.700	45.0	0.22		0.588
GYPSUM PLASTER,SAND AGGREGATE	0.467	5.600	105.0	0.20		0.179
CEMENT PLASTER,SAND AGGREGATE	0.417	5.000	116.0	0.20		0.200

Table 3-16 (cont.)

THERMOPHYSICAL PROPERTIES OF TYPICAL INSULATION AND BUILDING MATERIALS (IN ASCENDING ORDER BY DENSITY)

DESCRIPTION	THERMAL COND (K) (BTU/ HR-FT-F)	THERMAL COND (K) (BTU-IN/ HR-SQFT-F)	DENSITY (LB/CUFT)	SPECIFIC HEAT (BTU/LB-F)	RESISTANCE (SQFT-HR-F/ BTU)	RESISTANCE PER INCH (SQFT-HR-F/ BTU-IN)
ROOFING MATERIALS						
INSULATION,PREFORMED	0.030	0.360	10.0	0.25		2.778
WOOD SHNGLS,PLAIN AND PLASTIC FILM FCD			32.0	0.31	0.94	
ROOF GRAVEL OR SLAG	0.834	10.008	55.0	0.40		0.100
ASPHALT ROLL ROOFING			70.0	0.36	0.15	
ASPHALT SHINGLES			70.0	0.30	0.44	
BUILT-UP ROOFING	0.094	1.126	70.0	0.35		0.888
ASBESTOS-CEMENT SHINGLES			120.0	0.24	0.21	
SLATE			120.0	0.30	0.05	
SIDING MATERIALS						
ASBESTOS CEMENT SIDING,0.25IN,LAPPED					0.24	
ASPHALT INSULATION SIDING(0.5 IN BED)					1.46	
ASPHALT ROLL SIDING					0.15	
METAL CLAD SHEATHING,HOLLOW-BACKED					0.61	
METAL CLAD SHTH,INS-BD BACKED,0.375IN					1.82	
MTL CLAD SHTH,INS-BD BCKD,0.375IN,FOIL					2.96	
PLYWOOD,0.375IN,LAPPED					0.63	
WOOD SHINGLES,16IN, 7.5IN EXPOSURE					0.87	
WOOD SHINGLES,DOUBLE,16IN, 12IN EXPOSURE					1.19	
WOOD,BEVEL,0.5IN BY 8IN,LAPPED					0.81	
WOOD,BEVEL,0.75IN BY 10IN LAPPED					1.05	
WOOD,DROP,1IN BY 8IN					0.79	
WOOD,PLUS INS BACKER BOARD,0.3125IN					1.40	
HARDBOARD SIDING	0.124	1.488	40.0	0.28		0.672
ASBESTOS CEMENT SHINGLES			120.0		0.21	
ARCHITECTURAL GLASS	0.440	5.280	169.0	0.20		0.189
WOODS						
WOOD,FIR	0.063	0.756	26.0	0.57		1.323
WOOD,CYPRESS	0.056	0.672	29.0	0.65		1.488
WOOD,SOFT	0.065	0.780	32.0	0.65		1.282
WOOD,WHITE PINE	0.065	0.780	32.0	0.67		1.282
WOOD,HARD	0.085	1.020	38.0	0.57		0.980
WOOD,OAK	0.085	1.020	38.0	0.57		0.980
WOOD,YELLOW PINE	0.082	0.984	40.0	0.67		1.016

Table 3-17

THERMOPHYSICAL PROPERTIES OF INDUSTRIAL INSULATION

DESCRIPTION	MAX-SER TEMP (F)	NOMINAL DENSITY (LB/CUFT)	THERMAL CONDUCTIVITY (K) AT MEAN TEMP. F BTU-IN/HR-SQFT-F (-25)	(25)	(75)	(100)	(200)	(300)	(500)
BLANKETS & FELTS									
GLASS FIBER,ORGANICALLY BONDED	350	0.75	0.24	0.27	0.32	0.34	0.48		
GLASS FIBER,ORGANICALLY BONDED	350	1.0	0.23	0.25	0.29	0.32	0.43		
GLASS FIBER,ORGANICALLY BONDED	350	1.5	0.21	0.23	0.27	0.28	0.37		
GLASS FIBER,ORGANICALLY BONDED	350	2.0	0.20	0.22	0.25	0.26	0.33		
GLASS FIBER,ORGANICALLY BONDED	350	3.0	0.19	0.21	0.23	0.24	0.31		
FELT,SEMIRIGID,ORGANICALLY BONDED	400	3-8		0.24	0.25	0.26	0.27	0.35	
HAIR,FELT	180	10		0.26	0.29	0.30			
BLOCKS, BOARDS, & PIPE INSULATION									
ASBESTOS PAPER,LAMINATED	700	30				0.40	0.45	0.50	0.60
ASBESTOS PAPER,CORRUG & LAMIN 4PLY	300	11-13			0.54	0.57	0.68		
ASBESTOS PAPER,CORRUG & LAMIN 6PLY	300	15-17				0.51	0.59		
ASBESTOS PAPER,CORRUG & LAMIN 8PLY	300	18-20			0.47	0.49	0.57		
CALCIUM SILICATE	1,200	11-15				0.38	0.41	0.44	0.52
CELLULAR GLASS	900	8.5	0.30	0.32	0.35	0.36	0.42	0.49	0.70
CORK BOARD	180	5-6	0.24	0.25	0.26	0.26			
GLASS FIBER,ORG BND BLK & BOARD	400	3-10	0.19	0.22	0.25	0.26	0.33	0.40	
GLASS FIBER,ORG BND,NONPUNKING BIND	1,000	3-10				0.26	0.31	0.38	0.52
GLASS FIBER,ORG BND PIPE INSULATION	350	3-4		0.21	0.23	0.24	0.29		
GLASS FIBER,ORG BND PIPE INSULATION	500	3-10		0.22	0.25	0.26	0.33	0.40	
GLASS FIBER,INORG BONDED BLOCK	1,000	10-15				0.33	0.38	0.45	0.65
PHENOLIC FOAM PIPE INSULATION	275	2.0		0.21	0.23	0.24	0.29		
POLYSTYRENE,EXTRUDED SMOOTH SKIN	170	2.2	0.16	0.18	0.19	0.20			
POLYSTYRENE,EXTRUDED CUT CELL SURF	170	1.8	0.20	0.23	0.24	0.25			
POLYSTYRENE,MOLDED BEADS	170	1.5	0.19	0.21	0.23	0.24			
POLYISOCYANURATE,FOIL FACED R-11 EXP	250	2.0		0.12	0.14	0.15			
POLYURETHANE,R-11 EXPANDED,UNFACED	210	1.5-2.5		0.17	0.16	0.17			
RUBBER,ELASTOMERIC FOAM	220	4.5		0.20	0.22	0.23			
WOOL,FELT,PIPE INSULATION	180	20		0.28	0.31	0.33			
LOOSE FILL									
CELLULOSE	180	2.5-3.5			0.27	0.29			
GLASS FIBER,UNBONDED	1,000	1.65	0.20	0.25	0.29	0.31			
MINERAL FIBER,UNBONDED	1,000	4.8	0.22	0.24	0.28	0.29	0.36	0.44	0.65
PERLITE,EXPANDED	1,200	7.6	0.23	0.25	0.26	0.28	0.34		
VERMICULITE,EXPANDED	1,000	4-12	0.38	0.43	0.47	0.48	0.56	0.65	0.80

3.4 Vapor Barriers

Part of the successful performance of thermal insulation depends on keeping the insulation dry. One reason for this is that the conductivity of water is on the order of 10 times that of most insulations; thus, a wet insulation will lose most of its heat-retarding ability. If outdoor conditions are such that condensed moisture can freeze within the insulation, not only does this "conductivity ratio" increase from about 10 to about 40, but expansion of the freezing water may rupture cellular insulations, leading to cracks and eventual deterioration of the insulation itself. The same is true of insulation on pipes operating below freezing.

Note that the presence of gaseous water vapor within an insulation is normally not a problem. Only when the vapor condenses or freezes is insulation performance affected.

Vapor migration into insulation is caused by a difference in water vapor partial pressure in the air on opposite sides of a surface. In general, the indoor vapor pressure will be higher than the outdoor vapor pressure during the winter. Sources of indoor water vapor include kitchen and lavatory areas, as well as the occupants themselves. Cold outdoor air cannot hold as much moisture as warm indoor air (even at the same humidity, the water vapor pressure for cold air is less than for warm air; see psychrometric chart, Appendix E), so water vapor will migrate toward the outside, through insulated walls and roofs, if necessary.

Condensation will occur inside the wall or ceiling if the temperature at some point is equal to the dew point temperature corresponding to the indoor dry-bulb temperature and humidity. For this reason, it is usual to install a vapor barrier as close to the inside surface of a building as possible - the "warm-in-winter" side - to retard the flow of water vapor through the building surfaces. The exception to this rule occurs in humid climates where the vapor pressure inside a cooled building may be lower than that outside. Such climates occur in Hawaii and along the Gulf and Southern Atlantic coasts in the United States, and in other coastal areas within about 30° latitude of the equator worldwide [3]. For these buildings, the vapor barrier should be applied near the outside surface. For buildings that must be both heated and cooled during different times of the year, a vapor barrier may be applied near both surfaces. Alternatively, the insulation itself may be highly vapor resistant, such as cellular glass or reflective insulation [4].

Water vapor may enter building components at air infiltration sites. Air movement which carries water vapor with it is now recognized as a stronger mechanism of water vapor transfer than diffusion [3]. Control of infiltration will therefore help control vapor migration as well. Ventilation of a humid building space can help dispense excess moisture before it migrates into insulated building envelope components.

Vapor barriers are available in a variety of materials, including metal foils, laminated foils (such as foil-scrim-kraft (FSK) and vinyl-scrim-foil), metallized plastic films, sheets of vinyl or polyethylene (in various thicknesses), treated papers, coated felts, and mastics. Even paint has vapor barrier properties, and some are especially marketed for this purpose.

Vapor barriers do not completely stop the flow of moisture but act to slow it down significantly. By definition, a vapor barrier is a material that has a permeance of 1 "perm" or less. A perm is an English unit equal to 1 grain* of vapor transmitted per hour per square foot per inch of mercury difference in vapor pressure**; i.e.:

$$1 \text{ perm} = 1 \frac{\text{Grain}}{\text{hr ft}^2 \text{ in-Hg}}$$

Water vapor transmission is measured in grains per hour per square foot. Permeability is a material property, measured in "perm-inches", and is equal to permeance for a material that is 1 inch thick. Permeance, therefore, is a property of a certain thickness of a material. The two properties are related by:

$$\text{Permeance} = \text{Permeability/thickness}$$

The permeances of a number of materials are listed in Table 3-18 on the following page. Permeance values may be obtained by either the "wet-cup" or "dry-cup" test methods. Wet-cup values are usually about 5 times higher than dry-cup values, and the test method should be known when comparing different materials. These test methods are fully described in ASTM Standards (see Appendix A) and summarized in Reference [3].

The permeance of a material is greatly affected by punctures or holes. Proper care should be taken during installation to seal joints and edges, use proper fastening methods, and apply sufficient thicknesses. Stapling a vapor barrier into place, although a quick method, can increase its permeance by a factor of 20 [4]. Gluing is a preferable alternative to stapling; however, if the material is stapled, the staples should be covered with a moisture-resistant tape. As another example, one pinhole in every 4 square inches of an FSK laminate will double its permeance [3].

*A grain of water is a nominal "drop"; 7000 grains = 1 pound.

** Metric units for permeability and permeance do not seem to be agreed upon. In strict accordance with SI (International System) units, ASTM gives permeance in kilograms per Pascal-second-square meters (kg/Pa s m^2) and permeability in kilograms per Pascal-second-meter (kg/Pa s m). These units involve a factor of ten to the twelfth power when coverting to or from perms and perm-inches. ASHRAE [3] avoids this by using nanograms (ng) instead of kilograms. Reference [4] gives permeance in grams per millimeter of mercury-24 hours-square meter (g/(mm Hg)(24 hr)m^2) and permeability in gram-centimeters per millimeter of mercury-24 hours-square meter (g cm/(mm Hg)(24 hr)m^2). Factors for converting between these unit systems are given in Appendix D.

Table 3-18

WATER VAPOR PERMEANCE OF SELECTED MATERIALS

Material	Thickness (mils)	Permeance (Perms) Dry-cup	Wet-cup	Other
Plastic and metal foils and films(a)				
Aluminum foil	1	0.0		
Aluminum foil	0.35	0.05		
Polyethylne	2	0.16		
Polyethylene	4	0.08		
Polyethylene	6	0.06		
Polyethylene	8	0.04		
Polyethylene	10	0.03		
Polyvinylchloride, unplasticized	2	0.68		
Polyvinylchloride, plasticized	4	0.8-1.4		
Polyester	1	0.73		
Polyester	3.2	0.23		
Polyester	7.6	0.08		
Cellulose acetate	10	4.6		
Cellulose acetate	125	0.32		
	Weight (lb/100 sqft)			
Building paper, felts, roofing papers(b)				
Duplex sheet, asphalt laminated, aluminum foil one side	8.6	0.002	0.176	
Saturated and coated roll roofing	65	0.05	0.24	
Kraft paper and asphalt laminated, reinforced 30-120-30	6.8	0.3	1.8	
Blanket thermal insulation back up paper, asphalt coated	6.2	0.4	0.6-4.2	
Asphalt-saturated and coated vapor retarder paper	8.6	0.2-0.3	0.6	
Asphalt-saurated but not coated sheathing paper	4.4	3.3	20.2	
15-lb asphalt felt	14	1.0	5.6	
15-lb tar felt	14	4.0	18.2	
Single-kraft, double	3.2	31	42	

(a) These data are provided to permit comparisons of material. The values are intended for design guidance and should not be used as design or specification data. In the selection of vapor retarder materials, exact values for permeance or permeability should be obtained from the manufacturer of the materials under consideration or secured as a result of laboratory tests. A range of values shown in the table indicate variations among mean values for materials that are similar but of different density, orientation, lot or source. Values are summarized from ASHRAE [3].

(b) Low permeance sheets are used as vapor retarders. High permeance are used elsewhere in construction.

Table 3-18 (Con't)

WATER VAPOR PERMEANCE OF SELECTED MATERIALS

Material	Thickness (mils)	Permeance (Perms) Dry-cup	Wet-cup	Other
Liquid-applied coating materials				
Commercial latex paints (dry film thickness)(c)				
Vapor retarder paint	3.1			0.45
Primer-sealer	1.2			6.28
Vinyl acetate/acrylic primer	2.0			7.42
Vinyl-acrylic primer	1.6			8.62
Seimi-gloss vinyl-acrylic enamel	2.4			6.61
Exterior acrylic house and trim	1.7			5.47
Paint-2 coats				
Asphalt paint on plywood			0.4	
Aluminum varnish on wood		0.3-0.5		
Enamels on smooth plaster				0.5-1.5
Primers and sealers on interior insulation board				0.9-2.1
Various primers plus 1 coat flat oil paint on plaster				1.6-3.0
Flat paint on interior insulation board				4
Water emulsion on interior insulation board				30-85
	Weight (Oz/ft^2)			
Paint-3 coats				
Exterior paint, white lead and oil on wood siding		0.3-1.0		
Exterior paint, white lead-zinc oxide and oil on wood		0.9		
Styrene-butadiene latex coating	2	11		
Polyvinyl acetate latex coating	4	5.5		
Chloro-sulfonated polyethylene mastic	3.5	1.7		
	7.0	0.06		
Asphalt cut-back, 1.6 mm(1/16 in), dry		0.14		
4.8 mm(3/16 in), dry		0.0		
Hot melt asphalt	2	0.5		
	3.5	0.1		

(c) Cast at 0.25 mm (10 mils) wet film thickness.

The <u>resistance</u> to water vapor transmission is the reciprocal of the permeance of a material. Like thermal resistances, vapor transmission resistances of a series of materials may be added together to find the total resistance. The overall permeance of the assembly is the inverse of the total resistance.

In addition to maintaining low conductivity in insulation, vapor barriers can help prevent moisture-related structural damage (such as rotting of framing members) within buildings and can reduce paint blistering caused by moisture buildup behind the paint.

Special consideration must be given to vapor barriers applied to hot piping or equipment that is exposed to weather. Almost all insulations have some moisture content when they are first applied; this insulation will be dried out by a hot surface. Since vapor pressures will be quite high next to the hot surface, vapor will tend to migrate toward the outer jacket. If not allowed to escape, it can condense there and cause damage (especially to steel securements). Thus, the outer jacket or coating, while it must be a good liquid barrier to protect the insulation from rain and snow, should <u>not</u> be a tight vapor barrier [4].

Further general information on vapor and weather barriers is found in References [3,4,7-12]. Equations for the conductivities of moisture-containing polyurethane foam, polystyrene, perlite, and lightweight concrete are available in Reference [13]. Test wall sections containing cellulose fiber insulation have exhibited a 9% increase in thermal conductance by 8 weeks after installation [14]. Another similar test showed a 15% increase in conductance with a 10% increase in moisture content, but also indicated that thermal performance improved as the insulation dried out (15). Studies of moisture transfer through roof insulations found apparent permeabilities that were higher than ASTM wet-cup values by an average of 25% for polyurethane, 60% for extruded polystyrene, 220% for bead polystyrene, and 210% for pheolic foam [16]. Moisture problems are discussed with regard to suspended ceilings and 19th and 20th century frame buildings in References [17] and [18] respectively.

REFERENCES FOR SECTION 3.0

[1] Haggerty, Brian. "CPSC Ban of UF Foam Insulation Won't Stand Up In Court, Declare Industry Critics." Engineering Times, Vol. 4, No. 4, p. 13. April 1982.

[2] Loser, J.B., Moeller, C.E., and Thompson, M.B. Thermophysical Properties of Thermal Insulating Materials. Midwest Research Institute, Kansas City, MO. 1964.

[3] ASHRAE Handbook, 1981 Fundamentals. American Society of Heating, Refrigerating and Air-Conditioning Engineers, Inc. Atlanta, GA. 1981.

[4] Turner, W.C. and Malloy, J.F. Thermal Insulation Handbook. Robert E. Krieger Publishing Company, Malabau, FL, and McGraw Hill Book Company, New York, NY 1981.

[5] Bourne, J. G., et. al. Building Insulation Materials Compilation. Report No. CR-80.001, Naval Civil Engineering Laboratory, Port Hueneme, CA. January 1980.

[6] Horak, H.L., et. al. DOE-2 Reference Manual. Report Nos. LA-7689-M Ver. 211 and LBL-8706 Rev. 1. Lawrence Berkeley Laboratory and Los Alamos Scientific Laboratory. May, 1980.

[7] Department of Energy. Minimum Energy Dwelling (MED) Workbook - An Investigation of Techniques and Materials for Energy Conscious Design. SAN/1198-1. Dec. 1977.

[8] NAHB Research Foundation, Inc. Insulation Manual-Homes, Apartments. Rockville, MD. 1979.

[9] Latta, J.K. "Vapor Barriors: What Are They? Are They Effective?" Canadian Building Digest. p. 175. March 1976.

[10] Munawwar, S.M. "Concept of Vapor Barriors in Thermal Insulation." Journal of Thermal Insulation. Technomic Publishing Co., Inc. Vol. 4, pp. 288-306. April 1981.

[11] Turner, William C. and Johnson, John W. "Methods For Keeping Thermal Insulation Dry to Preserve Its Function For Energy Conservation." Journal of Thermal Insulation. Technomic Publishing Co., Inc. Vol. 2, pp. 67-86. Oct. 1978.

[12] Webber, Joseph F. "Cold Storage Insulation: The Vapor Barrior, Can We Get It Straight?" ASHRAE Journal. pp. 36-38. March 1979.

[13] Adams, Ludwig. "Thermal Conductivity of Wet Insulations." ASHRAE Journal. pp. 61-62. Oct. 1974.

REFERENCES FOR SECTION 3.0 (cont.)

[14] Burch, Douglas M., Contreras, Antonio, G. and Treado, Stephen J. "The Use of Low-Moisture-Permeability Insulation as an Exterior Retrofit System - A Condensation Study." National Bureau of Standards, Center for Building Technology, National Engineering Laboratory. Washington D.C. DE-79-3. No. 3, pp. 547-562.

[15] Tye, Ronald P. and Spinney, S.C. "A Study of the Effects of Moisture Vapor On The Thermal Transmittance Characteristics Of Cellulose Fiber Thermal Insulation." Journal Of Thermal Insulation. Vol. 2, pp. 175-196. April 1979.

[16] Hedlin, C.P. "Moisture Gains by Foam Plastic Roof Insulations Under Controlled Temperature Gradients." Journal of Cellular Plastics. pp. 313-319. Sept./Oct. 1977.

[17] Misselhorn, Donald J. "Some Problems With Insulation Over Suspended Ceilings." ASHRAE Journal. pp. 46-49. March 1979.

[18] Building Safety. "AIA Committee Warns Against Wall Insulation of Frame Buildings." and "Thermal Insulation Testing Criteria Announced." Building Standards. Special Issue, p. 5. Mar.-Apr. 1979.

4.0

Systems and Applications

This section describes many of the applications for thermal insulation. The plates, which start after page 4-28, show numerous methods of insulating various building sections and mechanical components. The techniques shown have all been used in actual construction. As many insulating techniques as possible are presented so the reader may be informed of the available alternatives.

Section 4.1, Building Systems, covers methods of insulating building sections. Many of the illustrations show typical wall, ceiling, roof, floor, and foundation applications for wood frame, masonry, and metal buildings. Window and door treatments are also presented, as are caulking and weatherstripping. Finally, the subjects of building "tightness", indoor air quality, and resulting health effects are discussed. Table 4-1 on the following page summarizes the applicability of various types of insulation to industrial or commercial and residential building components. The information contained in this table is based on several assumptions. The wall cavity insulations for retrofit applications were selected assuming that the cavity has been previously enclosed. Also, the flooring applications are for basement floors as opposed to between-story flooring. The engineer should compare his particular constraints with those used in typical building construction before using this table.

Section 4.2, Mechanical Systems, covers insulation systems for heating and air conditioning ducts, water and process piping, tanks, vessels, and other equipment. The remaining subjects (cold storage facilities, environmental chambers, and marine work) are briefly discussed in Section 4.3, Special Applications.

These plates illustrate suggested methods only and should not be used as shown in drawings or specifications until a thorough engineering investigation of actual design conditions and requirements is conducted. Some of the details may not be appropriate for specific functional, climatic, or geographical conditions. Also, material or system manufacturer's representatives should be consulted for guidance as to the applications and limitations of any particular product, especially in questionable or untried installations. These plates are not intended to circumvent design criteria which may disallow some of the installations shown.

4.1 Building Systems

A great deal of information is available concerning applications of thermal insulation to residential, commercial, and industrial buildings. Publications of several nationally-known organizations illustrate how insulation can be applied in residential wood frame buildings [1-3] and masonry buildings [4-8a]. Government reports [9-11a], privately published works [12-13], manufacturers' installation guides [14-18], and product literature are also sources of insulation installation techniques. State and local agencies, utilities, or other organizations may also provide such guides.

4.1.1 Walls

Residential wood frame walls (Plates 1 - 4) are most often insulated with mineral or glass fiber batts in new construction. Upon installation, batts with a kraft paper vapor barrier may be face stapled or inset stapled to the studs (Plate 1). A polyethylene sheet vapor barrier is recommended for use with unfaced batts (Plate 1), Occasionally, this sheet is also used with faced batts because some

Table 4-1

APPLICABILITY OF THERMAL INSULATION TO BUILDING ENVELOPES

INSULATION	LOOSE FILL INSULATION					RIGID INSULATION BOARDS				
COMPONENT	Cellulose	Glass Fiber	Mineral Fiber	Perlite	Vermiculite	Mineral/ Glass Fiber	Cellular Glass	Cellular Plastics	Wood Fiber	Composite Foam/Mineral
INDUSTRIAL/COMMERCIAL										
Roof/Ceiling										
Above Roof Deck	- -	- -	- -	- -	- -	N-R	N-R	N-R	N-R	N-R
Below Roof Deck	N-R	- -	- -	- -	- -	- -	- -	R	- -	- -
Walls										
In Cavities	R	- -	- -	N	N	- -	N	N	- -	- -
Sheathing or Siding	- -	- -	- -	- -	- -	N-R	N-R	N-R	N-R	- -
Floors										
Concrete Slab	- -	- -	- -	- -	- -	N	N-R	N	- -	- -
Wood or Steel Joists	- -	- -	- -	- -	- -	- -	N-R	N-R	- -	- -
RESIDENTIAL										
Roof/Ceiling										
In-Frame Cavities	N-R	N-R	N-R	N-R	N-R	- -	- -	- -	- -	- -
Above Roof Sheathing	- -	- -	- -	- -	- -	N	- -	N-R	N	N
Cathedral Ceilings	- -	- -	- -	- -	- -	N-R	- -	N-R	N-R	N-R
Walls										
In-Frame Cavities	R	R	R	R	- -	- -	- -	- -	- -	- -
Sheating or Siding	- -	- -	- -	- -	- -	N	- -	N-R	N	- -
Floors										
Wood Joisted	N-R	- -	- -	- -	- -	- -	- -	- -	- -	- -
Concrete Slab	- -	- -	- -	- -	- -	N	- -	N	- -	- -
Basement Wall										
Exterior	- -	- -	- -	- -	- -	N	- -	N	- -	- -
Interior	- -	- -	- -	- -	- -	- -	- -	N-R	- -	- -

N = Used in New Construction

R = Used in Retrofitting

- - = Not applicable

Table 4-1 (Con't)

APPLICABILITY OF THERMAL INSULATION TO BUILDING ENVELOPES

INSULATION	INSULATING BATTS OR BLANKETS		FOAMED-IN-PLACE INSULATION		SPRAYED-IN-PLACE		OTHER	
COMPONENT	Glass Fiber	Mineral Fiber	Urethane Foam	Urea-Based Foam	Cellulose	Mineral Fiber	Insulating Concrete	Reflective Insulation
INDUSTRIAL/COMMERCIAL								
Roof/Ceiling								
Above Roof Deck	- -	- -	N-R	- -	- -	- -	N-R	- -
Below Roof Deck	N-R	N-R	R	- -	N-R	N	- -	N-R
Walls								
In Cavities	N-R	N-R	N-R	- -	N	- -	- -	- -
Sheathing or Siding	- -	- -	- -	- -	- -	- -	- -	- -
Floors								
Concrete Slab	- -	- -	- -	- -	- -	- -	N-R	- -
Wood or Steel Joists	N-R	N-R	N-R	- -	N	- -	N-R	- -
RESIDENTIAL								
Roof/Ceiling								
In-Frame Cavities	N-R	N-R	- -	- -	N	- -	- -	N
Above Roof Sheathing	- -	- -	N-R	- -	- -	- -	- -	N
Cathedral Ceilings	N-R	N-R	- -	- -	N-R	- -	- -	- -
Walls								
In-Frame Cavities	N	N	N-R	N-R*	- -	- -	- -	N
Sheathing or Siding	- -	- -	- -	- -	- -	- -	- -	N
Floors								
Wood Joisted	N-R	N-R	- -	- -	N	- -	- -	N-R
Concrete Slab	- -	- -	- -	- -	- -	- -	- -	- -
Basement Wall								
Exterior	- -	- -	- -	- -	- -	- -	- -	- -
Interior	N-R	N-R	- -	- -	- -	- -	- -	N

N = Used in New Construction
R = Used in Retrofitting
- - = Not applicable
* = Check building codes for limitations.

contractors feel a more reliable seal can be obtained. Semirigid insulating sheathing can be added outside or inside the studs (Plate 2) to increase the overall wall thermal resistance (R-value). When insulating a wall containing water pipes, insulation should be placed between the pipes and the exterior wall, resulting in a thermal bridge between the pipes and inside space and thereby reducing the likelihood of freezing (Plate 2). The double-stud wall is finding applications in "superinsulated" dwellings (Plate 3).

Plates 1 - 4 also apply to retrofit construction, where the insulation may be a blown-in loose fill or a foamed-in-place plastic. Vapor barrier paints may be applied to the interior wall surface in lieu of installing a kraft paper or plastic sheet vapor barrier underneath the wallboard.

Other retrofit applications include residing, where an insulating sheathing and new siding are installed over old siding (Plate 3). Provision should be made for venting between the old siding and the sheathing, especially if the sheathing is foil faced, so that moisture migrating from the building interior does not build up between these layers and possibly result in wall deterioration. Vapor permeable sheathings (e.g., glass fiber) may also be used.

Masonry walls, both residential and commercial, are shown in Plates 5 - 11. The simplest retrofit to an existing building with concrete block walls is adding loose fill insulation or foam to concrete block cores through small holes (1 to 2 inches in diameter) drilled every 4 to 8 feet in the wall; the holes are patched after the job is completed (Plate 5). Concrete block cores are easily filled during construction in new walls.

The addition of foil backed wallboard on furring strips (Plates 6 - 7) or batt or board insulation (Plates 6 - 7) will result in a "new" interior finish. A technique becoming known as "outsulating" [19,20], in which rigid insulation is added to block wall exteriors, (Plate 7) is growing in popularity [19,20]. Computerized energy studies have shown that a masonry building insulated on the outside will consume less energy than if insulated on the inside with the same amount of insulation. This effect occurs because of the thermal storage capacity of masonry materials. The exact percentage savings depends on the climate and whether or not the thermostat temperature is set back at night.

Rigid insulation may be installed in new cavity walls (Plates 8-9). Existing cavities can be retrofitted with loose fill or foam (Plates 8-9). Solid combination walls (and also cavity walls) can be given new insulated interiors, as with concrete block walls (Plate 10). One innovative technique for new construction uses stacks of molded insulating foam blocks to form a cavity into which structural concrete is poured (Plate 11).

In commercial buildings (metal building walls are shown in Plates 12-13), steel studs are sometimes used instead of wood studs (Plates 10-12). Steel stud walls can also be "outsulated" (Plate 12). A number of techniques using prefabricated insulating panels, sometimes with factory-applied interior finishes, are available for commercial buildings (Plates 11,13).

4.1.2 Ceilings

Cathedral ceilings and ceilings that separate conditioned spaces from unheated attics are insulated using techniques very similar to those for insulating walls. Batt or loose fill insulation may be installed between rafters in flat ceilings (Plate 14); care must be taken to keep soffit vents clear (Plate 14) for venting moisture. Rigid insulation board or sheathing may be applied under rafters for extra thermal resistance (Plate 14). Extra-thick batts can extend above rafters (Plate 15), or additional batts or loose fill insulation may be added on top of the first layer (Plate 15). If additional batts are used, they should be installed perpendicular to the original batts unless roof trusses interfere.

Batts should be split at cross bridging and tucked under and over the bridging to maintain a continuous insulation layer (Plate 16).

Insulation should always be kept at least 3 inches away from recessed light fixtures or other electrical equipment that relies on convection with the surrounding air to prevent overheating (Plate 16). Covering fixtures with insulation could pose a possible fire hazard.

Cathedral ceilings may be insulated with batts (Plate 17) or batts plus rigid board (Plate 17) if the rafters are to be concealed. If exposed rafters are esthetically desirable, the spaces between them may be partially filled with rigid insulation. This insulation is subsequently covered with gypsum board or accoustical tile (Plate 17). Rigid insulation may also be applied above a cathedral ceiling; this method does not affect the interior (Plate 17). Sloped ceilings containing either metal beams or purlins instead of wood beams (Plates 18-20) are insulated with rigid board (Plates 18-19), batts (Plate 18), blankets, or combinations of these three insulation types (Plates 19-20). Sprayed-on insulation (Plate 20) may also be used on metal ceilings.

4.1.3 Roofs

Of the several types of commercial building roofs in use today, the built-up roof (BUR) has the longest history. The simplest BUR consists of a roof deck and several alternate layers of asphalt and roofing felt, topped with a protective surface such as gravel or rock agregate. The deck may be wood, concrete, or metal. Roofing felts are usually asphalt-saturated blankets of asbestos, glass fiber, or organic fiber. Basic uninsulated BURs are shown on Plate 21.

Thermal insulation in a BUR system greatly reduces the amount of heat loss or gain through the roof (Plates 22 - 23). Concern, however, has been expressed that increasing insulation thickness will increase the number of BUR failures because the membrane is subjected to more extreme temperatures. Also, it is feared that temperature differences between the top and bottom of the insulation will cause the insulation to deform due to differential expansion, thereby placing additional strain on the membrane. Roof membranes over insulation may reach 80°F above ambient temperature because of solar heating, or may fall to 20°F below ambient temperature because of radiative nighttime cooling [21]. At worst, however, the actual temperature increase for an insulated BUR surface over an uninsulated roof is only about 15°F [22]. Also, 75 to 80 percent of this temperature rise occurs after only 1 inch of insulation is added; additional insulation has only a minor effect on roof temperature [22].

The absorptivity (or color) of the roof surface has been shown to be more important than insulation thickness in determining roof surface temperature. Temperature increases of 25° to 30°F have been estimated to occur between a white roof and a black roof [21,22], regardless of whether or not the roof is insulated. Other sources confirm that fears of shortened roof life as a result of insulation are unfounded [23,24,25]. Other studies are in progress [26].

Because of the wide range of temperatures experienced by an insulated BUR, the insulation must have a coefficient of thermal expansion less than that of the membrane itself, or ridging and cracking of the membrane can occur. Reference [34] recommends the insulation coefficient of expansion be less than one-fourth that of the membrane.

Shear stresses between deck and insulation have caused the insulation to loosen from the deck in some cases. The adhesive or bitumen used should, therefore, have an acceptable shear strength affinity for the deck, insulation, and membrane materials used in the roof. Application temperature may affect these properties and should be carefully controlled.

Insulation materials for BUR systems include rigid glass fiber with asphalt kraft facings, urethane panels with asphalt felt, glass fiber, aluminum foil, or polyethylene-coated kraft paper facing, polystyrene board, polyisocyanurate board with asphalt-asbestos or foil facings, phenolic foam boards, and perlite boards. Various composite boards are also available and include urethane/perlite, perlite/urethane/perlite, glass fiber/urethane, and isocyanurate/perlite. In securing insulation to a steel deck with hot asphalt, about one day will be needed for the asphalt to cure sufficiently to prevent slippage [27]. Asphalt-based adhesives require several days to several weeks to cure, depending on the type of adhesive. Solvent-based mastics will damage polystyrene foam and should not be used as adhesives in this case.

Moisture build-up in BURs is of greater concern than insulation problems. Moisture-related problems include loss of strength and dimensional changes of roofing materials, corrosion, leaching, blistering, rotting of organic materials, and growth of algae, mold, and fungi [28]. Ponded water resulting from poor drainage of a roof can accelerate these problems. Water-saturated roofing materials and insulation will also experience an increase in the thermal conductance [29-32].

Some results of an NBS Study [30-32] on moisture in roofing materials are shown in Figures 4-1 and 4-2 on pages 4-8 and 4-9. To combat related problems, BUR maintenance program manuals are available [33,34]. In addition, ponding on flat roofs can be controlled with tapered insulation systems that provide suitable water drainage (Plate 23).

It has been predicted that nonconventional roofing systems will acquire a large share of the roofing market from the BUR in the 1980s [36]. One nonconventional roof system is the single-ply membrane roof (Plates 24 - 26). These systems use a single sheet of material in place of multiple layers of asphalt and felt. The membrane can be either fixed to the insulation with adhesive and a finish coat applied (an "attached" system), or weighed down with ballast, which is usually smooth stone applied at about 10 pounds per square foot (a "loose-laid" system). With a loose-laid roof, high winds, floating insulation, or workmen can displace the ballast, resulting in a loss of protection of the membrane from fire, wind, and ultraviolet radiation. Periodic inspections would identify any problems, however, and a loose-laid roof can be very successful. Membrane materials include polyvinylchloride (PVC), a PVC/nylon scrim reinforcing /PVC laminate, chlorinated polyethylene (CPE), a CPE/polyester reinforcing/CPE laminate, and an ethylene-bitumen- anthracite dust composite, with or without a glass fiber reinforcing laminate.

A method of retrofitting an existing BUR with insulation involves placing new single-ply membrane roof directly over an old BUR; two examples of this upside-down roof system (USD) are shown on Plate 23. There are several advantages of this scheme: (a) the BUR, if intact, still functions as the primary weather barrier; (b) increased roof insulation can be obtained without having to remove the existing roof; (c) the BUR membrane is protected from temperature extremes by the insulation, thereby allaying any remaining concerns of roof aging [35].

In silicone-urethane roofing systems, a sprayed-in-place urethane foam layer is applied to the roof deck and covered with one or two coats of sprayed-on silicone rubber (Plates 27 - 28). In one 3-year test, silicone-urethane roofs exhibited no appreciable signs of degradation [37].

Insulating perlite or vermiculite concrete is sometimes installed over polystyrene insulating board (Plates 29 - 30). It can be sloped to provide drainage. In addition, this material resists wind uplift, has a 1 to 3 hour fire rating (depending on the details of the roof), and provides a surface that resists traffic and equipment loads when cured [38].

Other metal building roof insulation systems use glass fiber batts and blankets draped over and/or suspended between metal joists or purlins (Plate 31). Industry experts estimate that 5 billion square feet of metal building roofs in the U.S. are underinsulated; such buildings have been identified as a sizable insulation retrofit market [39].

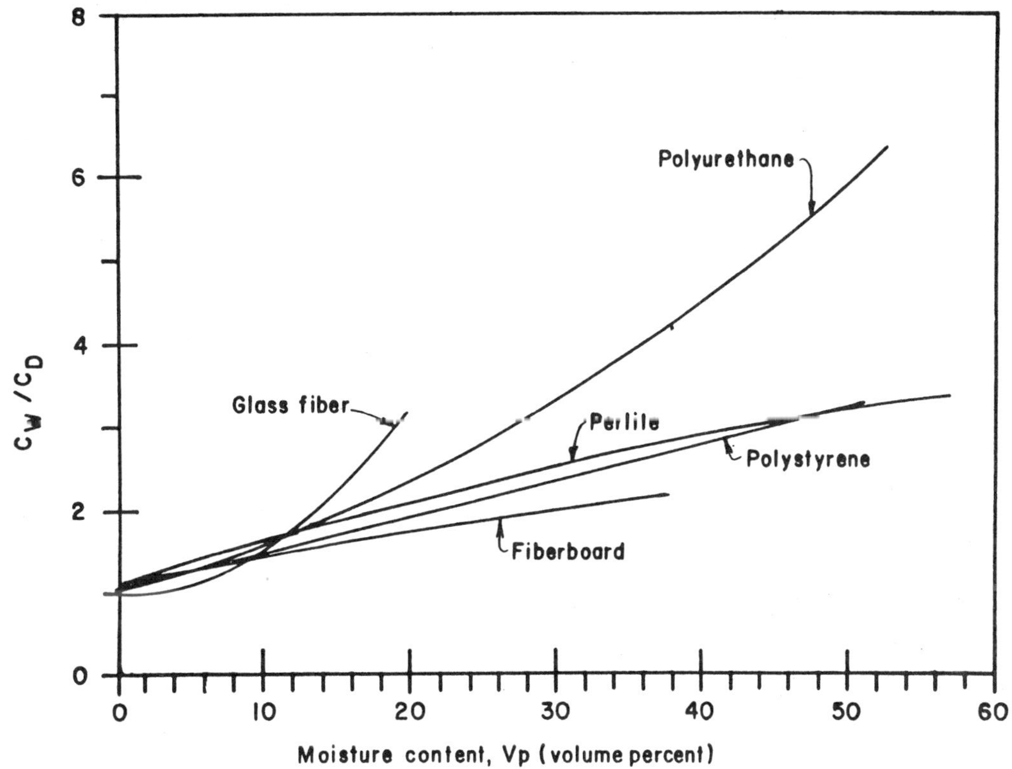

Figure 4-1 Ratio of "wet" to "dry" thermal conductance for BUR system specimens containing 1 inch insulation

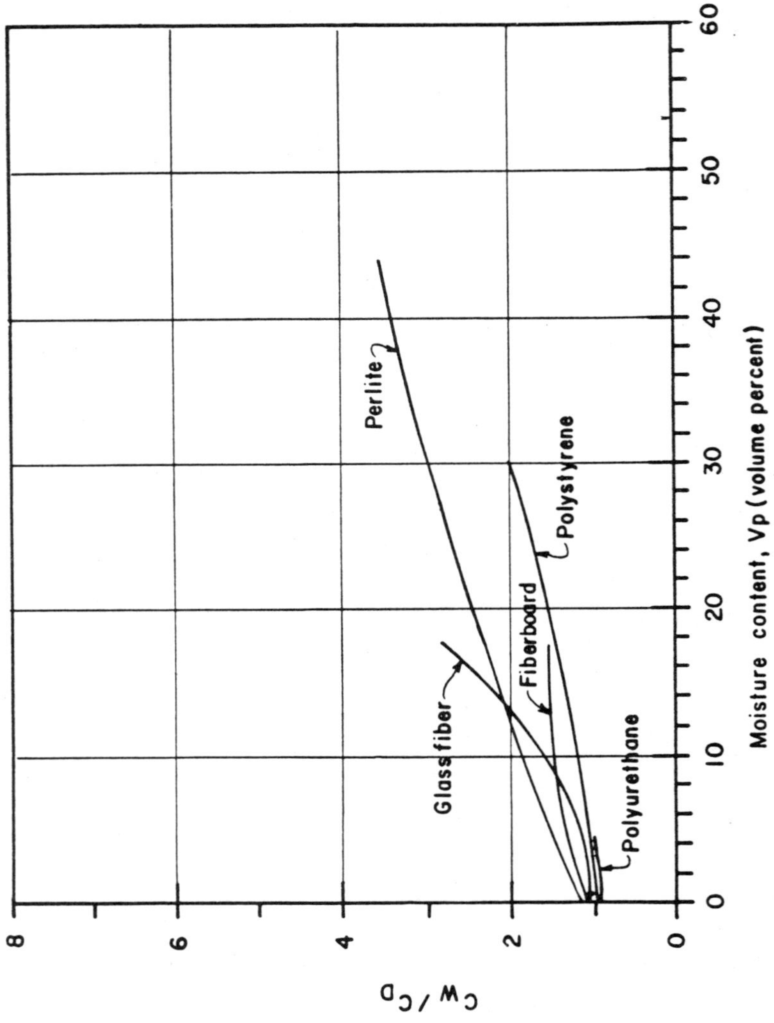

Figure 4-2 Ratio of "wet" to "dry" thermal conductance for BUR system specimens containing 2 inches of insulation

Two interesting studies have addressed dividing roof insulation into above-deck and below-deck portions. In one study, the placement of R-13 glass fiber blanket below a steel deck and the installation of 2-7/16 inches of rigid board above the deck resulted in an initial cost savings of 16 percent [40]. In the other [41], maximum heat load leveling for a concrete deck was shown to occur if one-half of the insulation is above deck and one-half is below; however, the load was relatively flat up to a division of one-fourth to three-fourths of the insulation, either above or below deck.

4.1.4 Floors and Foundations

Floors over unheated basements or vented crawl spaces may be insulated in much the same way as frame walls. Insulation, usually a glass or mineral fiber batt, is installed between the joists (wood or metal). One of several methods is used to keep the batt suspended (Plate 32). The vapor barrier is a facing on the batt and, as usual, should be installed toward the heated space. In new construction, the vapor barrier can be laid over the top of the joists before the subfloor is installed. Ordinary faced batts may be face stapled to the top of the joists (Plate 33).

Over unvented crawl spaces, batt insulation can be fastened to the stringer and header joists (Plate 34) and allowed to drape down the foundation wall. The batt, which should extend at least 2 feet inward from the wall, can be weighted down with rocks, tape, or scrap framing members. A polyethylene moisture barrier should be laid over exposed earth to prevent vapor migration from the ground. Overlapping edges should be taped. Rigid insulation board may be substituted for the batts (Plate 34). Insulating basement walls in these manners can reduce energy losses to 20 to 60 percent of uninsulated values, depending on insulation thickness [42].

Insulating the perimeter of a building will also reduce heat losses to the ground (Plates 34 - 35). Polystyrene foam and cellular glass have been used under concrete slab floors to decrease heat losses (Plate 35). One should note that concrete alone is <u>not</u> a good insulator and that under-slab insulation is often economically justified [43].

4.1.5 Windows

A wide variety of window treatments aimed at reducing heat loss are available. Storm windows are the best known, and these are now being applied to the interior of windows as well as the traditional exterior. Interior storm windows offer an advantage in ease of installation and reduction in thermal bridging and infiltration. Mounting systems include screw-attached vinyl moldings and magnetic strips (Plate 36). Storm window glazings need not be glass; acrylic or flexible plastic are often used.

Rolling shades provide some insulating value. Conventional shade fabrics can reduce heat loss through windows by 25 to 30 percent, while insulating fabrics can reduce heat loss by 35 to 45 percent. An additional 10 to 20 percent reduction can be achieved with track systems that seal the edges of the shade to the window [44].

Insulating shutters can be installed inside or outside a window, or even between the panes of a double-pane window. Exterior shutters are possibly the most desirable option in terms of preventing condensation and minimizing thermal stresses in the window glass [45].

Windows with excessive infiltration that cannot be sufficiently corrected by storm windows, caulking, weatherstripping (see Section 4.1.7), or other repairs are subject to two possible corrective actions: they may be replaced entirely with new wood or vinyl sash windows [46], or permanently covered with opaque, laminated insulating panels. Vinyl sash windows, or aluminum frame windows incorporating a thermal break (usually rigid vinyl or high-strength polyurethane) of an adequate thickness, may offer twice the resistance to heat flow as an ordinary aluminum frame window [47].

Not only do storm windows, rolling shades, and insulating shutters save building heat in winter, they can also reduce cooling loads in summer by intercepting direct sunlight that would otherwise be absorbed by the cooled space. Heat-absorbing or reflective sheets of glass or acrylic, will transmit less sunlight than ordinary window glass. Reflective plastic films may be applied to existing windows.

The amounts of solar energy admitted to a room through (a) a clear window, (b) a clear window with a reflective film, and (c) a window with heat-absorbing glass are compared in Figure 4-3 on the following page. Note that heat-absorbing glass and reflective films will also reject solar energy in winter months, resulting in increased heating requirements. A means of raising or removing the absorbing glass or film may therefore be beneficial [48]. Reference [48] also presents a method of estimating the potential energy savings from these materials.

Skylights, used to provide daylighting, retain heat less efficiently than an equivalent section of insulated ceiling or roof. This heat loss problem can be minimized if the skylights contain double or triple glazings and thermal breaks. Unwanted solar gain during the summer can be offset with translucent or reflective glazings. Models are also available that open to provide ventilation.

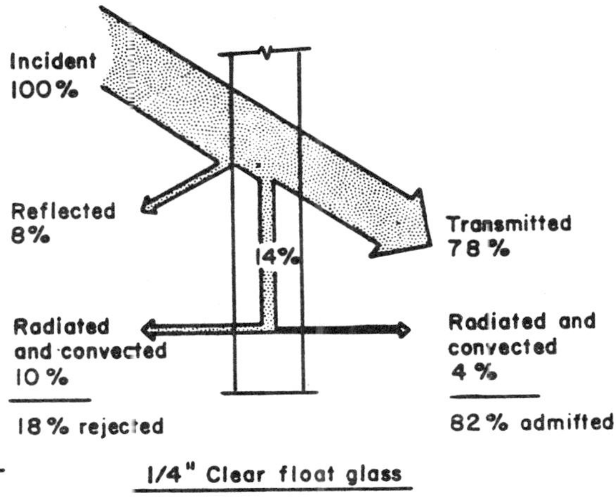

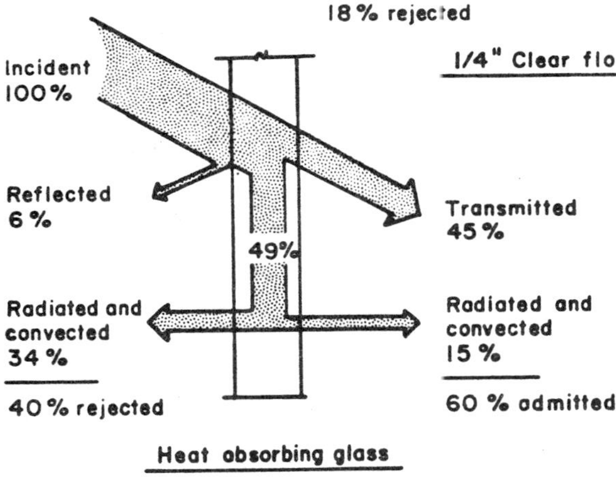

Figure 4-3
Solar energy admitted through three types of glass

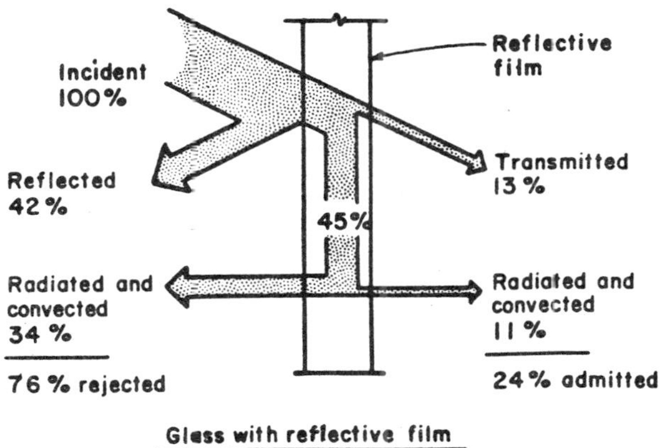

4.1.6 Doors

Storm doors can be added to residential entrances to decrease heat losses. On sunny winter days, the primary door can be opened to allow sunlight into the living space--an added benefit. A poor primary door, however, should be repaired or replaced before a storm door is added. Replacement entrance doors are available with foam insulation cores and fully weather-sealed perimeters. Sliding storm doors for existing glass patio doors are also available.

Commercial building entrance doors are available with insulating (double-pane) glass, thermal breaks, and full weatherstripping. For industrial buildings, traffic and access doors can be obtained with cores containing foam or loose-fill insulation.

Several manufacturers now produce insulated overhead garage doors for industrial and commercial buildings (Plate 39). These doors usually have cores containing 1-1/2 inches to 2 inches of rigid polyurethane foam. Again, door perimeters are fully sealed against adverse weather.

4.1.7 Caulking and Weatherstripping

Caulking and weatherstripping are used to reduce the rate of air infiltration into a building. Estimates of the cost of heating or cooling infiltrated air range from 17 percent to 67 percent of the annual energy cost of a dwelling. One study of 29 electrically heated homes showed that caulking and weatherstripping reduced infiltration an average of 41 percent [49].

In addition to saving energy, caulking and weatherstripping produce other benefits. By sealing a building against outside air that is dry in winter and humid in summer, these products can help maintain more comfortable humidity levels inside the building. Reducing drafts also add to comfort in a building. Furthermore, a dwelling which is effectively sealed against outside moisture is much less likely to experience structural degradation [50].

There are two basic methods of measuring infiltration. With all the doors and windows of the building closed, a fan can be installed in a window or doorway; the flow rate (or pressure rise) of the fan air stream is then measured. Alternatively, a tracer gas, such as hydrogen, helium, ethane, or sulfur hexafluoride, can be introduced into a building and its rate of dilution measured. Descriptions of these methods may be found in References [49] and [51] through [54]. Several analytical methods are describe in References [55] through [57]. The ASHRAE crack method [57] is based on an estimate of crack areas around windows and doors; however, this method is recognized as imprecise, and engineers and architects often multiply the estimate by a "geographical" or "experience" factor [49]. The air change method of estimating infiltration is often used with sufficient accuracy for sizing heating and cooling equipment.

A number of caulks are available. Oil-based caulks are not very expensive; however, they are the least durable (1 to 5 years) and have poor elasticity. Moderately priced acrylic latex caulks are more durable (2 to 10 years), have fair to good elasticity, clean up with water, and are paintable. Butyl rubber caulks are slightly more durable (5 to 10 years) than latex caulks and are about the same in elasticity and cost; however, they require solvent clean-up and are more difficult to apply. High-performance caulks include polysulfide, polyurethane, and silicone caulks. These are the most expensive caulks and are somewhat difficult to apply, but they have excellent elasticity and may last 20 years or more. A thorough comparison of caulking compounds is found in Reference [58]. Sealant manufacturers also publish design guides. The American Concrete Institute has published guides for concrete joint sealants [59,60].

Numerous applications for caulking exist, including those areas where siding intersects with windows, doors, brick veneers, chimneys, or foundation walls, and also at siding corners. Other areas are penetrations in outside walls and attic ceilings for water faucets, electric and phone lines, gas pipes, mail chutes, and dryer vents. Caulking may also be applied around air conditioners and skylights, and is especially effective when applied to cracks in siding, bricks, stucco exteriors, and foundation walls.

Whereas caulking is usually applied to stationary joints, weatherstripping is used for surfaces that move, primarily around doors and windows. Weatherstripping can also be used in some applications for caulking. Various forms of weatherstripping are illustrated in Figure 4-4 on the following page. Felts and adhesive-backed foams and EPDM weatherstrips are easily applied but low in durability. They should be used only where they will be compressed and not subjected to abrasion. Tubular or triangular gaskets, made from vinyl, PVC, thermoplastic or silicone rubber, may be nailed onto or inserted into routed grooves in surfaces that either butt or slide. Spring plastic or metal are especially effective for door frames; metal weatherstrips are quite durable. Nylon brush weatherstrips are used in many window and door frames. A variety of weatherstrips are compared in Reference [61].

To prevent air and moisture penetration, a number of door thresholds are used (Plates 37 - 38). The simplest of these is the saddle threshold (Plate 37). Sweeps can be added for better moisture protection; these are made of felt, rubber, and, more recently, flexible vinyl, or neoprene (Plate 37). Foam-filled vinyl tubular weatherstripping can be used as a sweep (Plate 37).

The interlocking saddle provides excellent protection against dust, drafts, and light (Plate 37). The bubble seal provides a good weather seal, but is subject to premature wear from traffic (Plate 38). With or without a rain guard, the door shoe (Plate 38) gives equal or better weather sealing than the bubble threshold and is not subject to traffic. A complete comparison of threshold types is given in Reference [62]; and less extensive comparisons appear in References [50] and [63].

Various types of duct tapes, flue tapes, and weather sealing tapes are available to combat infiltration and air leaks to nonconditioned spaces.

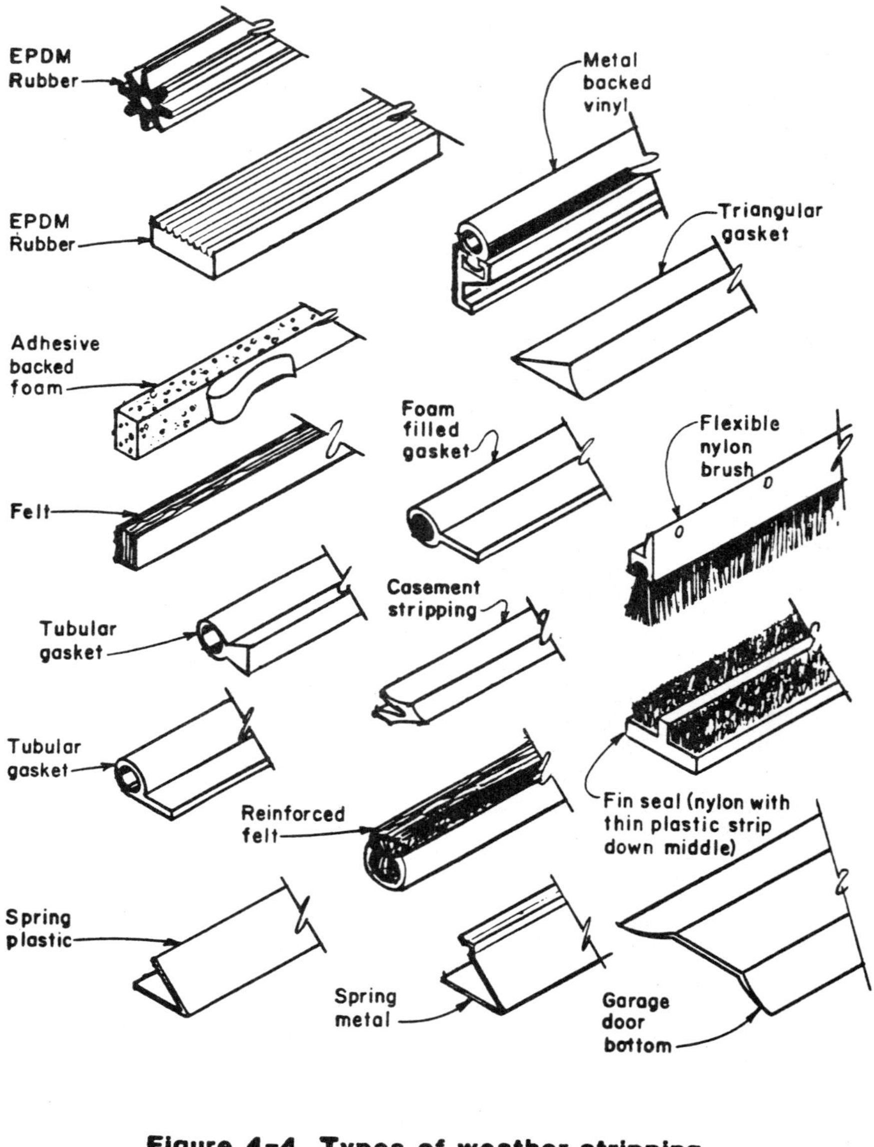

Figure 4-4 Types of weather stripping

An air infiltration barrier in the form of a 6-mil thick sheet of high-density polyethylene fibers, intended to be wrapped around the sidewalls of a building before the sheathing is put on, is now on the market. According to the manufacturer, this product will reduce sidewall infiltration up to 90 percent, and ceiling heat loss up to 20 percent; in addition, it is said to be highly permeable to moisture, thus preventing a condensation barrier [64] which would occur if a material such as polyethylene were used.

4.1.8 Indoor Air Quality and Health Effects

An aggressive infiltration reducing campaign can cause a dwelling to become "too tight" in extreme cases. Sealing some houses can eliminate their main source of ventilation, resulting in a build-up of indoor pollution.

Sources of indoor pollution include gas stoves, malfunctioning forced-air furnaces, and unvented gas space heaters; these produce carbon monoxide and oxides of nitrogen. Formaldehyde is present not only in urea-formaldehyde foam insulation, but in glues used in particle board and plywood as well as plastics used in interiors.

Radon, part of the radium-226 decay chain, is a gas which has four short-lived daughter products that attach themselves to airborne particles. These particles are retained in the lungs when inhaled, resulting in a radiation dose to the lungs as the radon daughters decay to lead. Since radium occurs naturally in small concentrations in rocks and soil, these materials, as well as concrete and brick, are radon sources. Several references [65-70] contain more extensive information on indoor pollutants.

Additional sources of indoor pollutants include paints, glues, cleaning products, aerosol sprays, and furniture polishes [50,66,67]. Smoking of tobacco injects a number of chemical compounds and particulates into the air; generation rates and allowable concentrations in indoor spaces are summarized in Reference [69].

Signs of a very tight house include condensation on the inside of double-glazed windows in winter (excessive moisture build-up), musty smells or lingering odors. Possible health problems indicating polluted indoor air are frequent headaches, watery eyes, nausea, and fatigue [50,70].

While most experts agree that tightening most existing houses to the point of creating indoor pollution problems would be difficult, there is concern over extremely air-tight older homes and some "energy-efficient" homes built since the mid-1970s. These houses may be tight enough to have infiltration rates as low as 0.2 air change per hour compared to about 1 air change per hour for the average American home and a minimum recommended rate of 0.5 air change per hour [70]. Infiltration safety guidelines (reflecting only the opinions of experts in the field) relating to the fan pressurization test method are shown in Figure 4-5 on the following page [49].

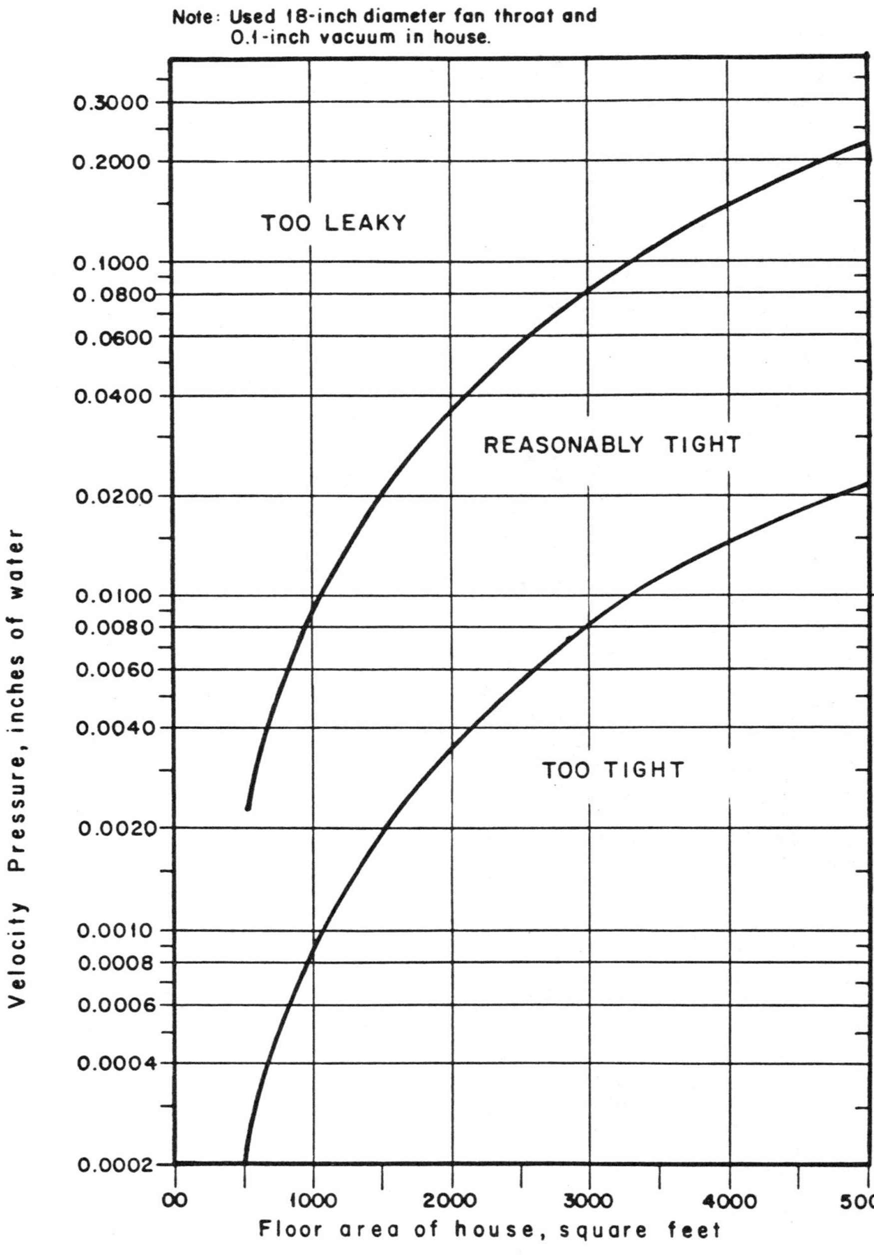

Figure 4-5 Infiltration safety guidelines

Acceptable indoor air quality and energy efficiency are not necessary incompatible goals. Often, "spot ventilation" - appliance exhaust fans or an opened window or door - will take care of temporary problems. Electrostatic precipitating air filters and high-efficiency fiber filters are effective at removing particulates from the air. Charcoal filters may be effective in removing polluting gases and vapors. The air-to-air heat exhanger is becoming more well-known as a method of providing ventilation while conserving energy [50,70]. Air economizer cycles, in which a control system allows outside air to be brought into a building, will help reduce pollution levels in both commercial and residential buildings [71,72].

4.2 Mechanical Systems

Mechanical systems which would benefit from being insulated include hot water, steam, and process piping, heating and cooling air ducts, tanks, vessels, furnaces, and so forth. Failure to insulate these systems is costing industry as a whole millions of dollars every day. For example, a study of 15 energy-intensive industries, conducted by the Thermal Insulation Manufacturer's Association (TIMA), found that, of 325 million linear feet of steam pipe used by the industries surveyed, 72 million feet (22 percent) is totally uninsulated. The rest is mostly underinsulated. Upgrading the insulation on these pipes to economic thicknesses would cost $6.2 billion, but would save the energy equivalent of 305,000 barrels of oil each day (worth $6.8 million per day - $2.5 billion per year - at $22.33 per barrel of oil equivalent), and would pay for itself in just 30 months [73]. After the payback period, insulation on these pipes would save industry additional billions of dollars every year.

This section describes and illustrates a number of techniques used to insulate piping, ducts, and equipment (Plates 40 - 45). It should serve as a good overview of insulation systems for mechanical components. The reader is referred to sources of more detailed information relating to various specific subjects.

4.2.1 Piping

Modern pipe insulation is basically available in the following forms:

- Preformed flexible foam or elastomeric rubber insulation (Plate 40), generally available only in small sizes, can be slipped over pipes before connecting pipes to fittings, or can be slit, slipped around pipes, and resealed with adhesive in retrofit situations.

- Rigid and semirigid preformed insulation comes in halves or slit for slipping around pipe (Plate 40). Insulation materials include glass and mineral fiber, cellular glass, calcium silicate, rigid polyurethane, polyisocyanurate and phenolic foams, and even perlite.

- Prefabricated piping systems of the pipe or pipes - within-a-pipe arrangement (Plate 41) may employ glass fiber, cellular glass, calcium silicate, and rigid urethane foam.

- Flexible blankets of glass or mineral fiber, and polyurethane may be cut to size and wrapped around pipes.

- Elastomeric tape may also be wound around pipe (Plate 44).

- Poured-in-place insulation may be used, but requires an enclosure around the pipe to hold it in place. The most frequent piping application for poured-in-place insulation is underground piping.

Pipe insulation jackets are used for a variety of purposes, including protection from weather, sunlight, chemicals, fire, physical abuse, and on cold piping, condensation. Jackets may also serve to physically hold insulation in place, although tape or metal bands often perform this function. Jacket materials include aluminum, stainless steel, paper and metal foil laminates, PVC, PVF, vinyl, vinyl laminated to polyester and foil, and glass fiber reinforced plastic. Various mastics and cements may also be used instead of other jackets.

Material compatibility is a concern with pipe insulation jackets and adhesives. On hot piping, water vapor migrates from the hot pipe toward the jacket, and may condense on the inside jacket surface. The condensate is usually alkaline and will corrode the jacket from the inside. Unprotected steel, aluminum, and galvanized steel jackets are most susceptible to this kind of corrosion. Aluminum and steel with anticorrosion coatings and stainless steel are less susceptible. Insulated pipes subjected to corrosive atmospheres may require stainless steel jackets.

Proper adhesives and sealants for jacketing are usually supplied by the jacketing manufacturer, but the user must ensure proper application of these products. In outdoor piping, special attention must be paid to sealing joints and seams in the jacketing. Allowance must also be made for thermal expansion of the piping to ensure jacket integrity as the piping expands and contracts. A leaky jacket allows water to pass through it, degrading the insulation. An additional problem with wet insulation is galvanic corrosion when the pipe and jacket are made of different metals. Galvanic corrosion can also occur between buried pipes if the ground is wet.

Flammability of mastics used as pipe insulation coatings is a concern in some applications. During a fire, a mastic with a relatively low melting point may drip onto other combustible materials, thus contributing to the spread of the fire, even though the mastic itself does not actually burn.

Insulation around pipe fittings is usually similar to that used on the adjacent pipe. Flexible insulation can be preformed into fitting covers that are simply slipped over the fitting and cemented into place (Plates 42-44). Fitting insulation may be fabricated out of rigid straight pipe insulation, secured with adhesive or wire, and covered with metal or plastic fitting covers or mastic (Plates 42-44). Blanket insulation may also be secured in the same manner (Plate 42). A thick coating of insulating cement, reinforced with wire mesh or glass fabric, is another alternative (Plates 42-44). Flexible tape, if used on the pipe, can be wrapped around fittings as they occur in the pipeline (Plate 44).

Summaries of pipe insulation considerations and techniques may be found in References [74] through [77]. References [76] and [77] also cover jacketing, finishes, and underground piping in more detail.

Reference [77] discusses heat-traced piping, thermal expansion and contraction considerations in high- and low-temperature piping, insulation attachment methods, and pipe supports and hangers for insulated pipes. Reference [78] describes design considerations for underground piping systems and related Federal Construction Council reports and specifications.

4.2.2 Equipment

Equipment insulation materials are much the same as those for piping. Rigid blocks or boards (glass or mineral fiber, polyurethane, polyisocyanurate, cellular glass, alumina fiber, or calcium silicate) may be applied directly to flat surfaces, or scored, beveled, or mitered (in the field or at the factory), and applied to curved surfaces (Plate 45). Semirigid blankets will also conform to curved surfaces (Plate 45) and are made from glass and mineral fiber. Flexible elastomeric insulation is attached to equipment surfaces with adhesive. Mineral fiber with binders and polyurethane and polyisocyanurate foams may be sprayed onto equipment (Plate 45). Enclosures around equipment may be filled with loose glass, mineral or alumina-silica fiber, or vermiculite or perlite. Insulating cements and mastics are also applied to mechanical equipment.

Jacketing is also applied to insulated equipment; sheet metal is the most common jacketing in larger applications (Plate 45). Semirigid and flexible blankets sometimes come faced with vinyl or other materials. Elastomeric sheets and sprayed-on insulation can be protected with sprayed-on coatings (Plate 45).

General information about equipment insulation may also be found in References [74], [75], and [77]. Again, Reference [77] is a good source of information about thermal expansion and contraction in insulated equipment and insulation attachment methods.

4.2.3 Ducts

Unlike pipe and equipment insulation, duct insulation appears to be limited to two materials: glass fiber and elastomeric rubber. Of these two materials, glass fiber is by far the more prevalent.

Existing rectangular metal ducts may be wrapped with a faced flexible blanket (Plate 46), or flexible elastomeric insulation may be applied with adhesive and covered by a protective coating if desired. Rigid faced board may also be applied (Plate 46). New ducts can be fabricated out of this product as well, without the use of a metal duct at all (Plate 46) [79]. Duct liner can be adhered to the inside of rectangular ducts (Plate 46). Existing round ducts are also wrapped with faced flexible blankets (Plate 47). New round duct may be either rigid or flexible fiberglass (Plate 47). Flexible round ducts contain an inside liner of either corrugated metal or two plastic films encasing a stiff wire helix.

Facing materials are more numerous than insulation materials themselves. Both flat and round duct insulation are available unfaced; one manufacturer offers a rigid unfaced round duct that contains an aluminum foil air barrier within the fiberglass insulation. Rigid and blanket insulations may come with aluminum foil or foil-scrim-kraft (FSK) vapor barriers; blankets also come with vinyl facing. Duct liner has a fire-resistant coating on the side that faces the air stream. Flexible round insulated duct facings include polyethylene, polyolefin, and reinforced metallized plastic films.

Heat losses from operating ducts have been investigated in a test program sponsored by TIMA [80]. Actual heat transfer coefficients (U-values) were generally higher than theoretical values. Duct construction technique was found not to be significant to thermal performance, <u>provided</u> the joints were sealed. More energy can be lost through air leakage from an unsealed duct than by heat transfer.

Duct liner, though mainly used for its noise-deadening properties, does have thermal advantages, and a higher density liner appears more beneficial than a greater thickness of low-density liner. U-values for duct liner increase as air velocity increases. Calculated U-values for ducts insulated with duct wrap, if based on actual installed thickness, were very close to test values. Duct wrap is often slightly compressed in actual practice, sometimes unavoidably, and TIMA found the U-value at 50 percent compression increased by 39 percent over the uncompressed value. Duct wrap has little effect on U-value if joints are sealed effectively.

Rigid duct board U-values were about 15 percent higher than predicted with minimal effects of air velocity. Flexible ducts with impervious liners performed very close to predictions. However, U-values for flexible ducts with pervious liners increase with air velocity; in fact, the U-value at an air velocity of 3000 fpm is twice the predicted U-value at that velocity. For sealing duct joints, pressure-sensitive aluminum tape is most often used; however, contractors have indicated that tapes are needed which can be applied at warmer and colder temperatures and which have improved longevity [81,82].

References [75-77] contain general information on duct insulation.

4.3 Special Applications

A few special applications for insulation do not fall into the Building or Mechanical Systems categories. One such application is commercial or industrial cold storage facilities.

Cold storage for food preservation originated in the use of natural caves which were unaffected by warmer temperatures above ground. Refrigerated, though uninsulated, caves are still operating in Scandinavian countries. Early refrigerated buildings used charcoal, wood shavings, sawdust, mineral wool, and granulated cork as insulating materials. Later, slab cork and polystyrene became common [83]. Polystyrene is still used in modern facilities, as are fiberglass and urethane foam. Further information on cold storage facility design may be found in References [76,77]. Suggested architectural specifications are contained in manufacturer's product literature.

The food industry has other insulation restrictions. Since food must be kept free of contamination, insulation should be nontoxic and free of dust and slivers. Vapor barriers and protective coatings have the same restrictions [77].

Environmentally controlled spaces, including computer rooms, clean rooms, and electronics laboratories, along with weather and altitude chambers, will also have definite requirements in terms of permissive finishes. Insulation materials for environmental spaces include fibrous materials, plastic foams, cellular glass, and reflective insulation (sometimes evacuated) [76].

Marine environments also have several specific requirements for insulation. High humidity requires that insulating materials and coatings resist fungus growth. Shipboard applications should be considerate of clearances and vermin-proofing [77].

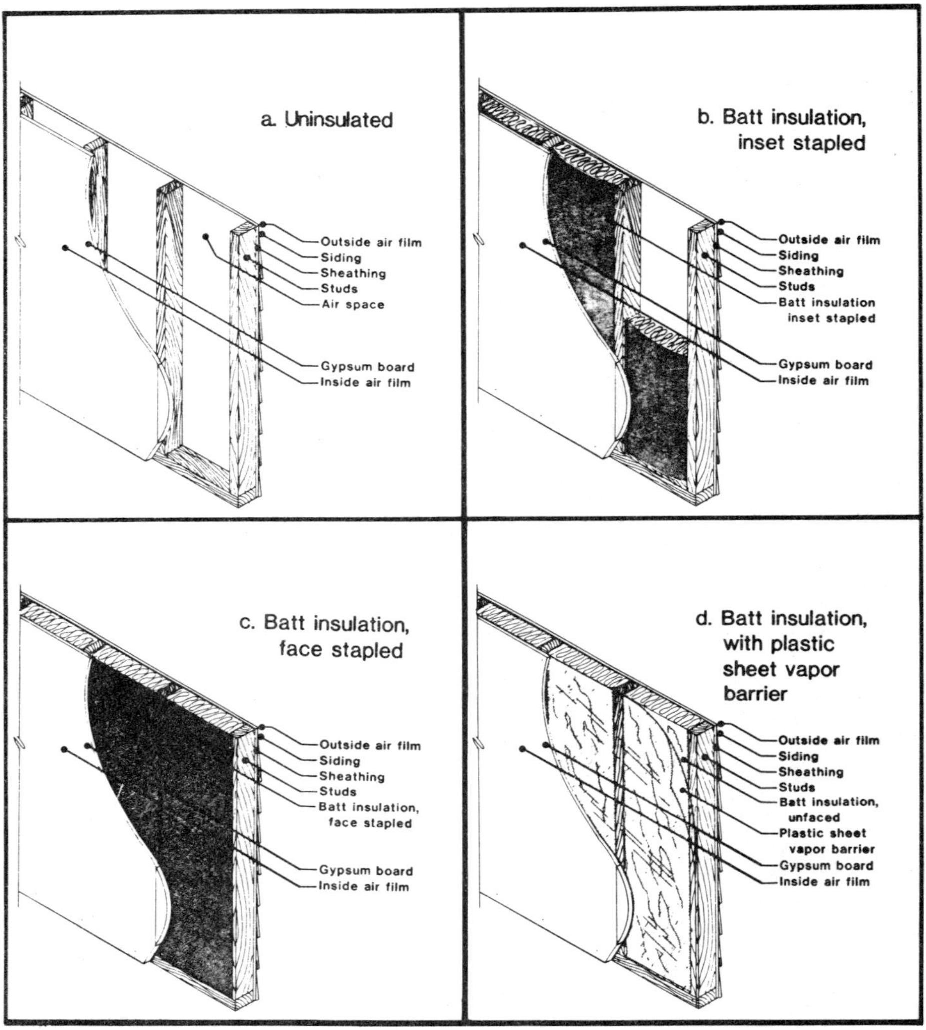

PLATE 1. Frame Walls – A

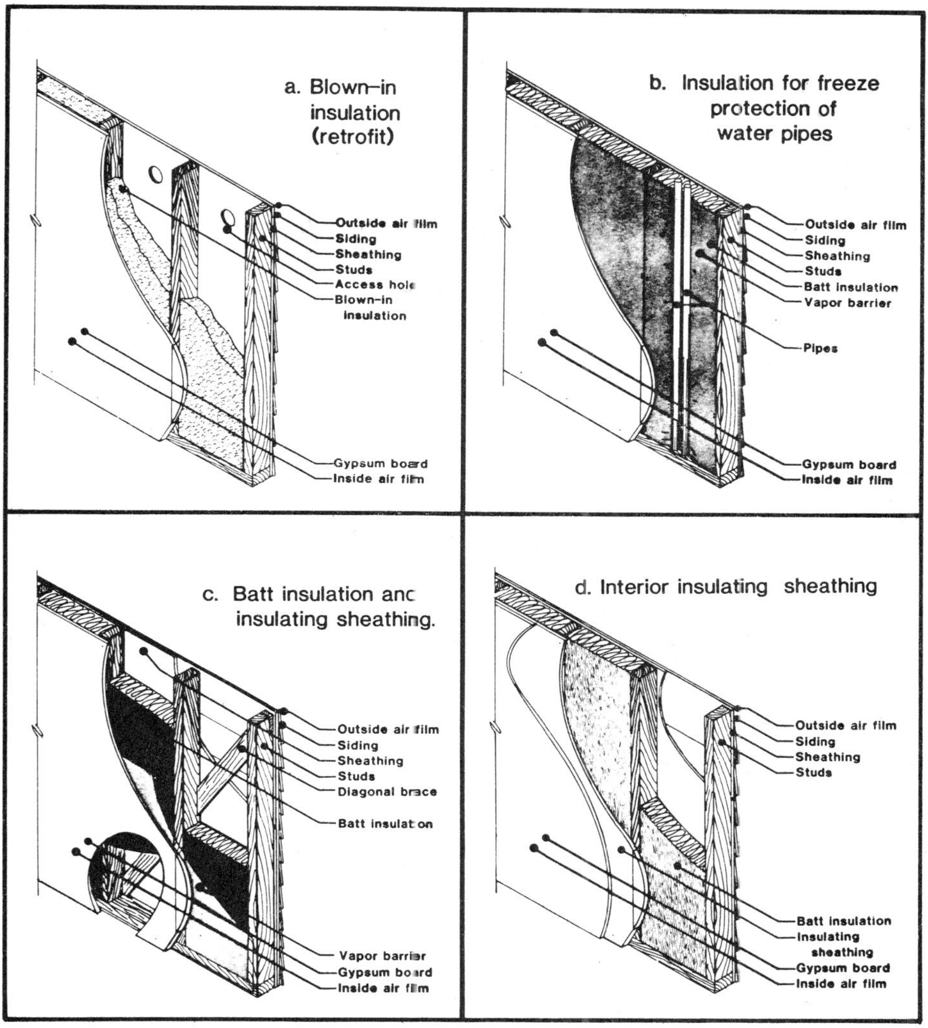

Plate 2. Frame Walls – B

2a: Access hole is plugged when insulation job is completed.

2b: Alternative method: pack space behind pipes with loose insulation, or cut and fit a piece of insulation to the space.

2c: Vapor barrier may be an integral batt facing or foil backing on gypsum board.

2d: Foil facing on insulating sheathing acts as vapor barrier. Other facings may require separate vapor barrier.

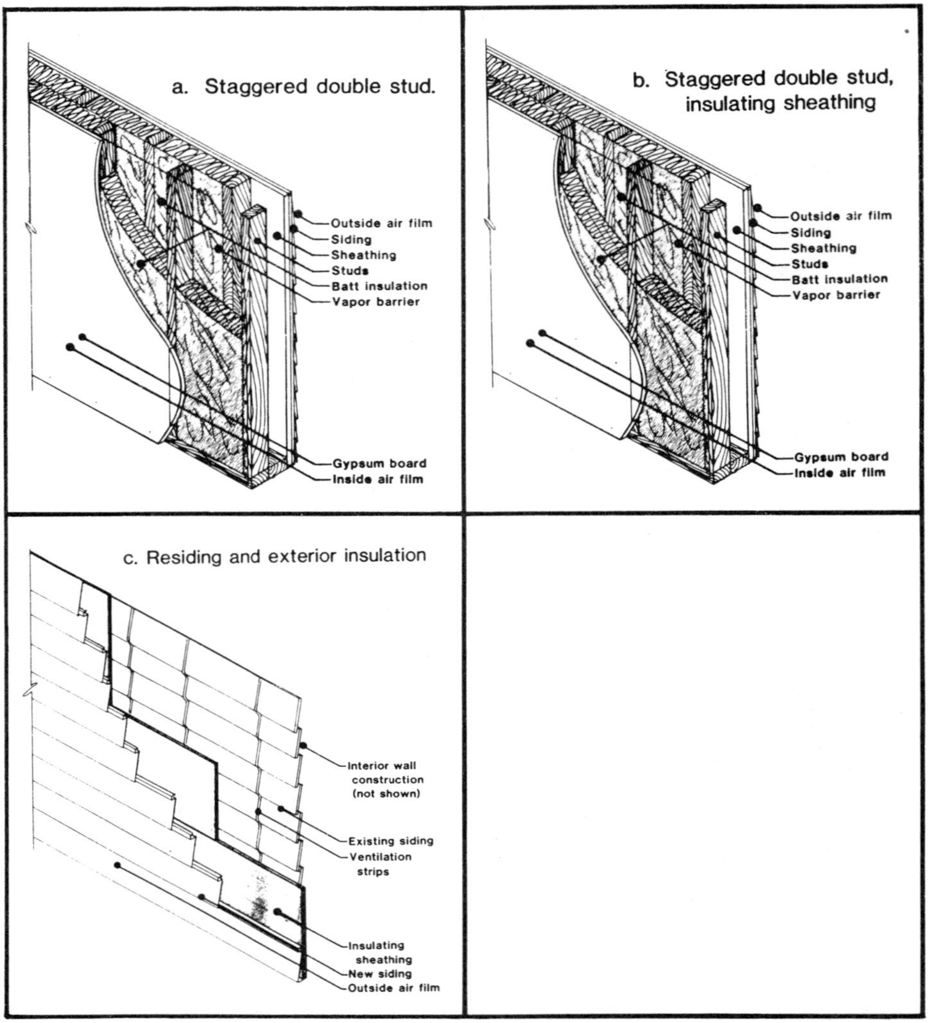

Plate 3. Frame Walls – C

3a,b: Vapor barrier may be an integral batt facing or foil backing on gypsum board.

3c: Ventilation required between existing siding and insulating siding sheathing if sheathing facing is a vapor barrier.

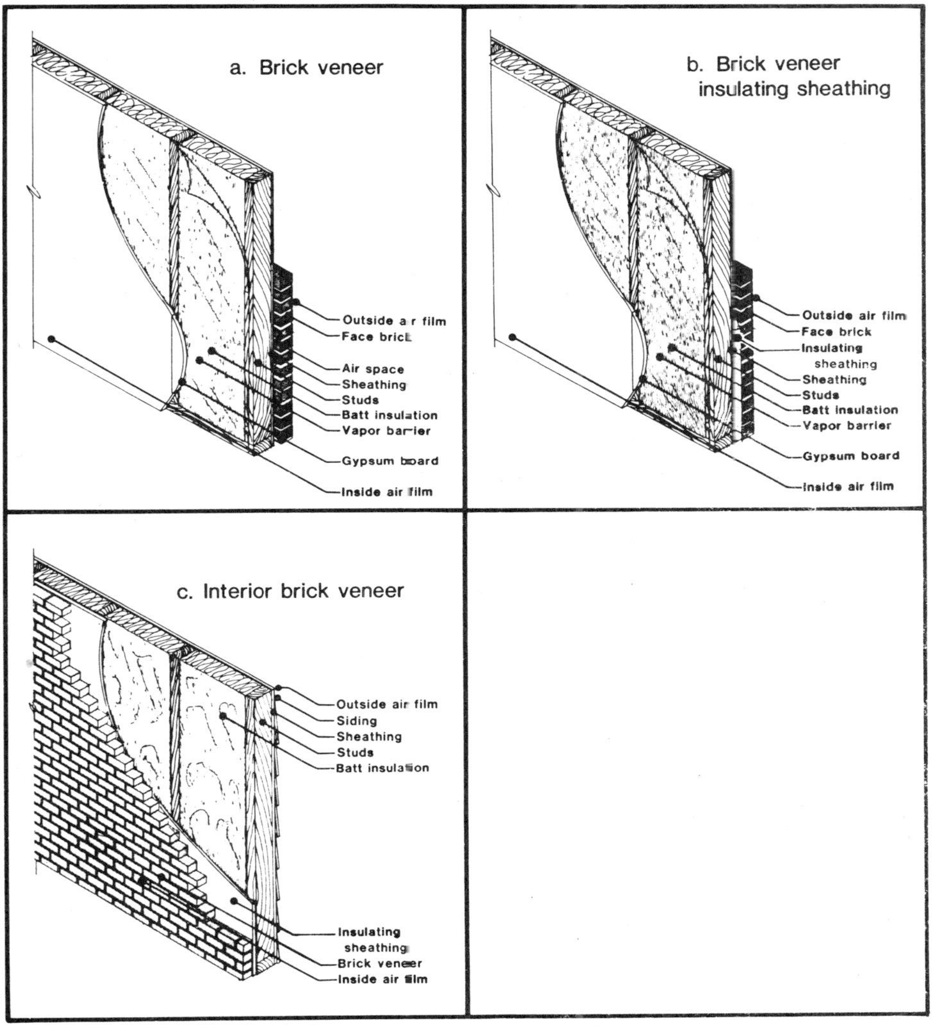

Plate 4. Frame Walls – D

4a,b: Vapor barrier may be an integral batt facing or foil backing on gypsum board.

4c: Foil facing on insulating sheathing acts as vapor barrier. Other facings may require separate vapor barrier.

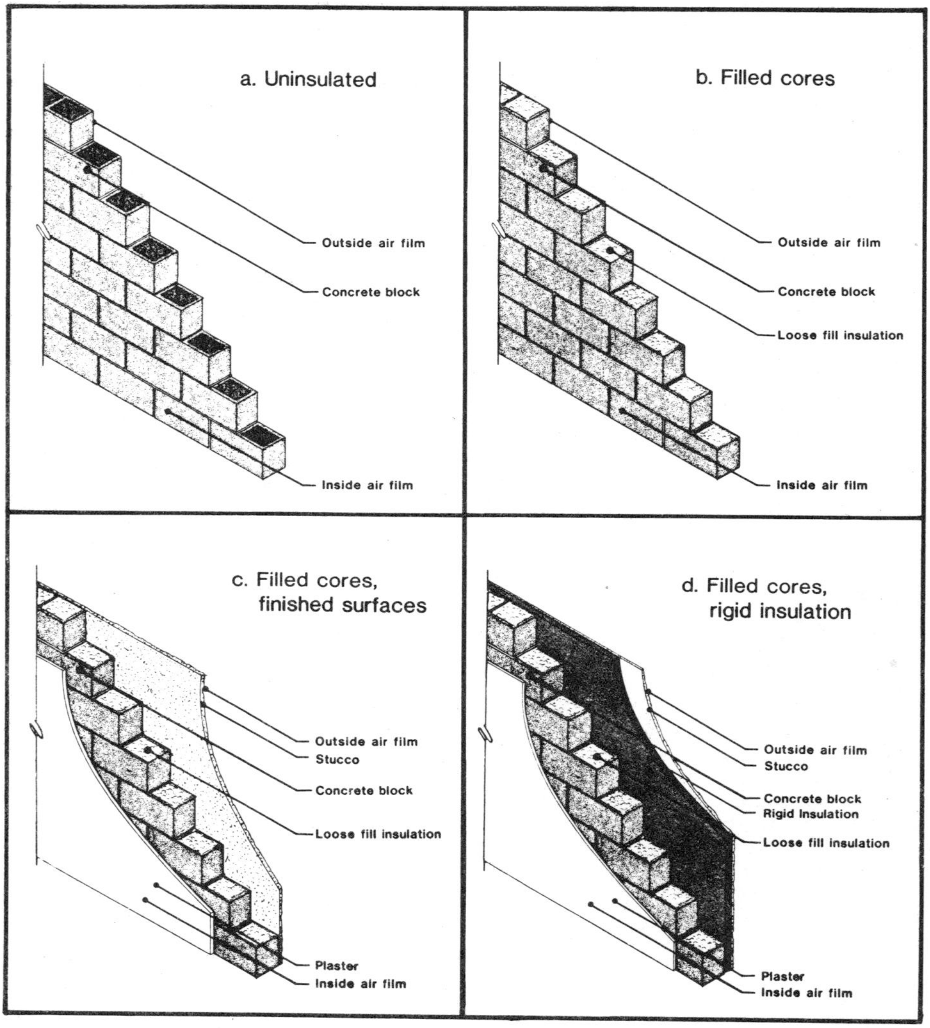

Plate 5. Masonry Walls – A

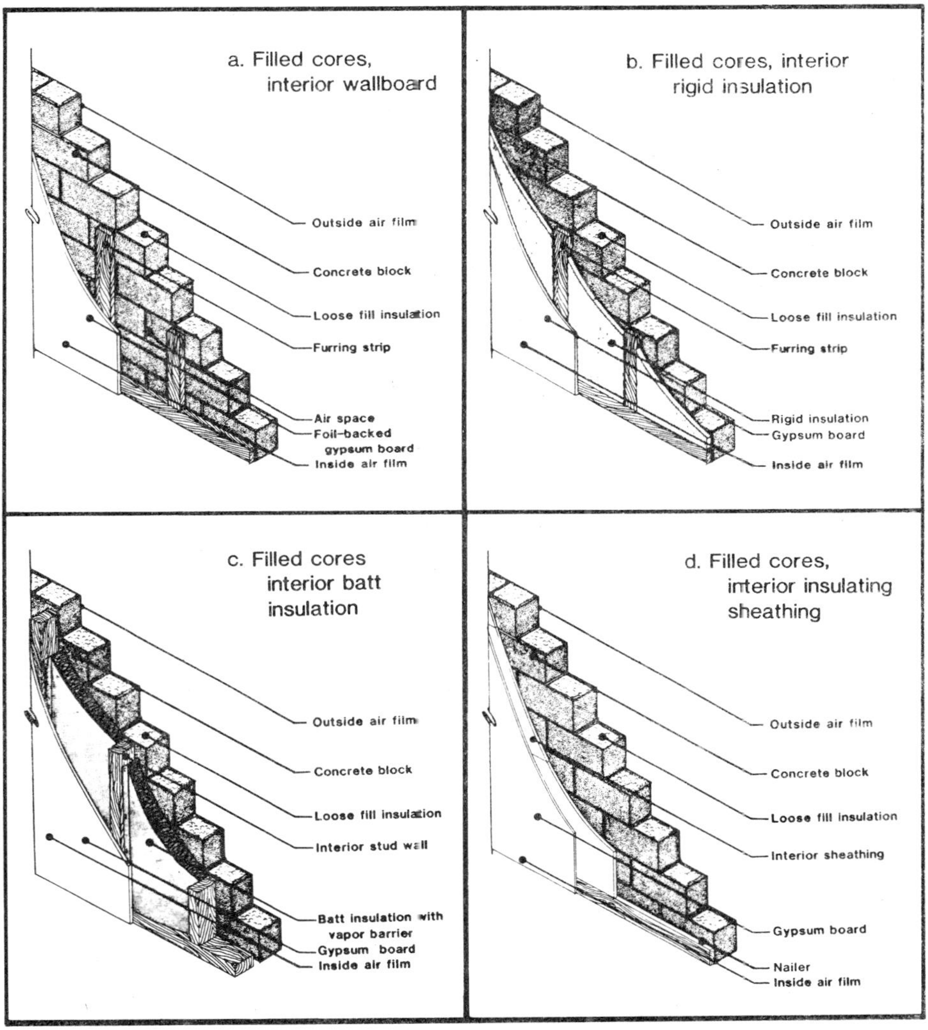

Plate 6. Masonry Walls – B

6c: With unfaced batts, vapor barrier may be either polyethylene sheet or foil backing on gypsum board.

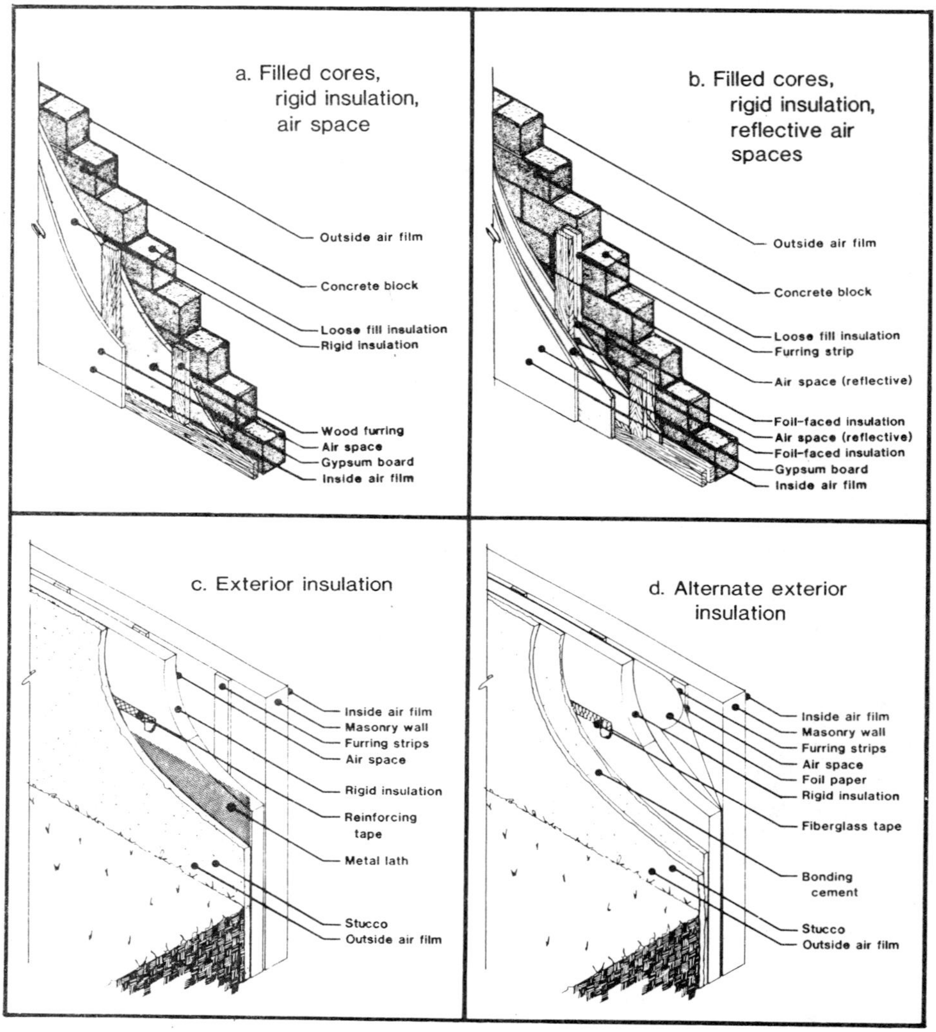

Plate 7. Masonry Walls – C

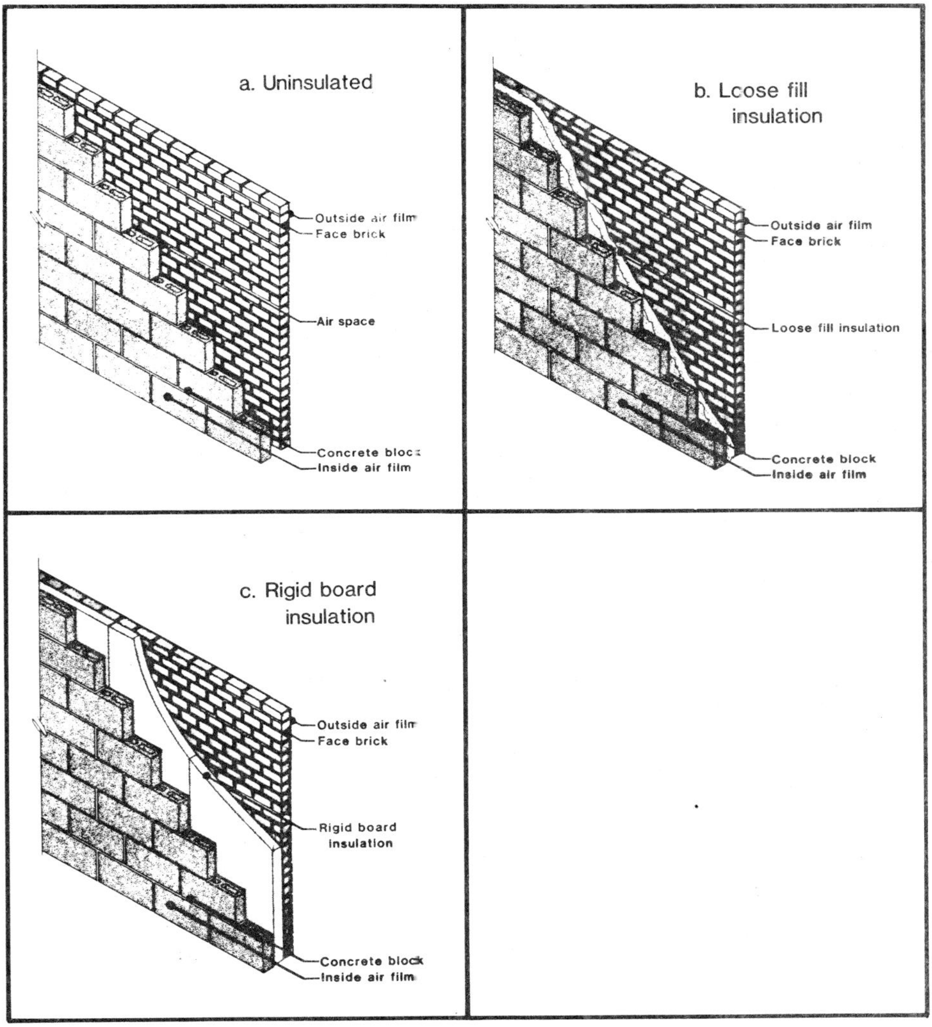

Plate 8. Masonry Cavity Walls – A

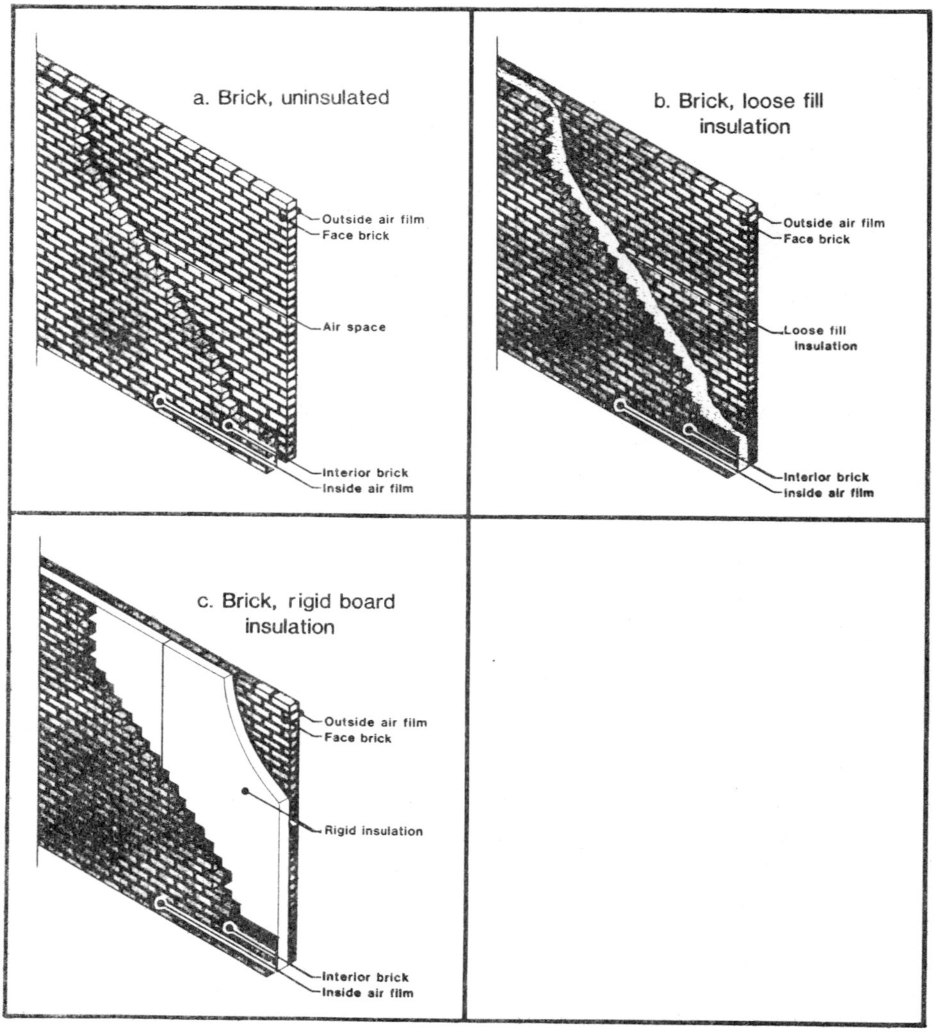

Plate 9. Masonry Cavity Walls – B

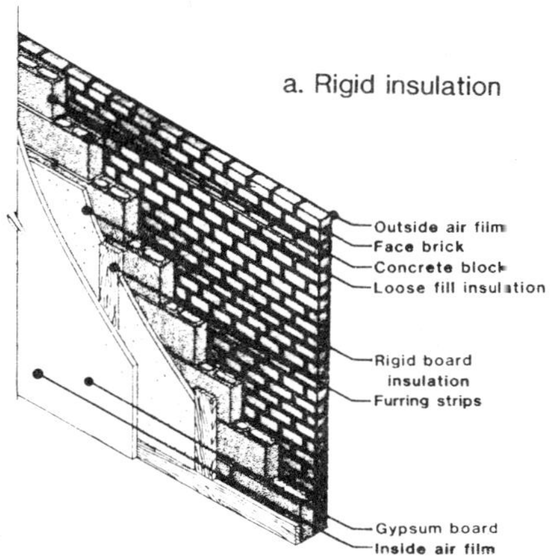

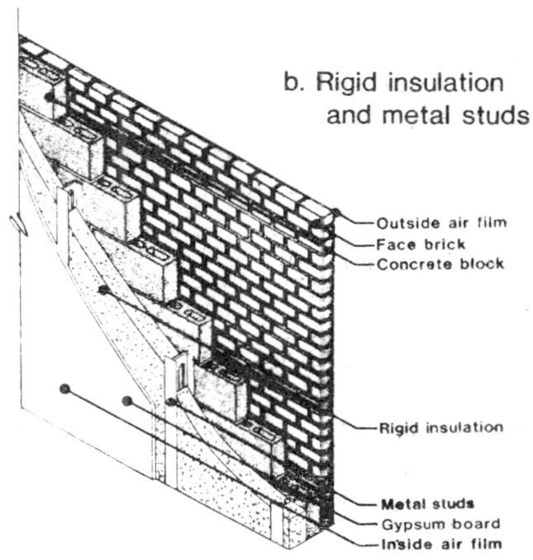

Plate 10. Masonry Combination Walls

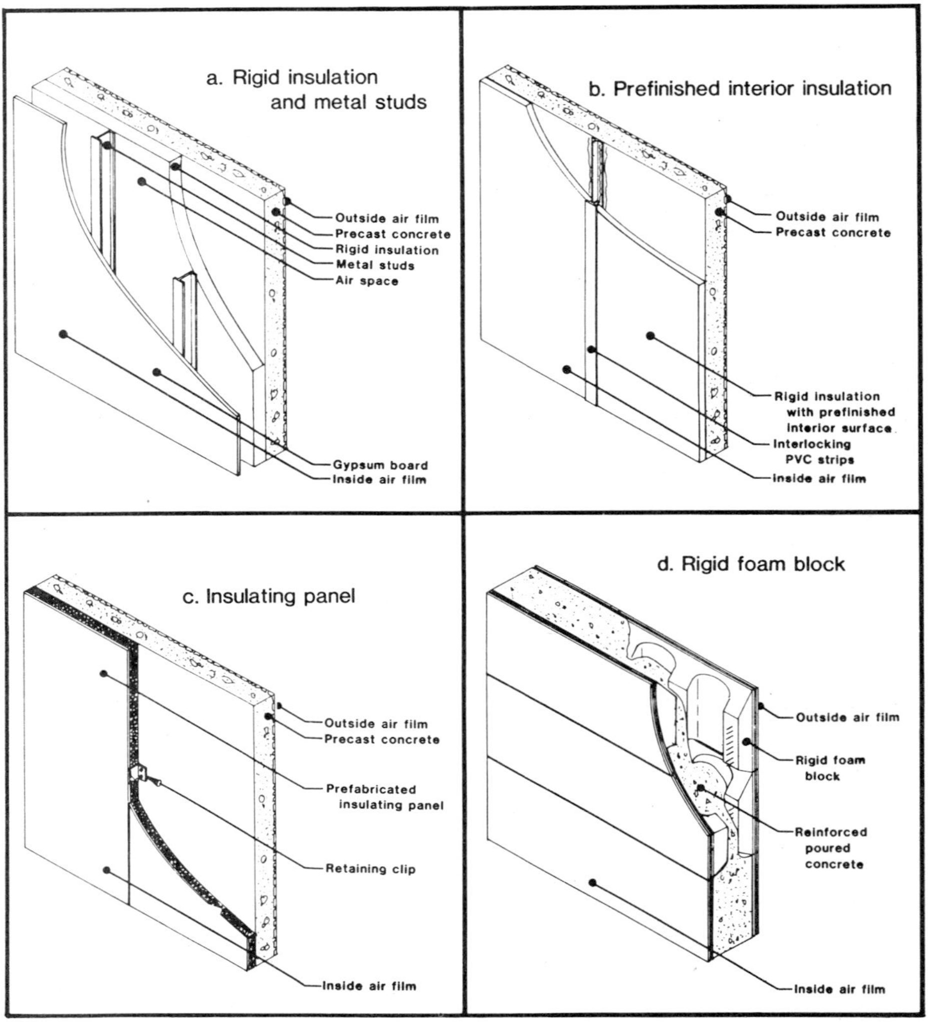

Plate 11. Concrete Walls

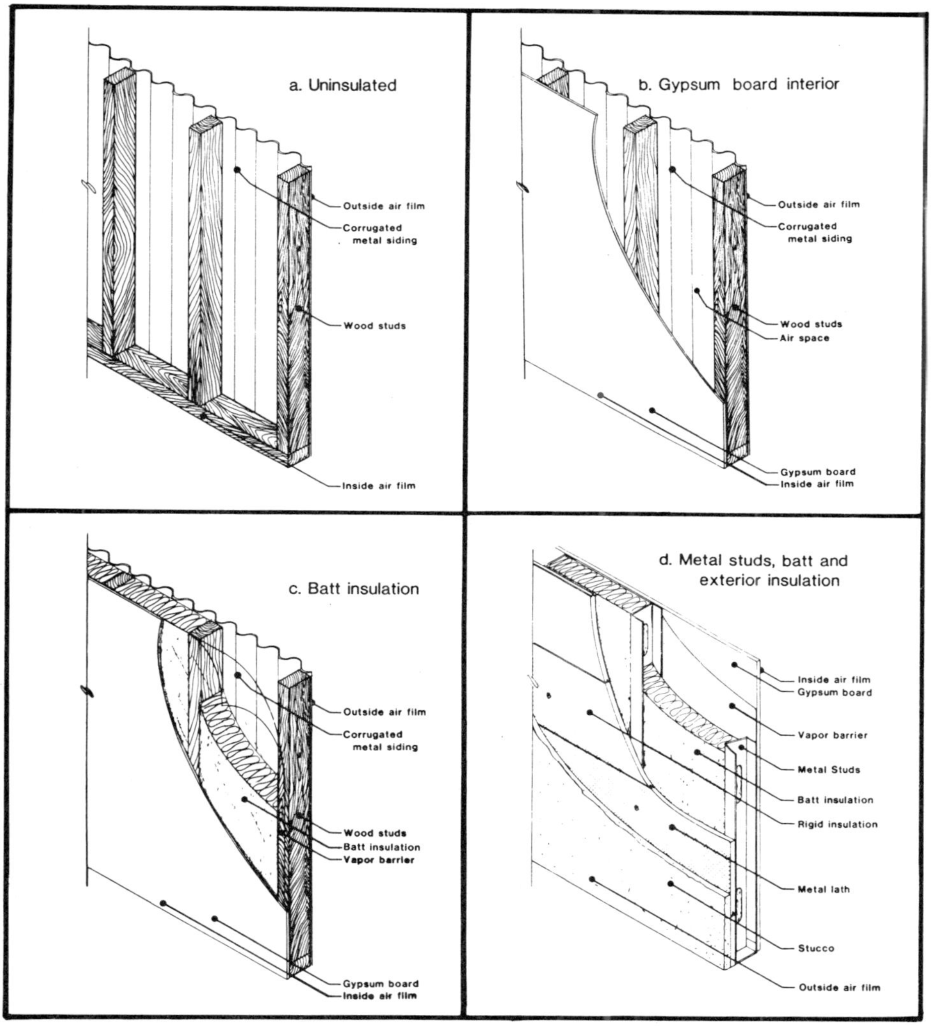

Plate 12. Metal Building Walls – A

12c,d: Vapor barrier may be an integral batt facing or foil backing on gypsum board.

12d: Exterior finish may be any finish that can be anchored through insulation to metal stud, including siding, masonry, or architectural panel.

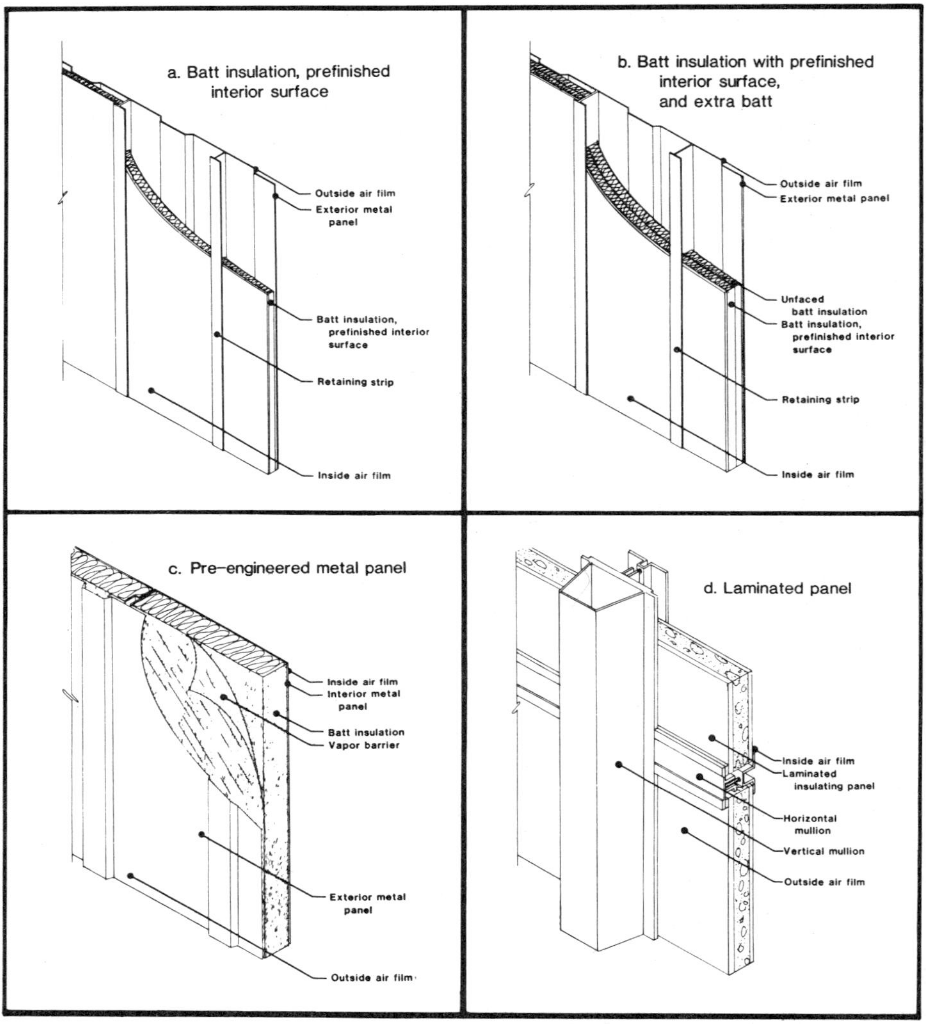

Plate 13. Metal Building Walls – B

13c: Vapor barrier may be an integral batt facing or foil backing on gypsum board.

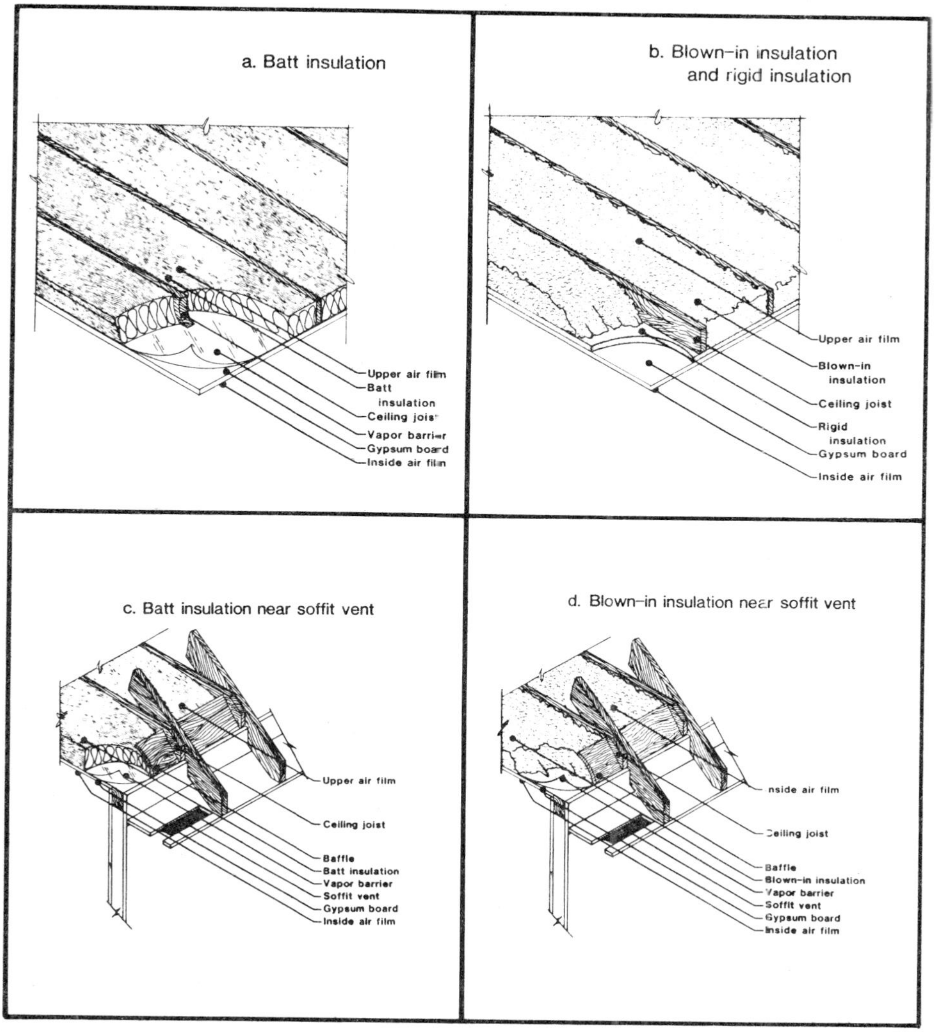

Plate 14. Frame Ceilings - A

14b,c: Free air space of at least 1 inch thickness should exist between roof sheathing and insulation.

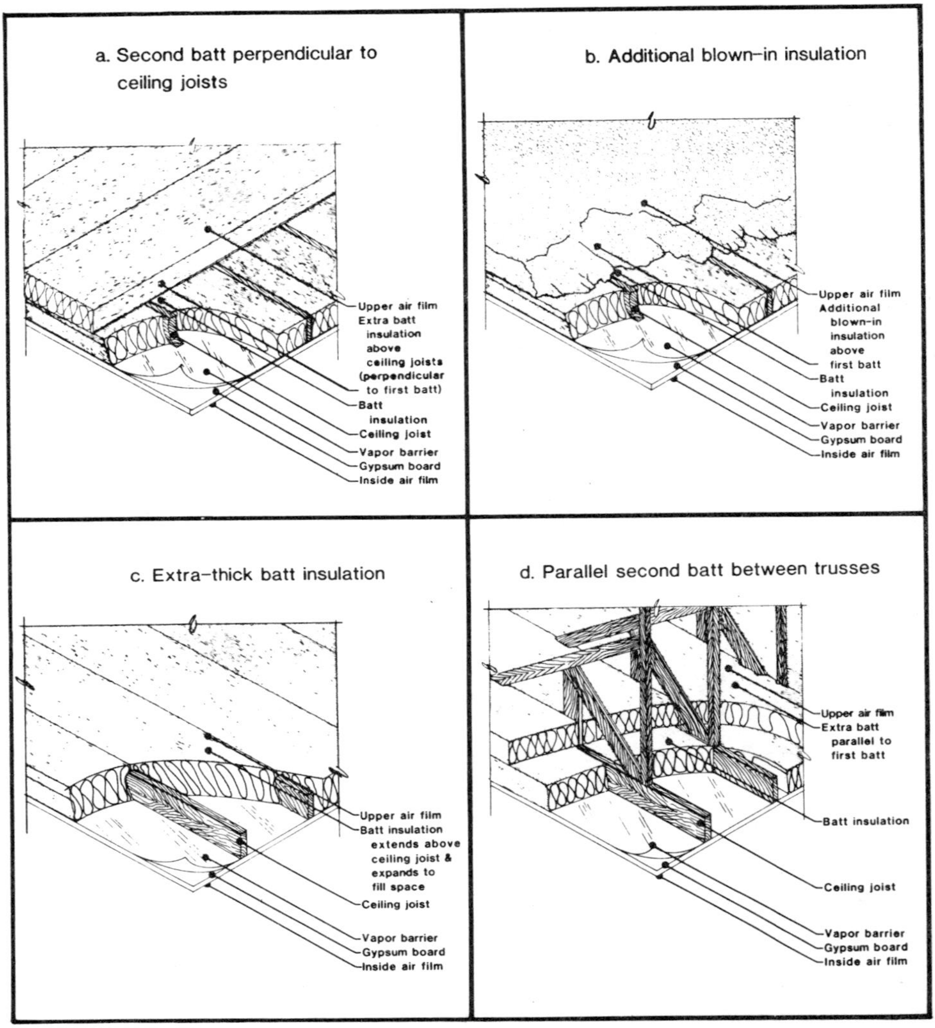

Plate 15. Frame Ceilings – B

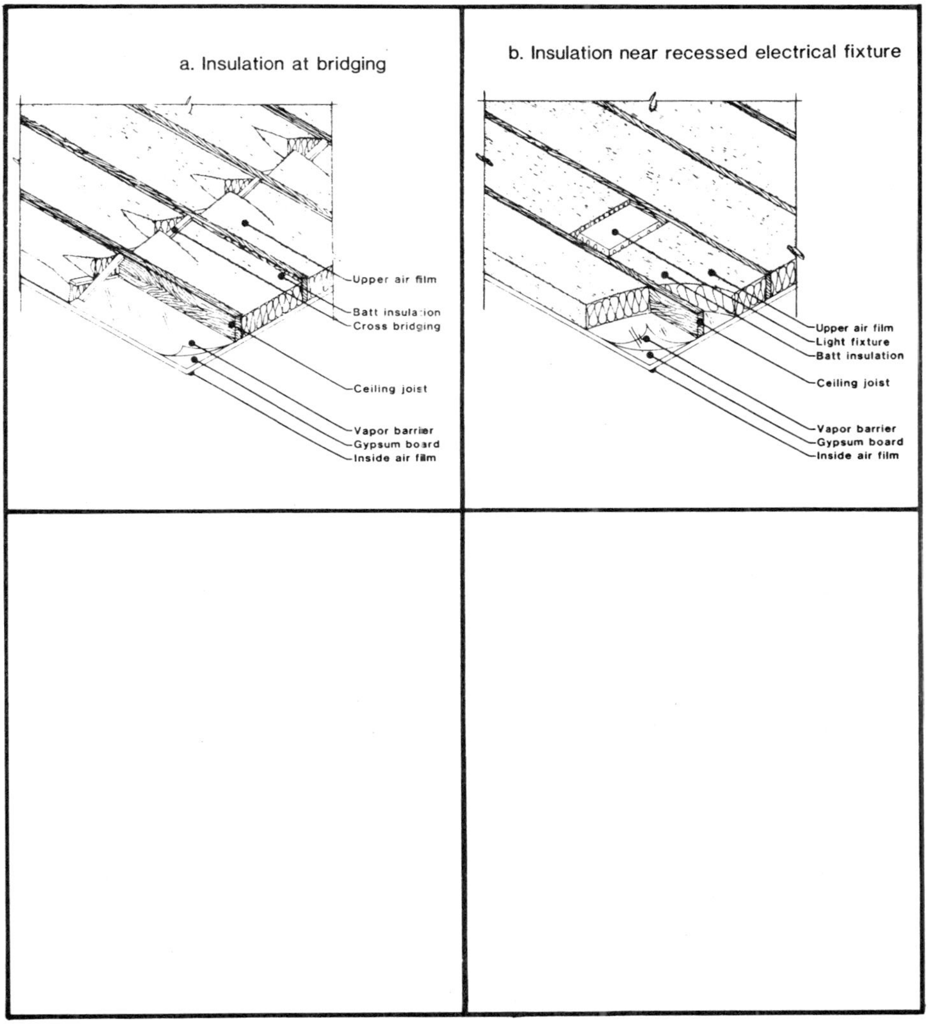

Plate 16. Frame Ceilings – C

16b: Insulation must be kept at least 3 inches away from recessed electrical fixtures. Use fireproof baffle with loose fill insulation.

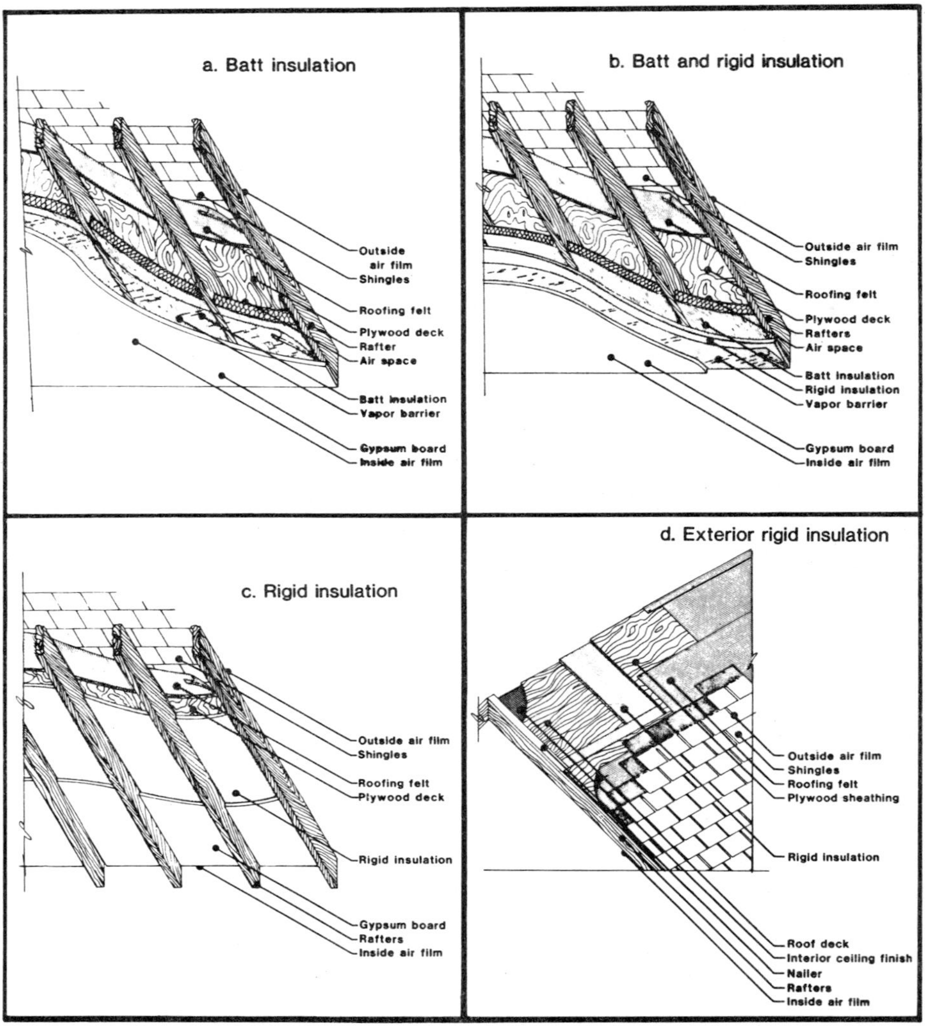

Plate 17. Cathedral ceilings

17a,b: Vapor barrier may be foil facing on rigid insulation or foil backing on gypsum board.

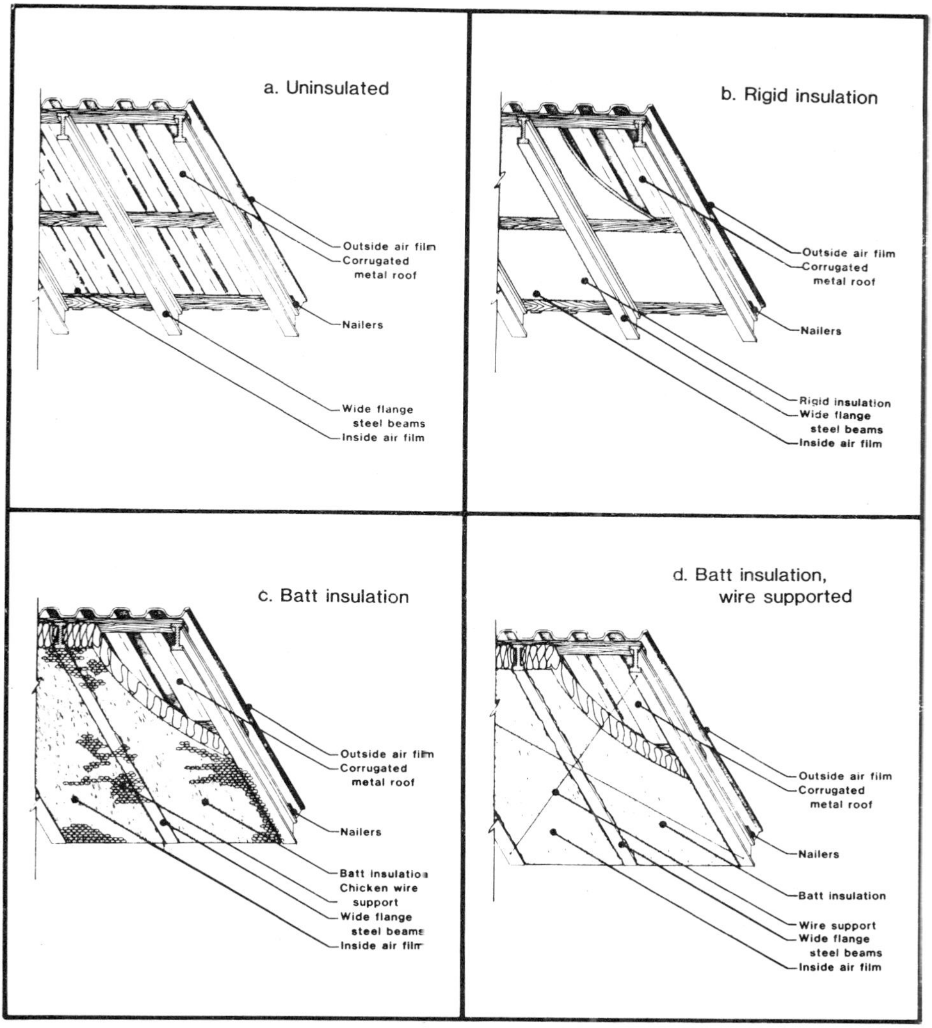

Plate 18. Metal Building Ceilings – A

18b: Fire hazard ratings may limit the use of foam insulation in some applications.

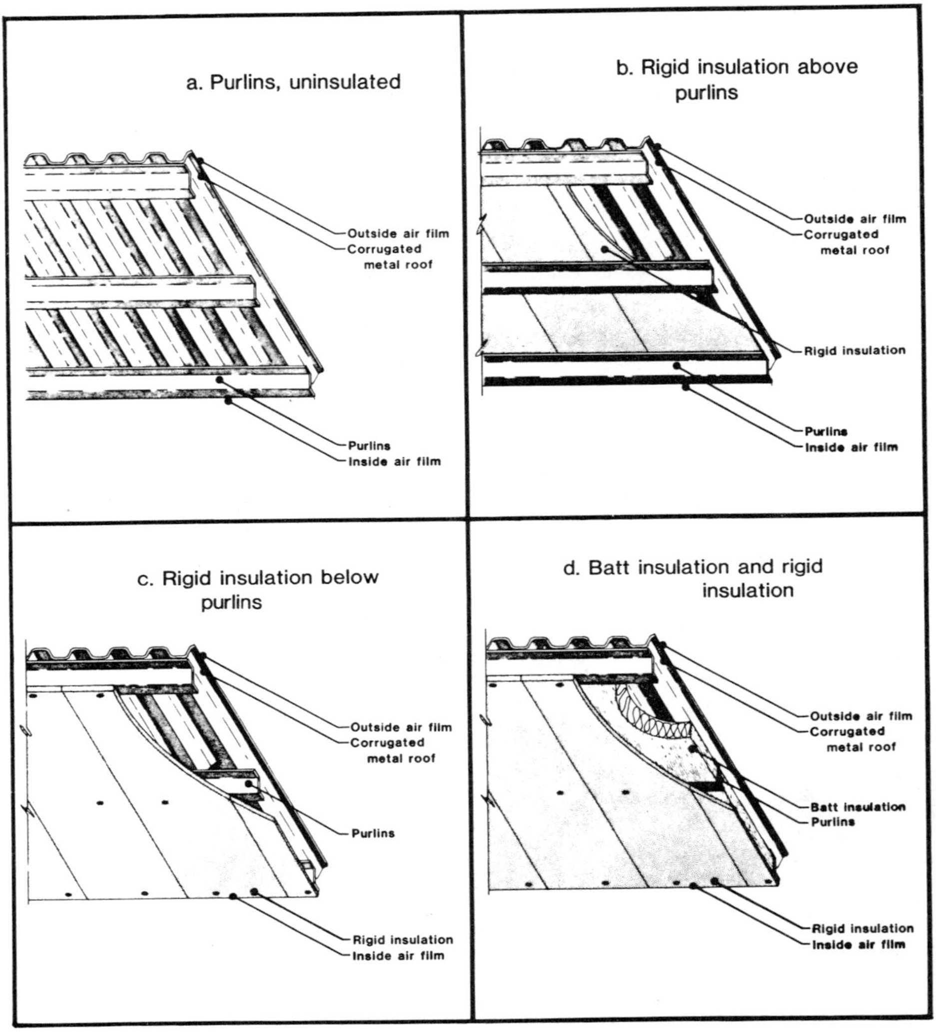

Plate 19. Metal Building Ceilings – B

19b–d: Fire hazard ratings may limit the use of foam insulation in some applications.

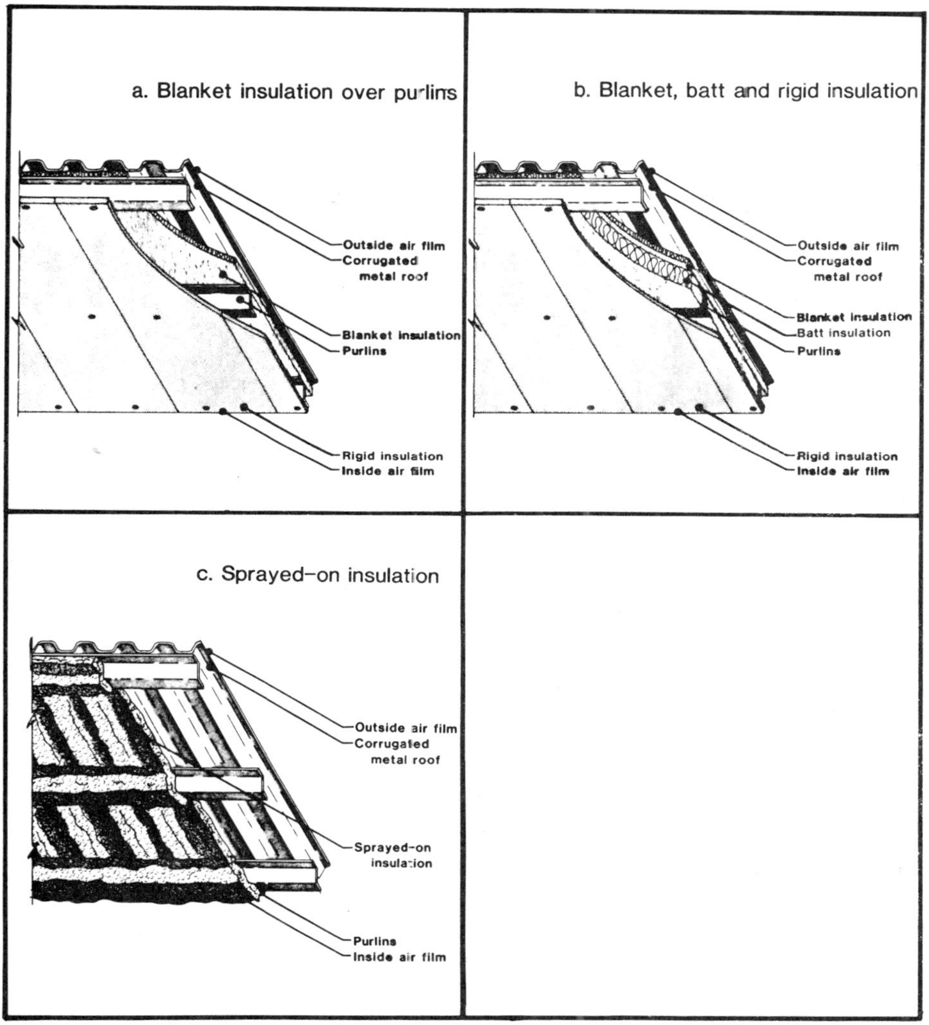

Plate 20. Metal Building Ceilings – C

20a–c: Fire hazard ratings may limit the use of foam insulation in some applications.

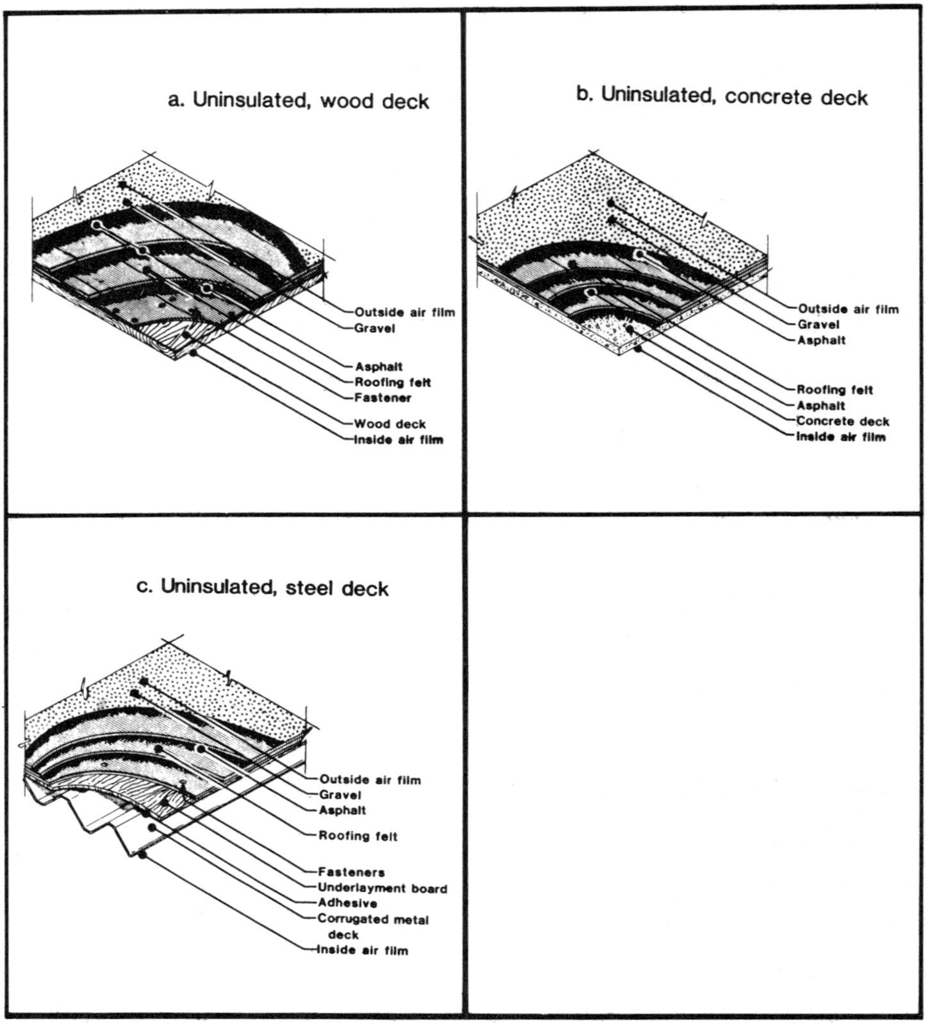

Plate 21. Built-Up Roofs - A

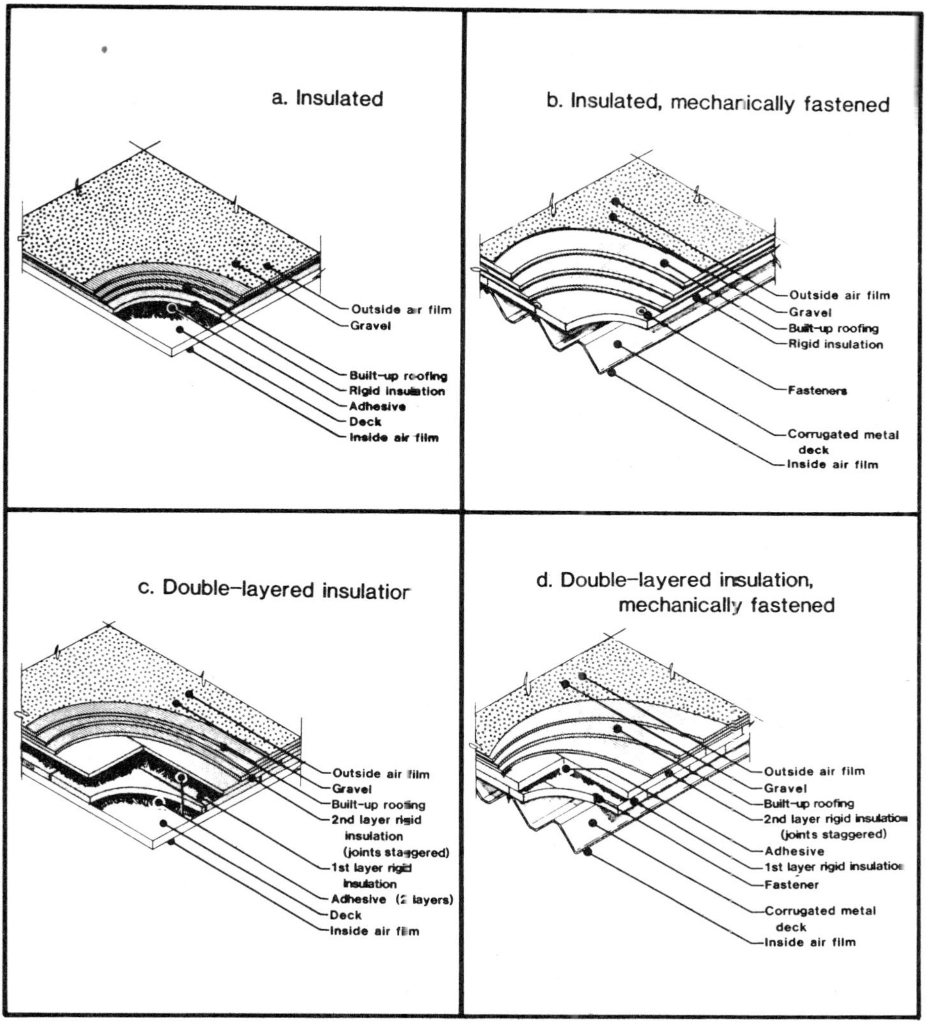

Plate 22. Built-Up Roofs - B

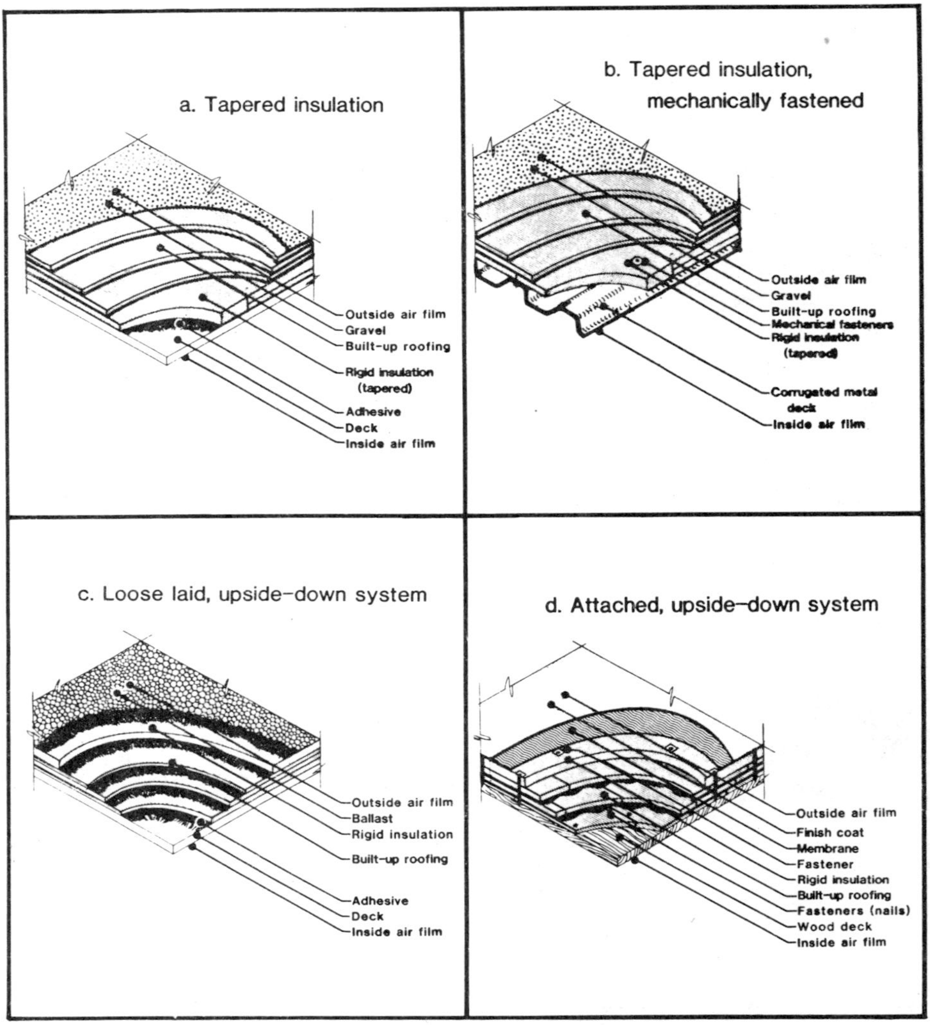

Plate 23. Built-Up Roofs - C

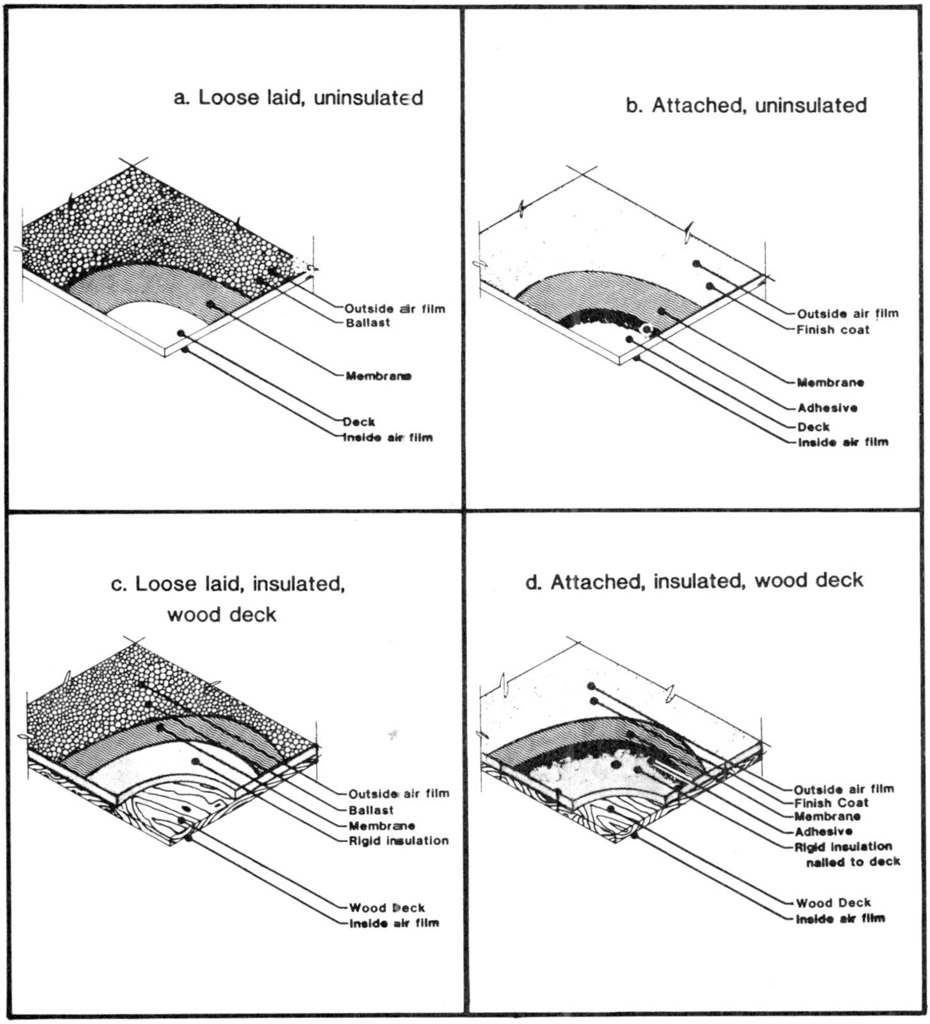

Plate 24. Single-ply Membrane Roofs – A

24a,c: Ballast is smooth stone 10 lb/sq ft. Roof deck must be sufficiently strong to support this load.

24b,d: May be used with roofs that cannot support ballast. Finish coat is UV-resistant.

24d: Deck may be lightweight nailable concrete.

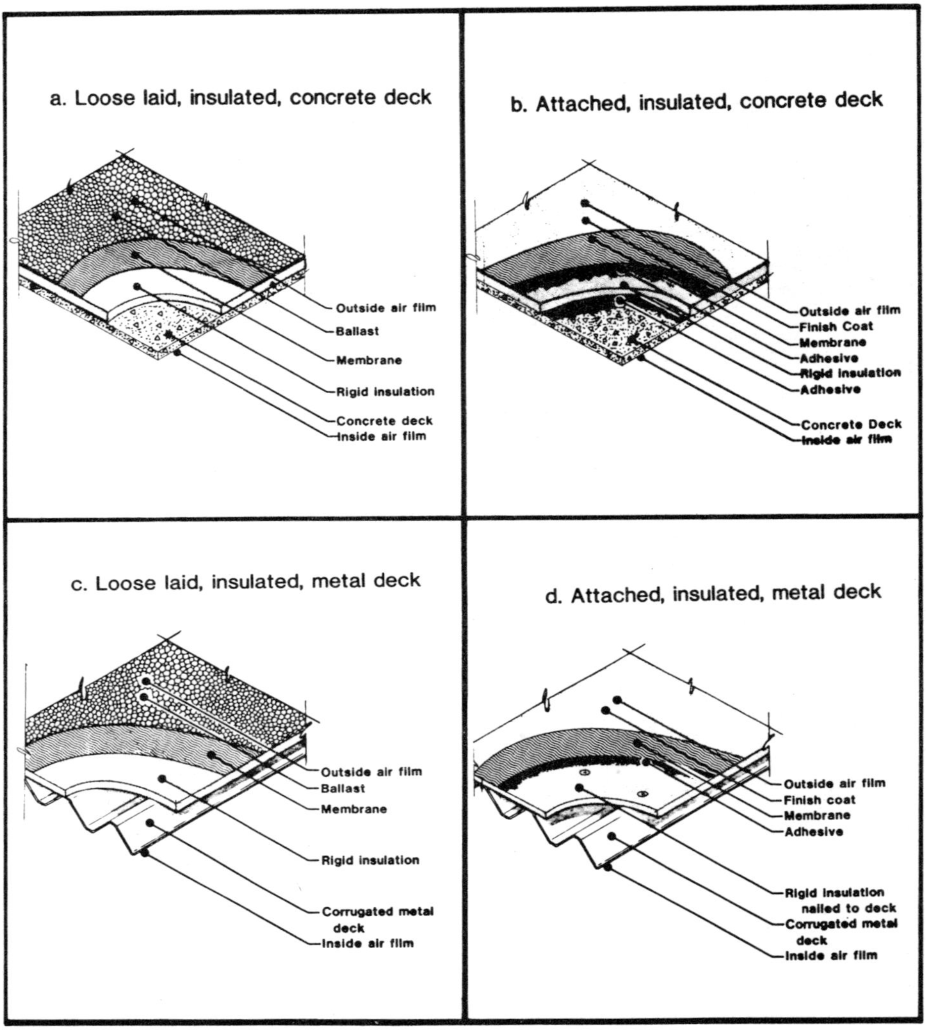

Plate 25. Single-ply Membrane Roofs – B

25a,c: Ballast is smooth stone, 10 lb/sq ft. Roof deck must be sufficiently strong to support this load.

25b,d: May be used with roofs that cannot support ballast. Finish coat is UV-resistant.

a. Alternate insulation attachment method

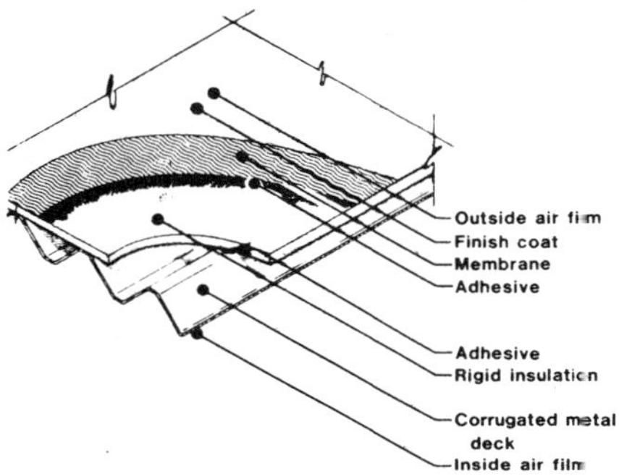

Plate 26. Single-ply Membrane Roofs – C

26a: May be used with roofs that cannot support ballast. Finish coat is UV-resistant.

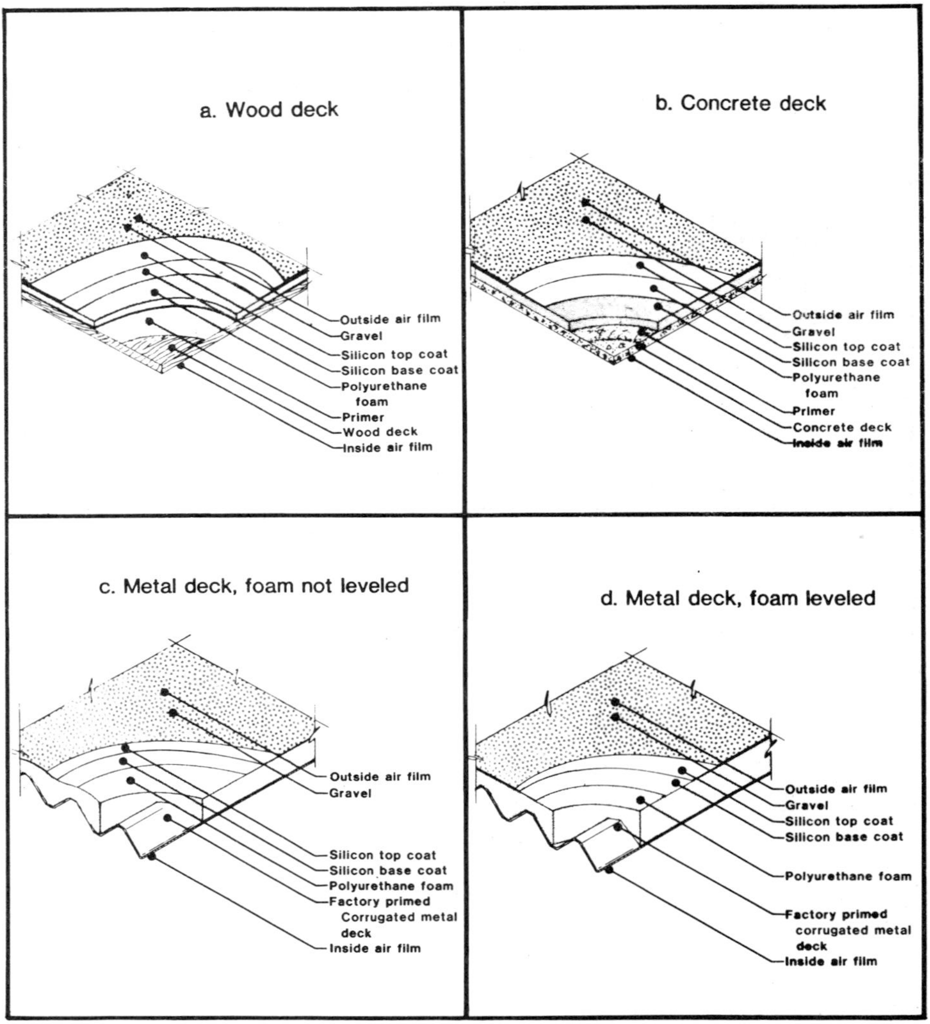

Plate 27. Polyurethane Roofs – A

27a,b: Primer is chlorinated rubber.

27a-d: Polyurethane foam has 3 lb/cu ft density.

27c: Water must drain from finished surface by proper slope of steel deck or control of foam thickness.

27d: Smooth surface may be achieved by filling deck flutes and shaving smooth, or by filling flutes with boardstock.

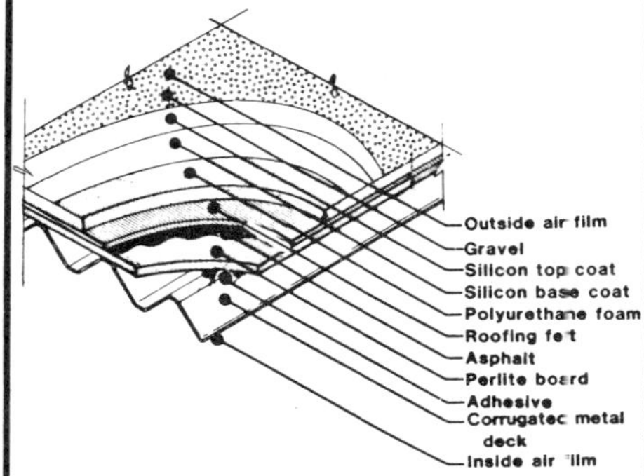

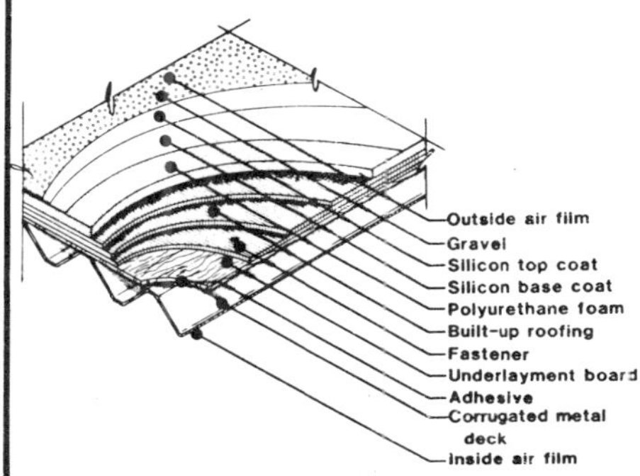

Plate 28. Polyurethane Roofs – B

28a: Approved fasteners may be substituted for adhesive.

28a,b: Polyurethane foam has 3 lb/cu ft density.

28b: This is a reroofing application.

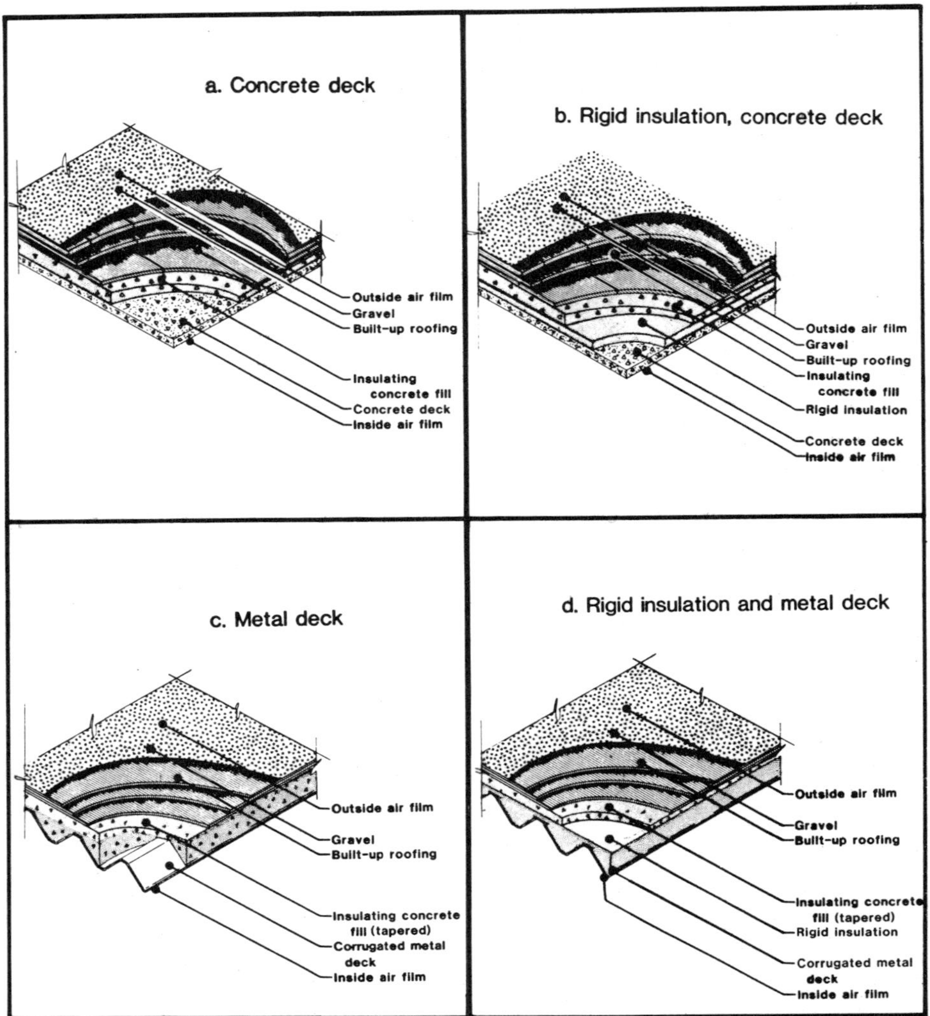

Plate 29. Lightweight Concrete Roofs – A

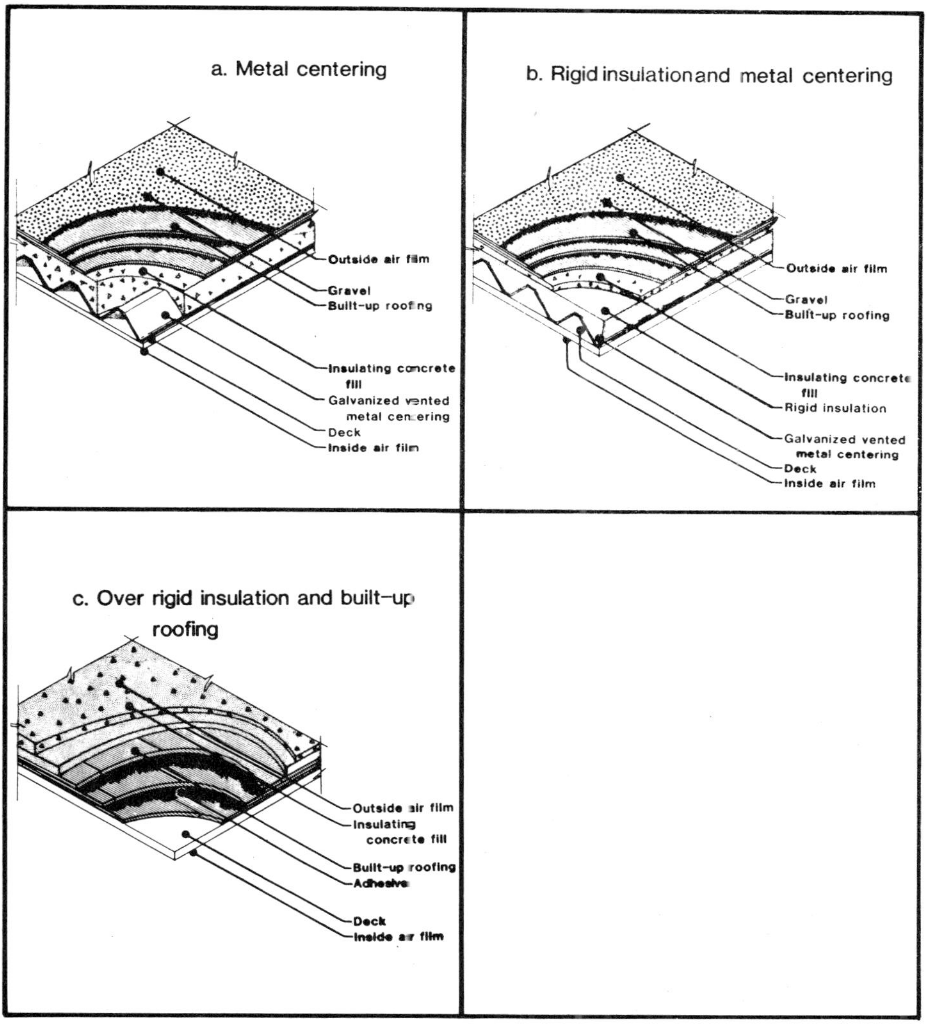

Plate 30. Lightweightweight concrete roofs – B

30c: This is a reroofing application.

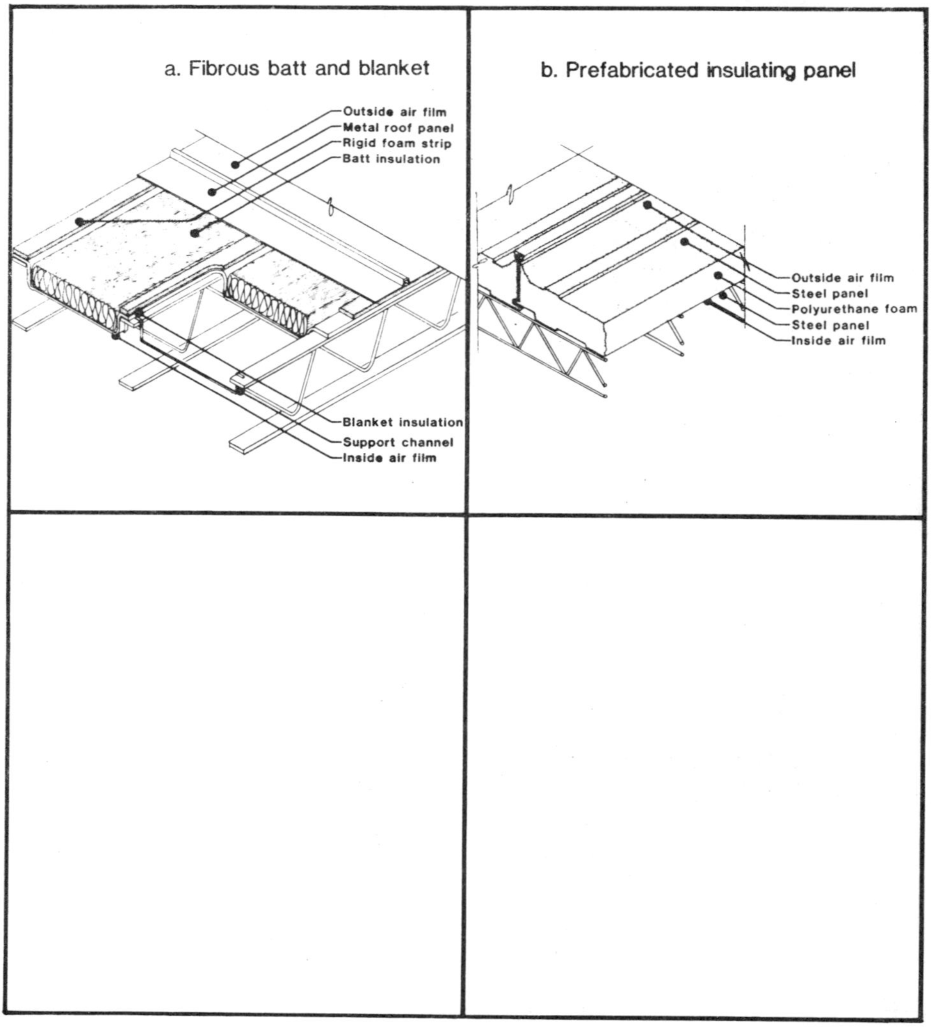

Plate 31. Steel Panel Roofs

31a: Rigid foam strips are optional.

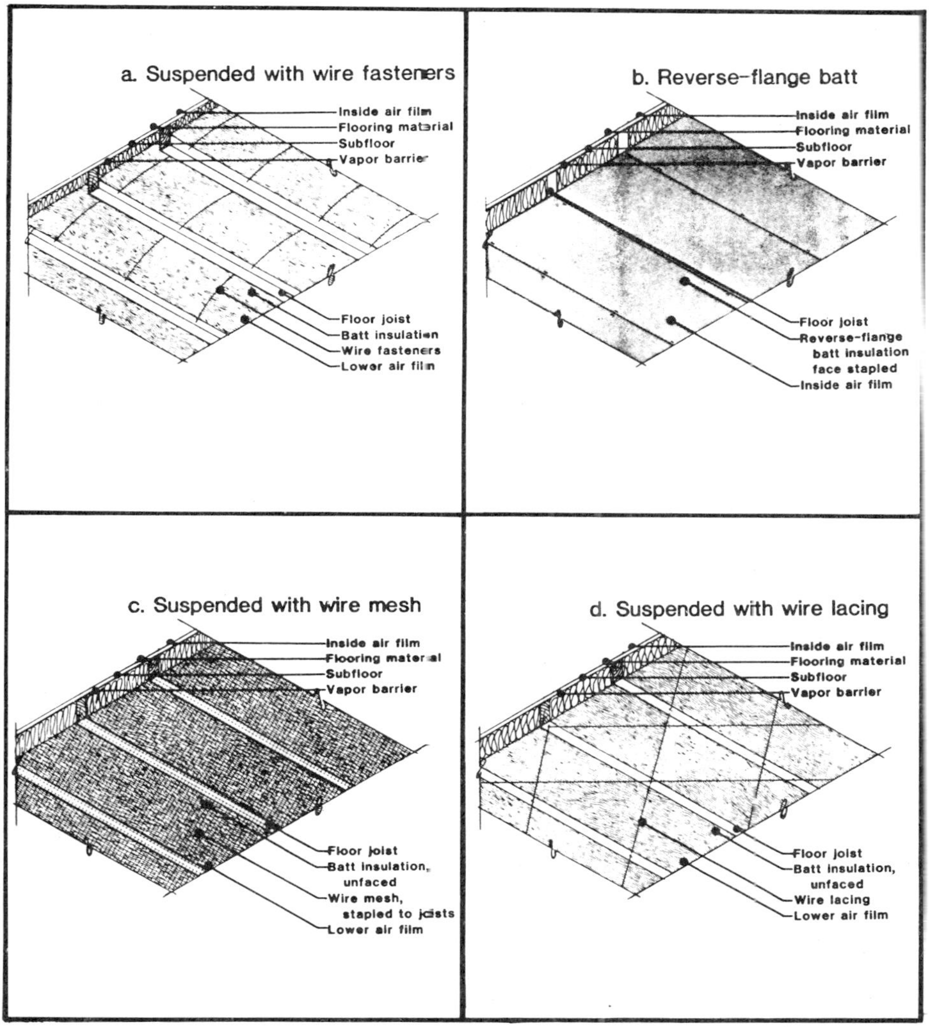

Plate 32. Floors - A

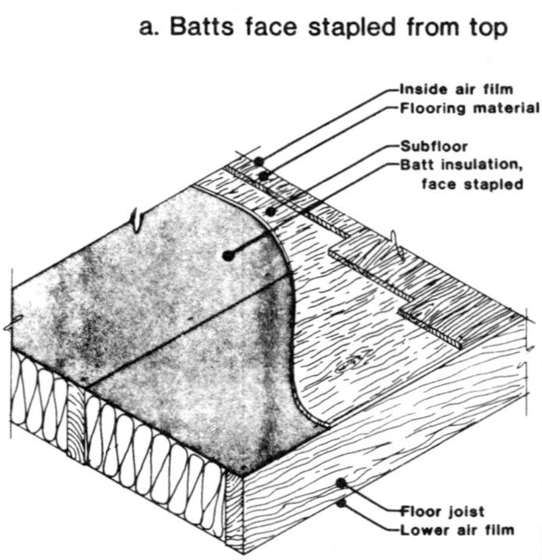

Plate 33. Floors – B

33a: New floor application.

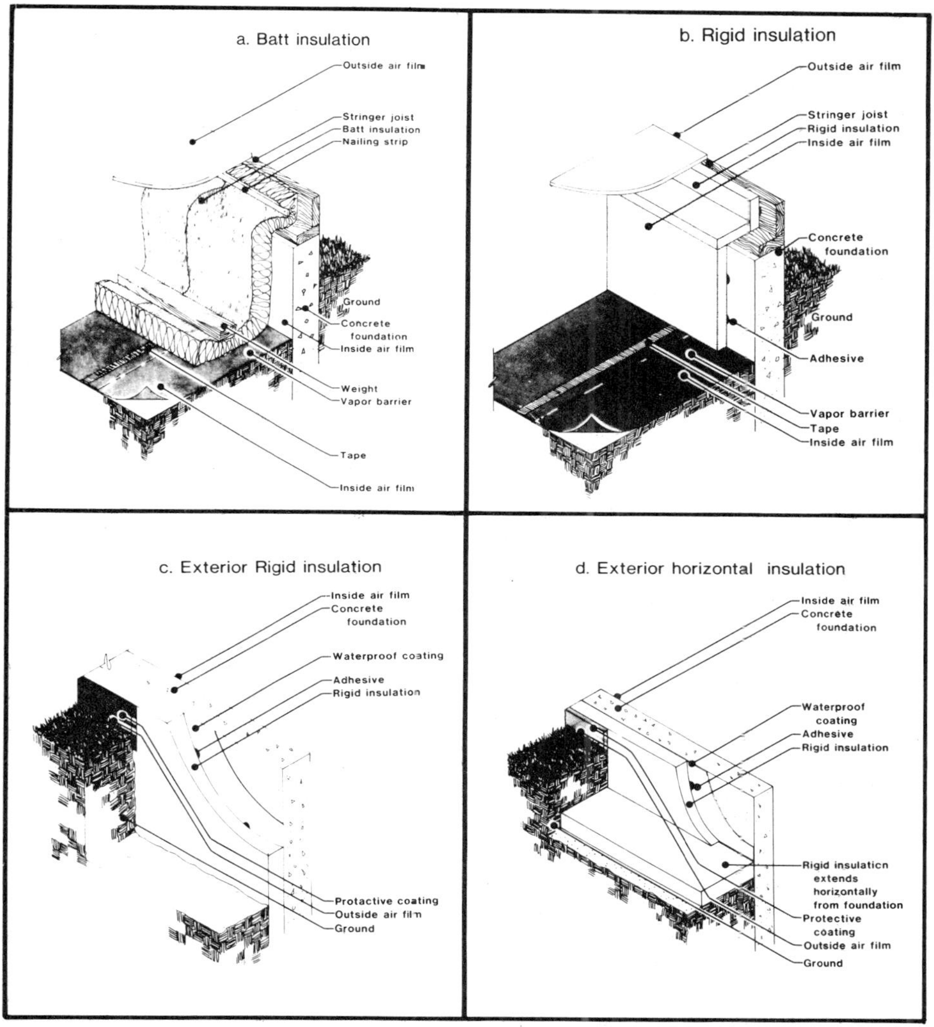

Plate 34. Foundations – A

34b: Fire hazard ratings may limit the use of foam insulation in some applications.

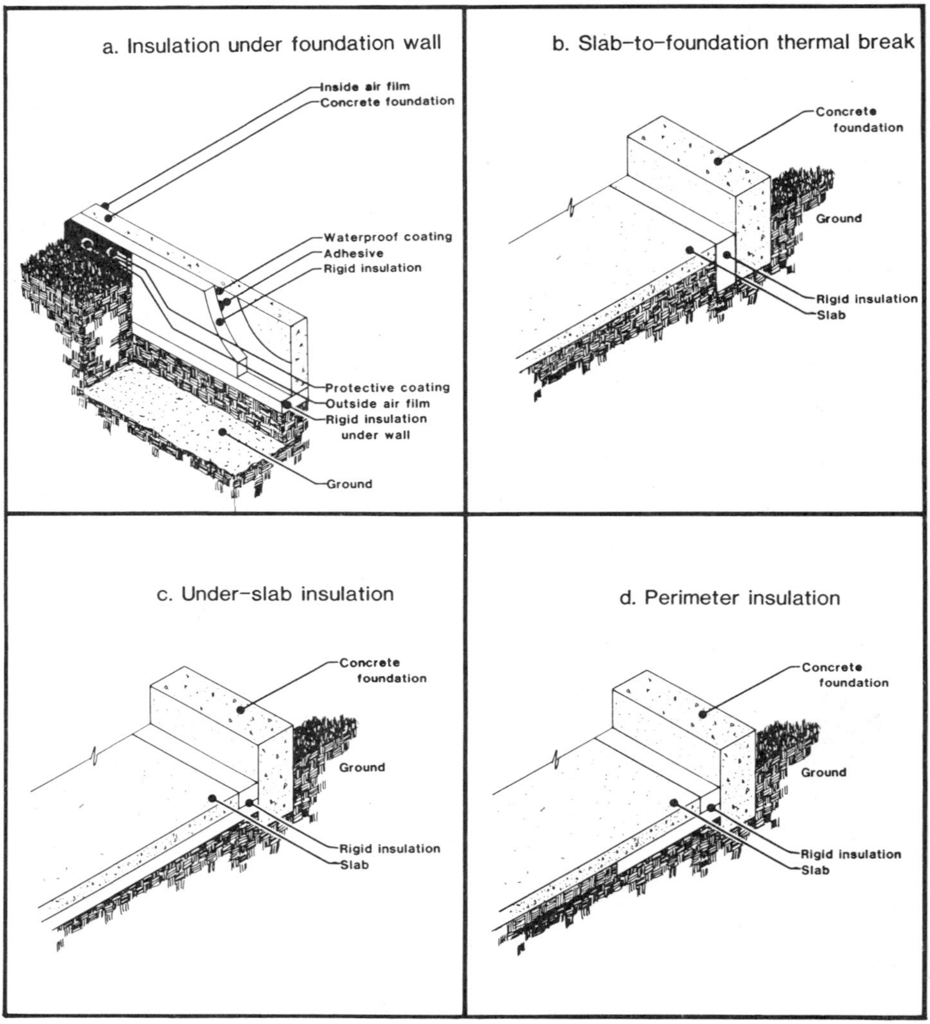

Plate 35. Foundations - B

35d: Insulation should extend under slab at least 24 inches from foundation.

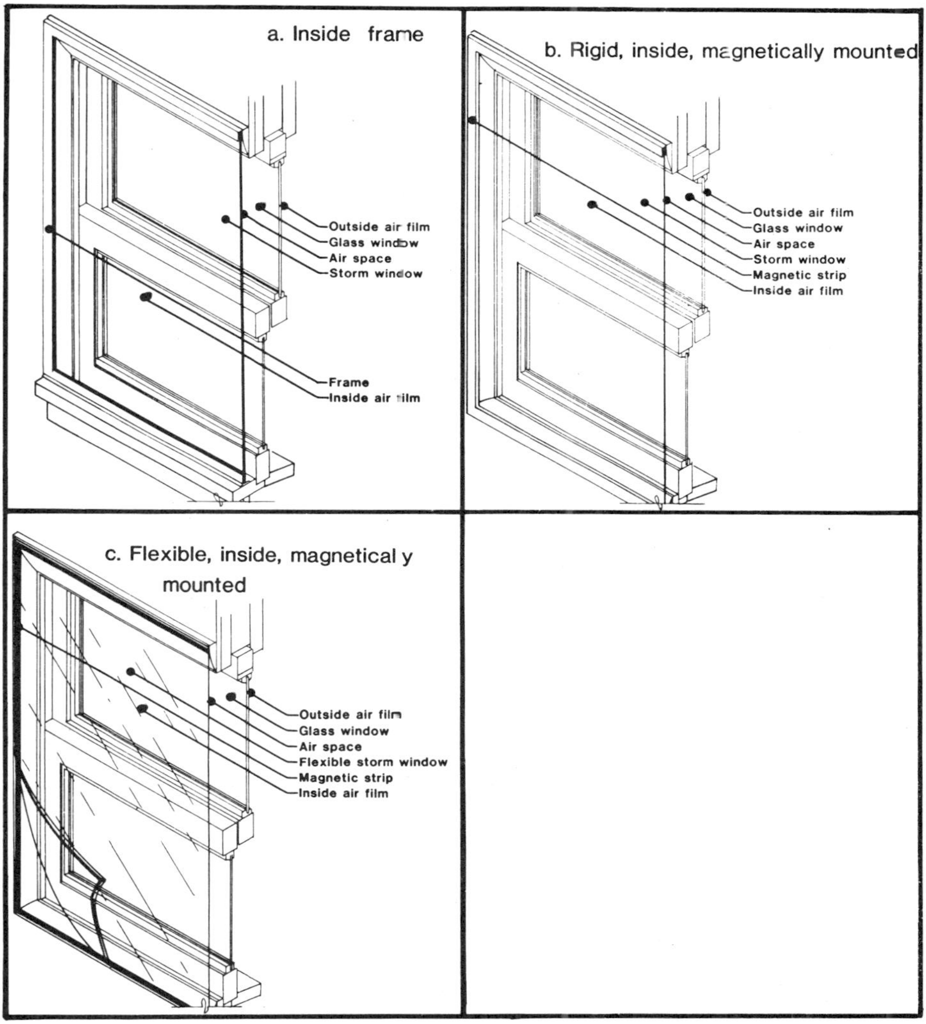

Plate 36. Storm Windows

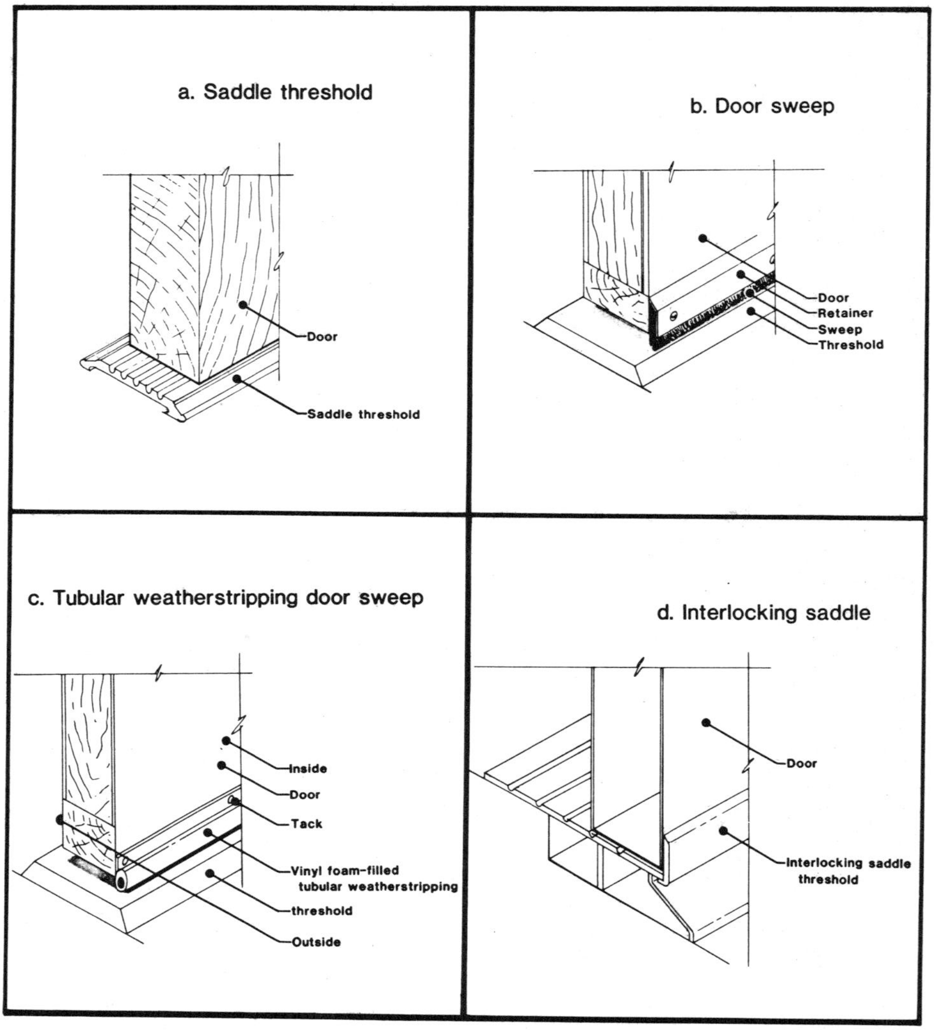

Plate 37. Thresholds – A

37b,c: Sweep mounted on outswinging door is more aesthetic, more effective against leaks than sweep mounted on inswinging door.

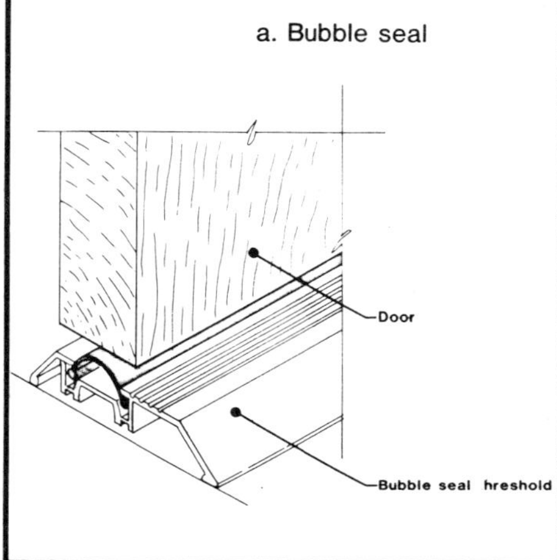

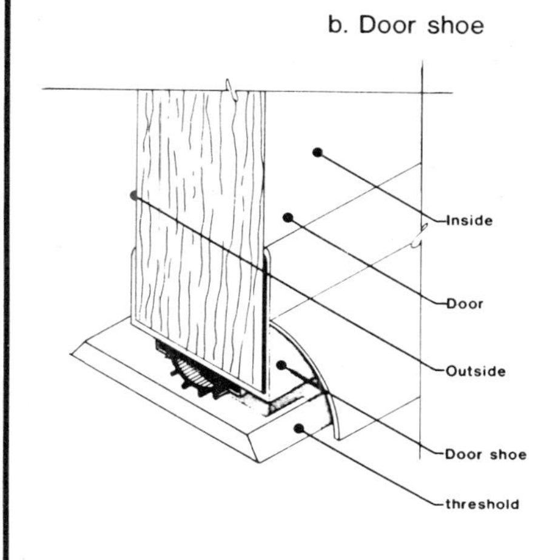

Plate 38. Thresholds – B

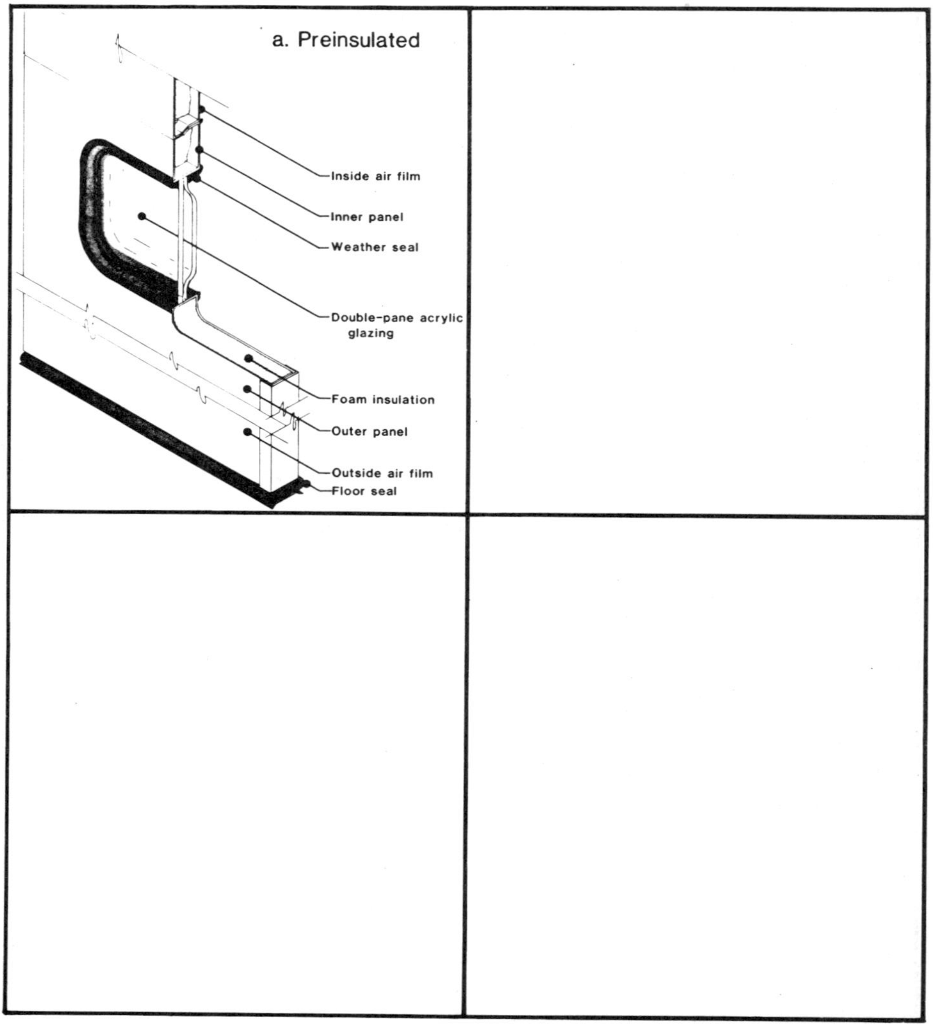

Plate 39. Overhead doors

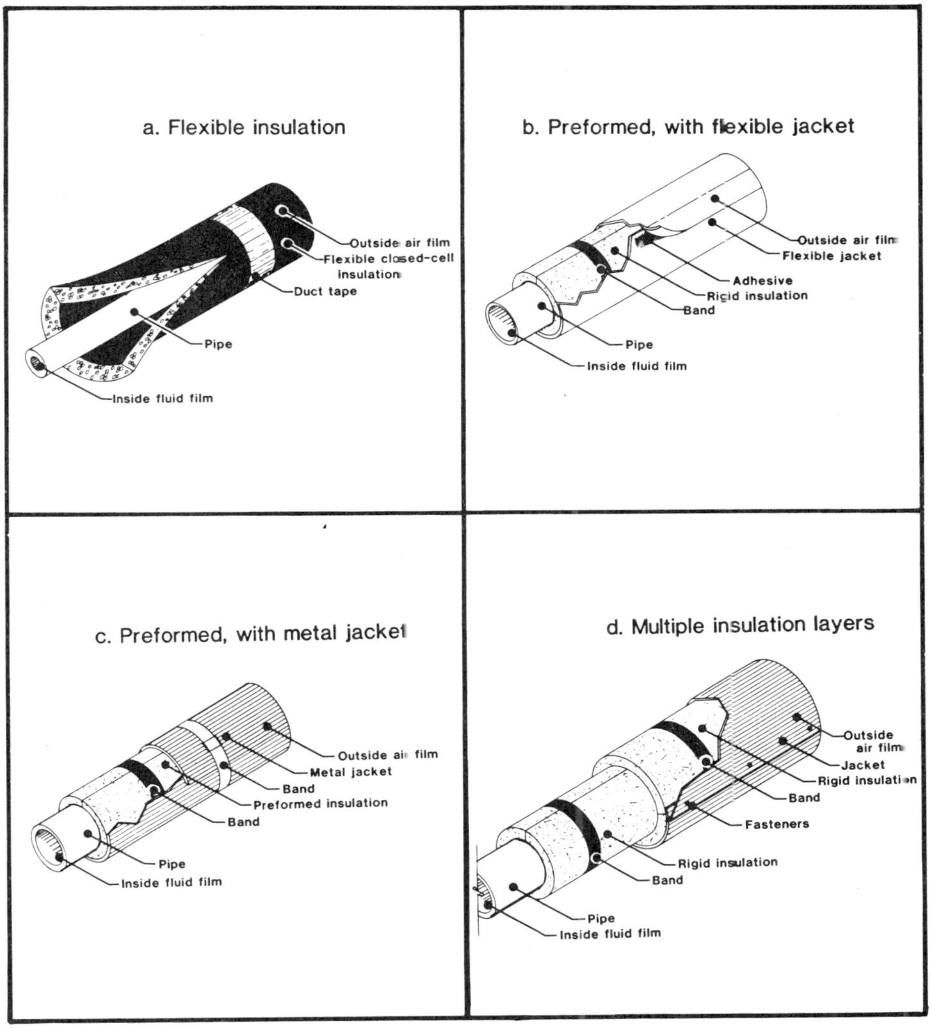

Plate 40. Piping – A

40a,b: Fire hazard ratings may limit the use of foam insulation in some applications.

40b–d: Bands or, alternatively, wire or tape may not be necessary with some insulation types.

40c: Metal jacket may be secured with screws or rivets instead of metal bands.

40d: Insulation joints are staggered.

40d: Fasteners may be screws or rivets, metal bands may be substituted.

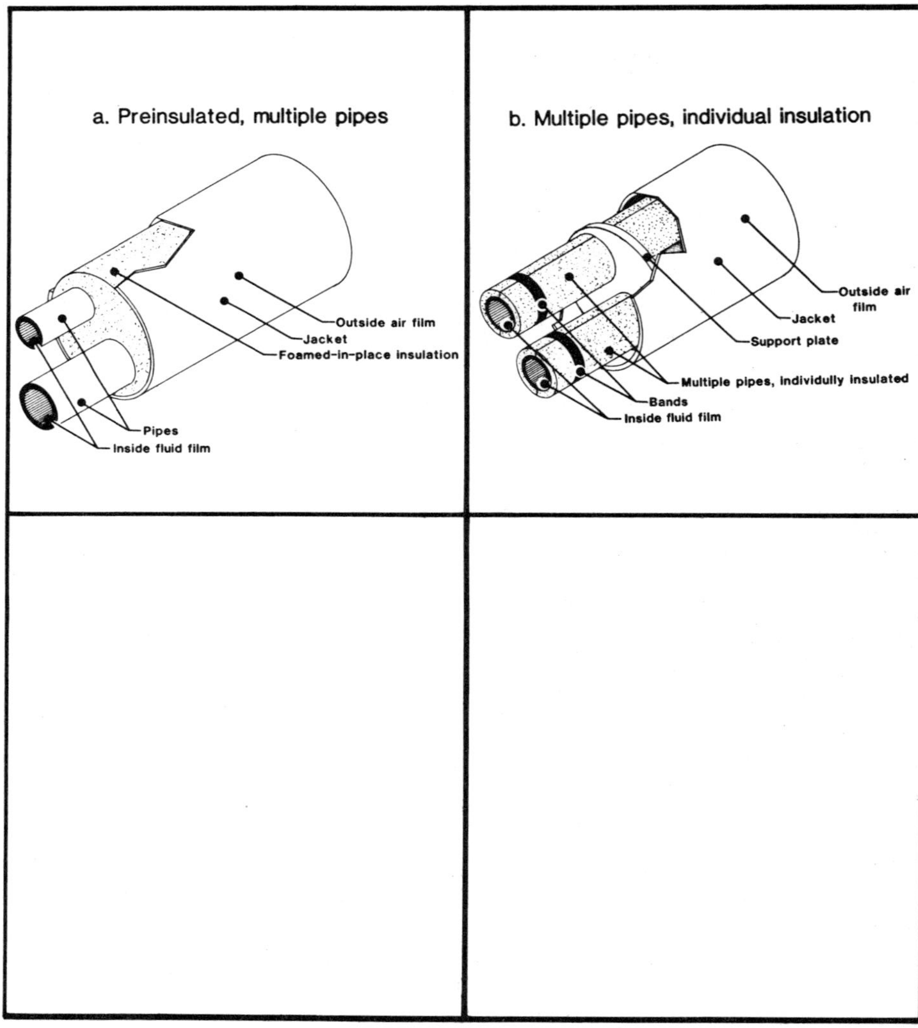

Plate 41. Piping – B

41b: Bands or, alternatively, wire or tape may not be necessary with some insulation types.

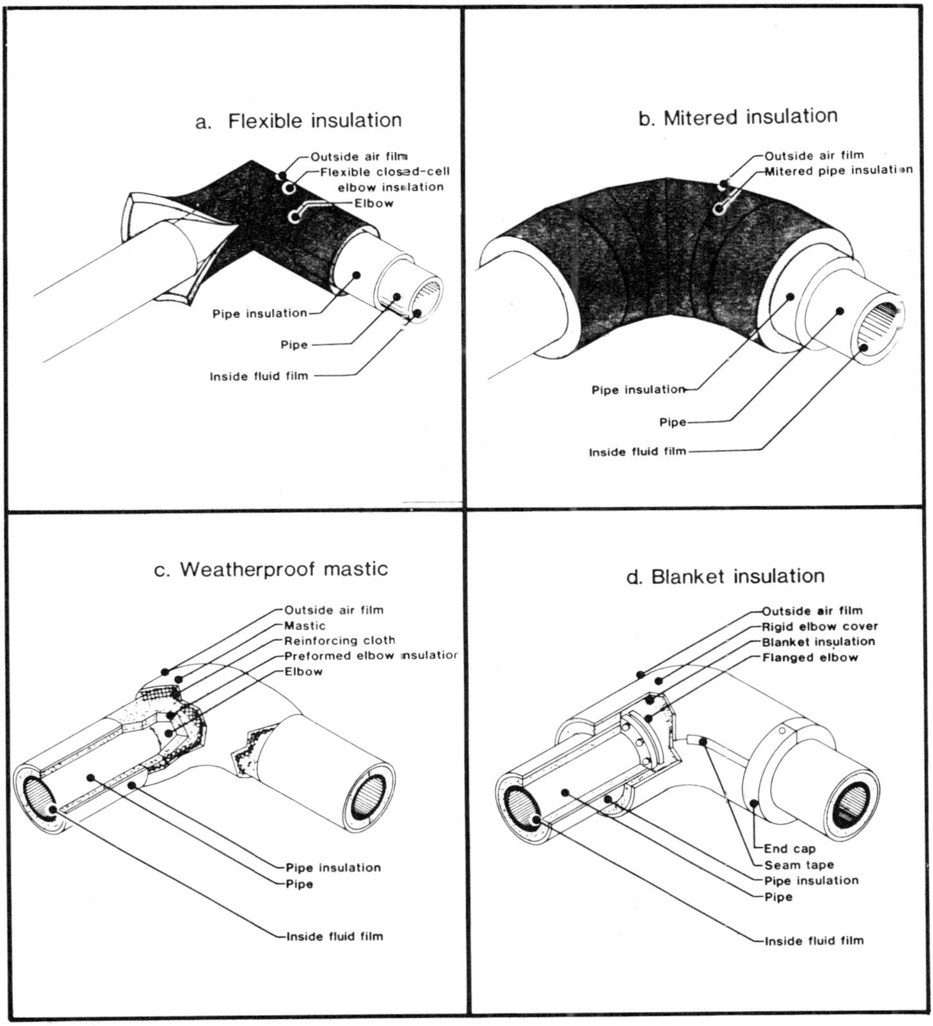

Plate 42. Elbows

42a: Fire hazard ratings may limit the use of foam insulation in some applications.

42b: Metal or plastic jacket may be added if applicable.

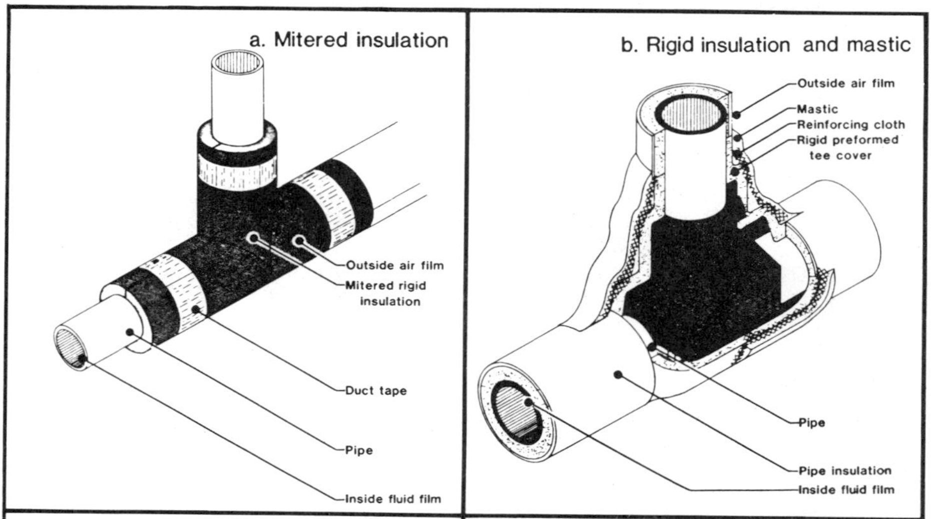

Plate 43. Tees

43a: Metal or plastic jacket may be added if applicable.

43a: Fire hazard ratings may limit the use of foam insulation in some applications.

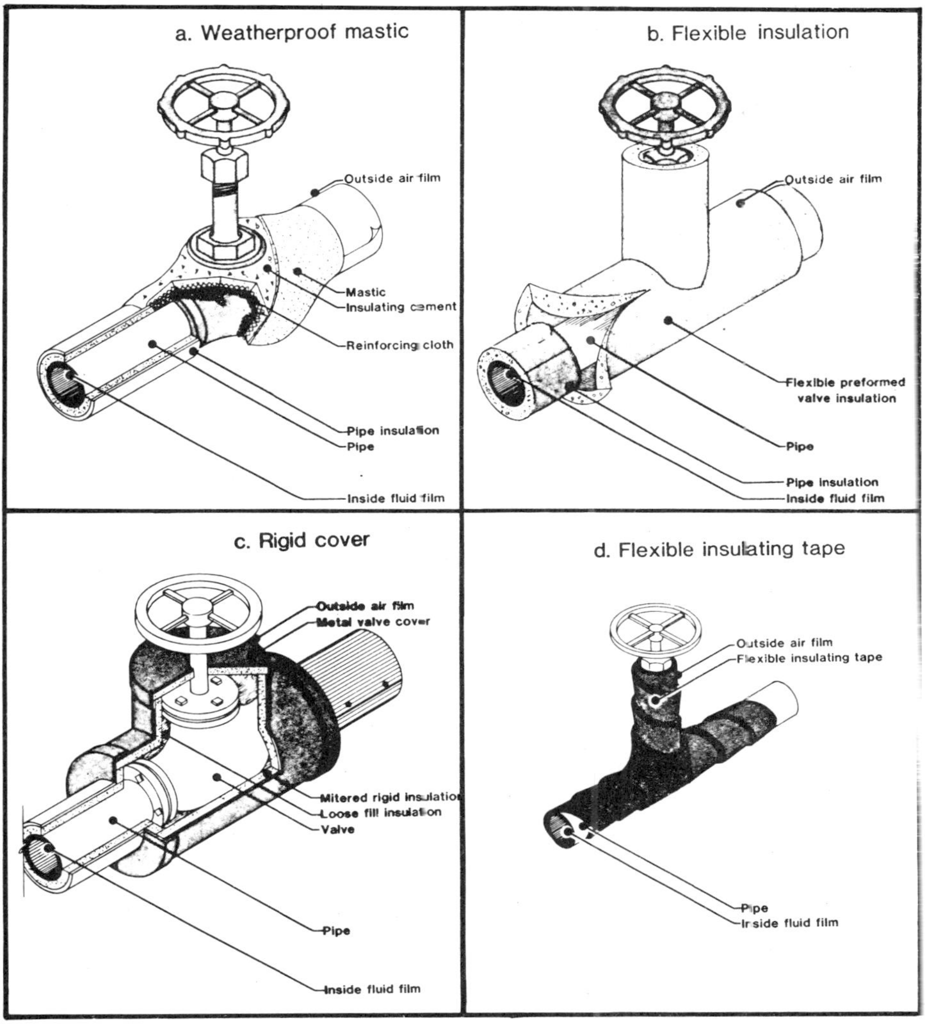

Plate 44. Valves

44b,d: Fire hazard ratings may limit the use of foam insulation in some applications.

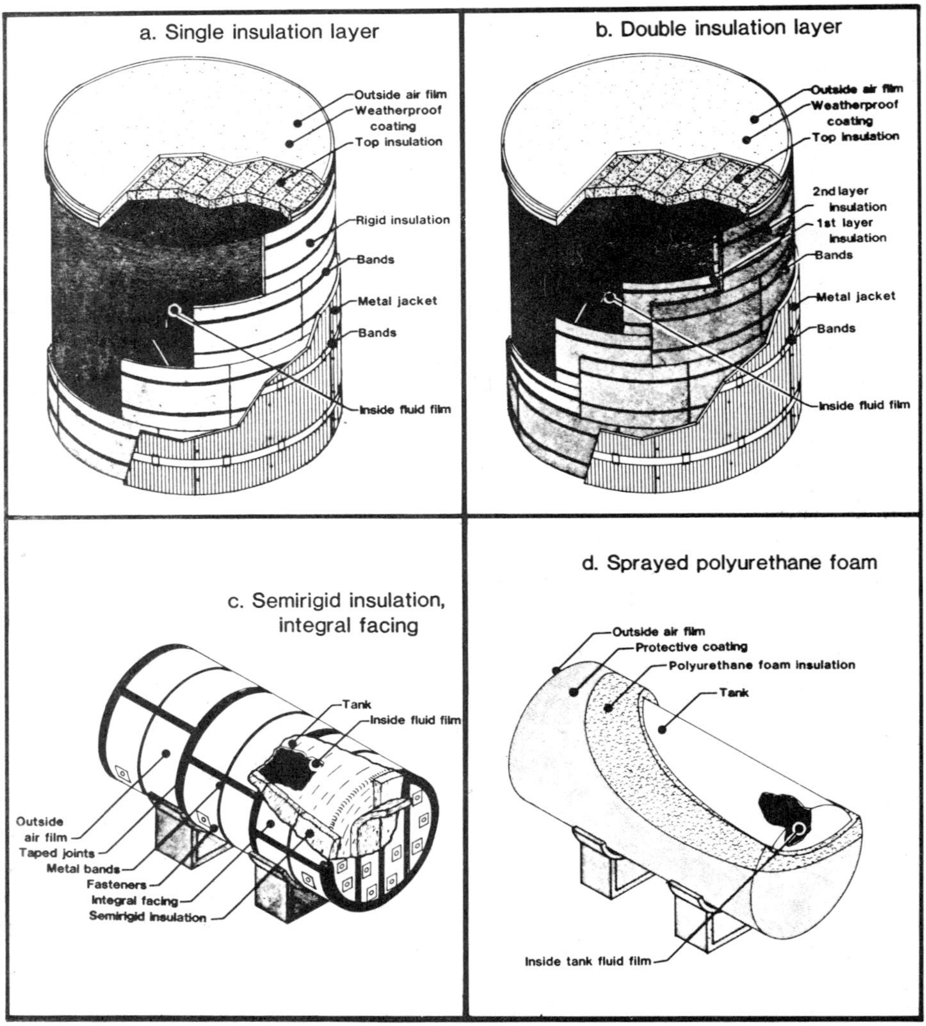

Plate 45. Tanks

45a,b,d: Fire hazard ratings may limit the use of foam insulation in some applications.

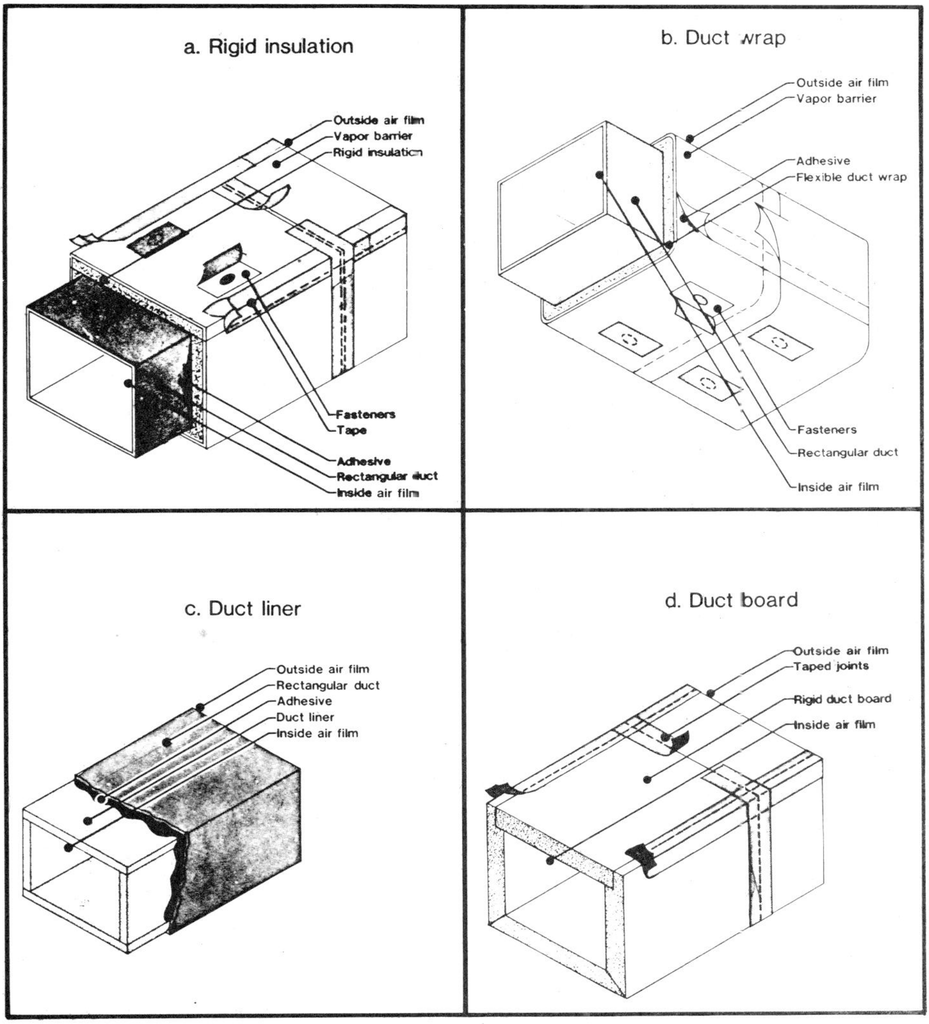

Plate 46. Rectangular Ducts

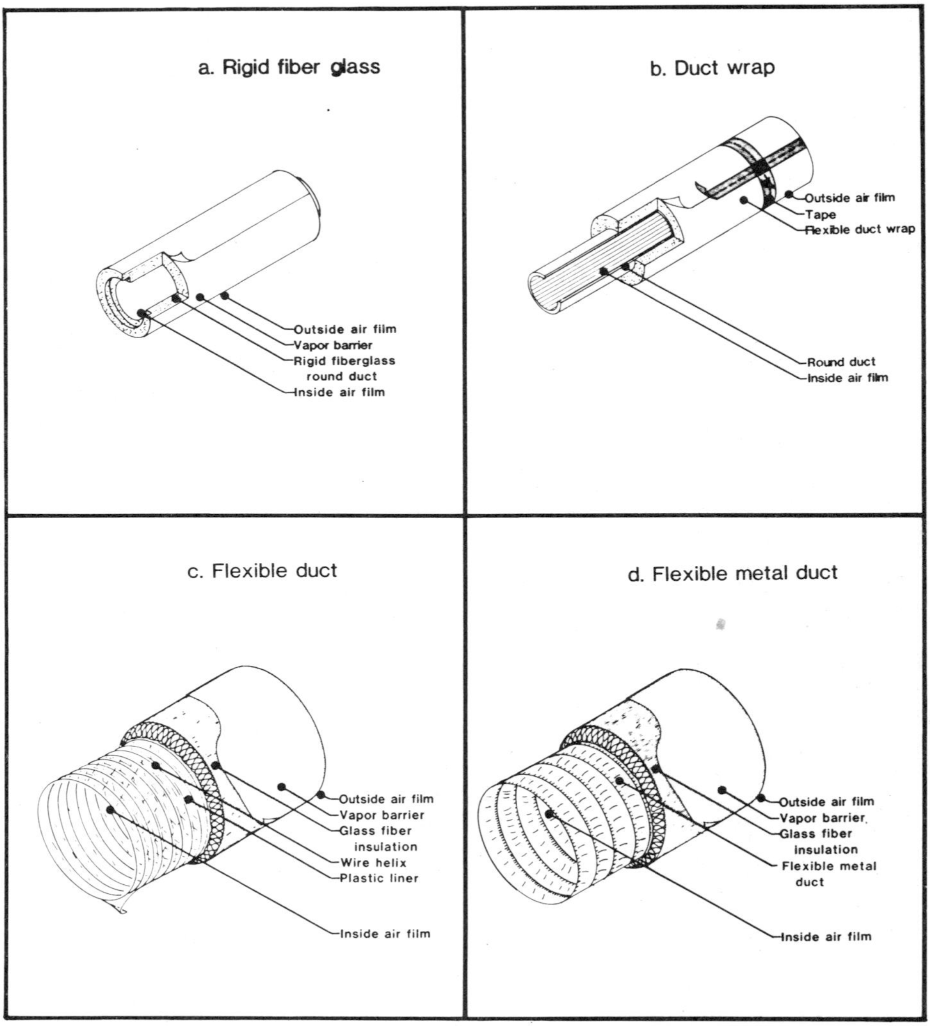

Plate 47. Round ducts

REFERENCES FOR SECTION 4.0

[1] ASHRAE Handbook, 1981 Fundamentals. American Society of Heating, Refrigerating and Air-Conditioning Engineers, Inc. Atlanta, GA. 1981.

[2] NAHB Research Foundation, Inc. Insulation Manual-Homes, Apartments. Rockville, MD. 1979.

[3] Mineral Insulation Manufacturers Association, Inc. "Mineral Fiber Batts and Blankets - Fiber Glass and Rock Wool - Standard for Installation in Residential and Other Light-Frame Construction." Summit, NJ. June 1981.

[4] Structural Clay Products Institute. "SCR Insulated Cavity Walls." Technical Note No. 21. McLean, VA. 1968.

[5] Portland Cement Association. "The Concrete Approach to Saving Energy." Skokie, IL. 1974.

[6] Brick Institute of America. "Heat Transmission Coefficients of Brick Masonry Walls." Technical Notes on Brick Construction. No. 4 Revised. Aug/Sept. 1974.

[7] National Concrete Masonry Association. "Thermal Comfort in Housing with Concrete Masonry." NCMA-TEK 26-A. Herndon, VA. 1979.

[8] National Concrete Masonry Association. "Testing the Thermal Performance of Concrete Masonry Walls." NCMA-TEK 112. Herndon, VA. 1979.

[8a] Portland Cement Association. "Special Considerations for the Selection of Tilt-Up Concrete Sandwich Panels." Skokie, IL. 1975.

[9] Department of Energy. Minimum Energy Dwelling (MED) Workbook - An Investigation of Techniques and Maerials for Energy Conscious Design. SAN/1198-1. Dec. 1977.

[10] Burch, D.M., and Hunt, C.M. "Retrofitting an Existing Wood-Frame Residence for Energy Conservation - An Experimental Study." NBS Building Science Series 105. National Bureau of Standards. Washington, D.C. July 1978.

[11] Steven Winter Associates, Inc. Passive Solar Construction Handbook. SSEC/SP-41187. Southern Solar Energy Center (U.S. Dept. of Energy). Atlanta, GA. Feb. 1981.

[11a] Rossiter, Walter J., Jr. and Mathey, Robert G. (eds.) "Criteria for Retrofit Materials and Products for Weatherization of Residences." NBS Technical Note 982. National Bureau of Standards. Washington, D.C. Sept. 1978.

REFERENCES FOR SECTION 4.0 (cont.)

[12] Turner, W.C. and Malloy, J.F. Thermal Insulation Handbook. Robert E. Krieger Publishing Company, Malabau, FL, and McGraw Hill Book Company, New York, NY. 1981.

[13] Watson, Don A. Construction Materials and Processes. McGraw-Hill Book Company. New York, NY. 1978.

[14] Johns-Manville Corporation. "Fiber Glass Building Insulation Installation Manual." Denver, CO. April 1981.

[15] Owens-Corning Fiberglass Corporation. "High-R Sheathing Application Instructions." Toledo, OH. Aug. 1979.

[16] The Celotex Corporation. "Application Instructions, Energy-Saving Insulation." Form 1669 Rev. D. Tampa, FL. July 1981.

[17] BASF Wyandotte Corporation. "EPS - Expanded Polystyrene Insulation." Wyandotte, MI. 1979.

[18] Dow Chemical U.S.A. How To Insulate bulletin series. Midland, MI. 1980, 1981.

[19] Holland, Elizabeth. "Insulation Moves Outside." Solar Age. Vol. 11, No. 11, pp. 22-27. Nov. 1981.

[20] National Concrete Masonry Association. "Concrete Masonry in Passive Solar Buildings." NCMA-TEK 90. Herndon, VA. 1977.

[21] Cullen, William C. "Solar Heating, Radiative Cooling, and Thermal Movement - Their Effects on Built-Up Roofing." NBS Technical Note 231. National Bureau of Standards. Washington, D.C. April 1964.

[22] Rossiter, Walter J., Jr. and Mathey, Robert G. "Effect of Insulation on the Surface Temperature of Roof Membranes." NBSIR 76-987. National Bureau of Standards. Washington, D.C. Feb. 1976.

[23] Richards, David E. and Mirra, Edward J. "Does More Roof Insulation Cause Premature Roofing Membrane Failure or Are Roofing Membranes Adequate?" Journal of Thermal Insulation. Vol. 1, pp. 171-180. Jan. 1978.

[24] Griffin, C.W. "Plug the Energy Leaks in Your Roof." Journal of Thermal Insulation. Vol. 1, pp. 206-214. Jan. 1978.

[25] Gumpertz, Werner. "Thermally Efficient Roofs Put a Lid on Heat Loss." Specifying Engineer. Vol. 42, pp. 107-109. Aug. 1979.

[26] Roofing, Siding and Insulation. "Blistering: Still Looking For Answers." p. 68. Feb. 1980.

[27] Roofing, Siding and Insulation. "Attaching Insulation to Steel Decks." pp. 65-67. Feb. 1980.

REFERENCES FOR SECTION 4.0 (cont.)

[28] Busching, Herbert W., Mathey, Robert G., Rossiter, Walter J., Jr., and Cullen, William C. "Effects of Moisture in Built-Up Roofing - A State-of-the-Art Literature Survey." NBS Technical Note 965. National Bureau of Standards. Washington, D.C. July 1978.

[29] Tobiasson, Wayne and Ricard, John. "Moisture Gain and Its Thermal Consequence for Common Roof Insulations." U.S. Army Corps of Engineers, Hanover, N.H. Reprinted From: Proceedings 5th Conference on Roofing Technology. April 1979.

[30] Knab, Lawrence I., Jenkins, David R., and Mathey, Robert G. "The Effect of Moisture on the Thermal Conductance of Roofing Systems." NBS Building Science Series 123. National Bureau of Standards. Washington, D.C. April 1980.

[31] Knab, Lawrence I., Jenkins, David R. and Mathey, Robert G. "The Effect of Moisture on the Thermal Conductance of Roofing Systems." Roofing, Siding, and Insulation. Special report, pp. 61-66, Part I. July 1980, and pp. 62-67 Part II. Sept 1980.

[32] Knab, Lawrence I., Jenkins, David R. and Mathey, Robert G. "The Effect of Moisture on the Thermal Conductance of Roofing Systems." Roofing, Siding, and Insulation. Special report, pp. 61-66, Part I. July 1980, and pp. 62-67 Part II. Sept 1980.

[33] The Roofing Industry Educational Institute. "Roof Maintenance." Englewood, CO. Feb. 1981.

[34] Department of the Air Force. "Built-Up Roof Management Program." AFM 91-36. Washington, D.C. Sept. 1980.

[35] Epstein, K. A. and Putnam, L. E. "Performance Criteria for the Protected Membrane Roof System." Journal of Thermal Insulation. Vol. 1, pp. 149-167.

[36] Roofing, Siding, and Insulation. "Hot Build-Up to Falter in the 80's." p. 58. Feb. 1980.

[37] Brady, Sam A. "Protecting Polyurethane Foam Roofing Insulation." Metal Building Review. p. 54-58. July 1977.

[38] Milanese, Robert. "The Super Insulating Perlite Roof Deck." Roofing, Siding, and Insulation. pp. 94-98. Oct. 1980.

[39] Roofing, Siding and Insulation. "Hidden Market: Insulating Metal Roofs." pp. 74-76. Oct. 1980.

[40] Morgenroth, Dan E. "A Cost Effective Solution To Achieve Thermal Performance of Roofing Systems." ASHRAE Journal. p. 54. March 1981.

[41] Sodha, M.S., Kaushik, S.C., Tiwari, G.N., Goyal, I.C., Malik, M.A.S. and Khatry, A.K. "Optimum Distribution of Insulation Inside and Outside the Roof." Building and Environment. Vol. 14, pp. 47-52. Feb. 1979.

REFERENCES FOR SECTION 4.0 (cont.)

[42] Shipp, Paul H. and Broderick, Thomas B. "Comparison of Annual Heating Loads for Various Basement Wall Insulation Strategies Using Transient and Steady State Models." Presented at the DOE-ORNL/ASTM Conference on Thermal Insulation, Materials, and Systems For Energy Conservation in the '80s. Clearwater Beach, FL. Dec. 1981.

[43] Turner, W.C. and Malloy, J.F. Thermal Insulation Handbook. Robert E. Krieger Publishing Company, Malabau, FL, and McGraw Hill Book Company, New York, NY. 1981.

[44] Grasso, Maureen M. and Buchanan, David R. "Roller Shade System Effectiveness in Space Heating Energy Conservation." ASHRAE Transactions. No. 2520. GIA-39.

[45] Quirouette, R.L. "Insulated Window Shutters." Building Practice Note No. 17. National Research Council of Canada. Ottawa, Canada. June 1980.

[46] Shepherd, P.B. "Performance Evaluation of Vinyl Replacement Windows." FESA-TS-70-78-D-0002, Task Order 10. Johns-Manville Sales Corp., Denver, CO. Prepared for the U.S. Army Facilities Engineering Support Agency, Fort Belvoir, VA. Jan. 1980.

[47] Hastings, S. Robert and Crenshaw, Richard W. "Window Design Strategies to Conserve Energy." NBS Building Science Series 104. National Bureau of Standards. Washington, D.C. June 1977.

[48] Bishop, Ronald R. "Performance Evaluation of Solar Films and Screens." Johns-Manville Sales Corp., Denver, CO. Prepared for the U.S. Army Facilities Engineering Support Agency, Fort Belvoir, VA. May 1979.

[49] Shepherd, P.B. and Gerharter, J.E. "Techniques for Control of Air Infiltration in Buildings." Johns-Manville Sales Corp., Denver, CO. Prepared for the U.S. Army Faciliites Engineering Support Agency, Fort Belvoir, VA. April 1979.

[50] Department of Energy. "Find and Fix the Leaks-A Guide to Air Infiltration Reduction and Indoor Air Quality." DOE/CS-0006. Washington, D.C. May 1981.

[51] Harrje, David T., Dutt, Gautam, S. and Beyea, Jan E. "Locating and Eliminating Obscure But Major Energy Losses in Residential Housing." Princeton University, Princeton, N.J. DE-79-3, No. 1.

[52] Harrje, David T. "Building Envelope Performance Testing." ASHRAE Journal. pp. 39-41. March 1981.

[53] Liptak, Bela G. "Savings Through CO_2 Based Ventilation." ASHRAE Journal. pp. 38-41. July 1979.

[54] Sandberg, Mats. "What is Ventilation Efficiency?" Building and Environment. Great Britain. Vol. 16, No. 2. pp. 123-135. 1981.

REFERENCES FOR SECTION 4.0 (cont.)

[55] Peterson, Joel E. "Estimating Air Infiltration into Houses." ASHRAE Journal. pp. 60-62. Jan. 1979.

[56] Tucker, Henry W. "Simplified Determination of Air Infiltration." ASHRAE Journal. pp. 44-47. July 1979.

[57] ASHRAE Handbook, 1981 Fundamentals. American Society of Heating, Refrigerating and Air-Conditioning Engineers, Inc. Atlanta, GA. 1981.

[58] "Exterior Caulking Compounds." Consumer Reports, Vol. 46, No. 10, pp. 579-581. Oct. 1981.

[59] American Concrete Institute. "Use of Epoxy Compounds with Concrete." ACI 503R-73. Detroit, MI. Sept. 1973.

[60] American Concrete Institute. "Guide to Joint Sealants for Concrete Structures." ACI 504R-77. Detroit, MI June, 1977.

[61] "Weather Stripping." Consumer Reports, Vol. 46, No. 10, pp. 576-578. Oct. 1981.

[62] Pemko Manufacturing Company. "What You Should Know About Interior and Exterior Weatherstrip and Thresholds." Emeryville, CA. 1970.

[63] Federal Energy Administration. "Project Retro-Tech Home Weatherization Manual." Conservation Paper No. 28C. Washington, D.C. May 1976.

[64] Henning, G.N. "Energy Conservation With Air Infiltration Barriers." Presented at the DOE-ORNL/ASTM Conference on Thermal Insulation, Materials, and Systems for Energy Conservation in the '80s. Clearwater Beach, FL. Dec. 1981.

[65] Hadley, John. "Energy Conservation and Indoor Air Quality." ASHRAE Journal. p. 35 March 1981.

[66] Hollowell, Craig D., Berk, James V. and Traynor, Gregory W. "Impact of Reduced Infiltration and Ventilation on Indoor Air Quality." ASHRAE Journal. pp. 49-53. July 1979.

[67] Hollowell, Craig D., Berk, James V. and Traynor, Gregory W. "Impact of Reduced Infiltration and Ventilation on Indoor Air Quality in Residential Buildings." ASHRAE Transactions. PH-79-10, No. 2. 1979.

[68] Kusuda, T., Hunt, C.M. and McNall, P.E. "Radioactivity (Radon and Daughter Products) As a Potential Factor in Building Ventilation." ASHRAE Journal. pp. 30-34. July 1979.

[69] Woods, James E. "Ventilation, Health and Energy Consumption: A Status Report." ASHRAE Journal. pp. 23-27. July 1979.

REFERENCES FOR SECTION 4.0 (cont.)

[70] "Can You Make a House Too Tight?" Consumer Reports. Vol. 46, No. 10, pg. 582. Oct. 1981.

[71] Woods, James E., Maldonado, Edwardo, A.B. and Reynolds, Gary L. "How Ventilation Influences Energy Consumption and Indoor Air Quality." ASHRAE Journal. pp. 40-43. Sept. 1981.

[72] Nelson, Lorne W. and Tobias, James R. "Energy Savings in Residential Buildings." ASHRAE Journal. pp. 38-45. Feb. 1974.

[73] Heating/Piping/Air Conditioning, "Steam Pipe Insulation Could Save 305,000 Barrels of Oil a Day." p. 27. Oct. 1980.

[74] Midwest Insulation Contractors Association. "Commercial and Industrial Standards." Omaha, NE. 1979.

[75] Roose, Robert W. "Thermal Insulation Guide for Ducts, Equipment, and Piping." Ch. 31 in Roose, Robert W. (ed.) Handbook of Energy Conservation in Mechanical Systems in Buildings. Van Nostrand Reinhold Company. New York, NY. 1978.

[76] ASHRAE Handbook, 1981 Fundamentals. American Society of Heating, Refrigerating and Air-Conditioning Engineers, Inc. Atlanta, GA. 1981.

[77] Turner, W.C. and Malloy, J.F. Thermal Insulation Handbook. Robert E. Krieger Publishing Company, Malabau, FL, and McGraw Hill Book Company, New York, NY. 1981.

[78] Roose, Robert W. and Pannkoke, T.E. "Thermal Insulation for Buried Piping." Ch. 33 in Roose, Robert W. (ed.) Handbook of Energy Conservation in Mechanical Systems in Buildings. Van Nostrand Reinhold Company. New York, NY. 1978.

[79] Sheetmetal and Air Conditioning Contractors' National Association, Inc. "Fibrous Glass Duct Construction Standards." 5th edition. Vienna, VA. 1979.

[80] Thermal Insulation Manufacturers Association. "The Thermal Performance of Operating Duct Systems." Mt. Kisco, NY. Nov. 1980.

[81] Heating/Piping/Air Conditioning. "Energy Conservation and HVAC Contractors." pp. 25-26. Nov. 1980.

[82] Heating/Piping/Air Conditioning. "Glass Fiber Ductwork." pp. 57-67. July 1980.

[83] Wallis, F.A. "Evaluation of Cold Storage Building Techniques." Insulation Outlook, pp. 24-27.

5.0

Insulation Economics

When confronted with an opportunity to insulate a building or mechanical system, two very basic but important questions are likely to be asked: What will insulation cost? How much money will the insulation save? The answer to these questions depends on exactly how much insulation is added. As it turns out, some thickness of insulation will be the economic optimum for any given application and will result in the lowest summation of both energy and insulation cost.

5.1 Definitions of Basic Economic Terms

Before giving a further explanation of economic analysis as specifically applied to thermal insulation, it is worthwhile to introduce a few general terms and methods of economic analysis. Perhaps the simplest economic evaluation method is simple payback (N_s), calculated as the initial investment divided by the resulting savings (or return) in the first year (see Equation 5-1).* The number obtained is the number of years needed to recover the investment assuming all costs and the yearly energy savings remain constant. While this method may be acceptable for investments yielding simple paybacks of 1 year or less [1], it has the following drawbacks for longer-term investments [2]:

- Simple payback assumes zero financing and alternative investment interest rates.
- Simple payback ignores inflation and elements of cost which escalate at a rate greater than the rate of inflation (e.g., energy). It treats dollars saved in future years the same as dollars saved today.
- The timing of cash flows within the payback period is neglected.
- Cash flows beyond the payback year are neglected.
- Simple payback does not provide for income taxes, depreciation, or investment credit.
- Simple payback does not provide for return on debt or equity capital.

The discounted payback method is an improvement over simple payback as it accounts for the discount rate available to the investor as well as the expected inflation rate, if any, in the annual savings. The discount rate is defined as the interest rate a sum of money could earn in an alternative investment. The discounted payback period (N_{dp}), in years, is given by equations 5-2 and 5-3 [3].

---

*Equations for all economic concepts discussed in this section are listed at the end of the section, starting on page 182.

Discounted payback still does not account for cash flows beyond the payback period or for tax-related items. An investment having a shorter discounted payback period than another often does not result in the greatest benefit over the long term. A concept that reveals the long-term benefit is present worth. A present worth analysis essentially determines the value, in current dollars, of all the savings gained over a period of years (often the lifetime of the installation) resulting from an investment. The present worth (PW) of future savings may be found using any of several equivalent equations, listed as Equations 5-4 through 5-9.

The equations for present worth do not have to be limited to future savings but are applicable to future costs as well. In Equations 5-4 through 5-9, f is the expected rate of increase in fuel costs; however, using values for expected rates of increase in operational costs, maintenance costs, or costs tied to general inflation would give the present worth of future expenditures for these quantities.

A cash outlay for an investment is already equal to its present worth and no further calculation is necessary. Alternatively, an investment may be financed through a short-term loan or as a part of a long-term mortgage (as with a new building or plant, where the mortgage covers the total costs of construction). In these cases, the present worth of all future expenditures attributable to the initial cost of the investment must also be determined.

The above ideas on present worth form the basis of life cycle cost (LCC) analysis, the generally accepted method of evaluating an investment which generates savings in future years. A proper LCC analysis should consider not only operational costs (which may be tied to an inflation rate for electricity as well as fossil fuel) and maintenance costs, but should also consider the effects of an investment on property taxes, income taxes (interest paid on a loan to finance the investment is deductible from income), and insurance rates. If applicable, depreciation and salvage value must also be considered. A life cycle cost analysis allows one to find the total net present worth of an investment by subtracting the present worths of all future costs associated with the investment from the present worths of all future savings (see Equation 5-10). If the result is not positive, the investment will cost more than it will save.

The benefit/cost ratio (B/C) is another measure sometimes used to evaluate an investment. The same quantities are used to find both the B/C ratio and total net present worth, but rather than subtracting the costs from the savings, a ratio of savings to cost is formed (see Equation 5-11).

The rate of return (ROR) is essentially the interest rate (i) earned by the investment in insulation and is tied to some period of analysis. This period may be equal to the expected life of the project, or it may be the number of years in which the investment is required to be paid back. One might think of the investment as a loan and the annual savings as the payment being returned. The ROR is simply the interest rate giving the correct ratio of payment to principal (i.e., savings to investment) for the number of years being considered in a standard loan analysis. If the savings are the same year after year, ROR may be determined from Equations 5-12 and 5-13. If the savings increase year after year, as will savings gained by not using some amount of energy, Equations 5-14 and 5-15 should be used.

The ROR can be thought of as the discount rate which makes the total net present value of the investment equal to zero. If this discount rate were available to the investor, the investor might be indifferent to making the investment. If the ROR turns out to be higher than the available d, the investment is attractive. The rate of return on investment (ROI), or real rate of return, is the amount by which the ROR is greater than the discount rate (see Equation 5-16).

To decide which of the above concepts should be used in an economic analysis, a designer must be aware of any specific economic criteria established by a particular client. The Department of Defense, for example, has published several documents describing its requirements [4-6] (Appendix C of Reference [6] cites additional references). These documents provide official economic guidance for this agency.

5.2 Optimizing Insulation Thickness

ASHRAE Standard 90A-1980 specifies recommended minimum insulation thicknesses for pipes, ducts, service water heaters, space heating boilers, and water storage tanks. These minimum thicknesses may not be the most economic choices, however, for any particular application. For this reason, the designer of insulation systems should be aware of economic techniques used to choose optimum levels of insulation.

The first method of determining the most economical thickness of thermal insulation was presented by L. B. McMillan in 1926 [7]. More recent methods have not changed the basic concepts; however, they have made these concepts easier to use. How an economic optimum insulation thickness occurs is best illustrated in Figure 5-1 on the following page. As the thickness of insulation is increased, the cost of the insulating job will also increase as a result of both material and labor expenditures. As pointed out in Section 2.0, however, the thermal energy lost through the insulation decreases as insulation is added, and so does the cost of providing that energy.

Because each additional unit increment of insulation saves less energy than the one before it, the cost of lost energy versus insulation thickness is a downward sloping curve approaching zero for very thick insulation. Since the insulation cost itself rises with increasing thickness, there exists some point at which more would be spent for an increment of insulation than could ever be saved by that increment. By adding the insulation cost curve to the lost energy cost curve, a total cost curve having a distinct minimum is found. The minimum occurs at the thickness of insulation for which the incremental insulation cost equals the incremental savings in lost energy cost. The costs referred to in the previous paragraph are not just first-year costs. Although the insulation may be paid for immediately, it may be expected to last for many years--and will, of course, save energy all the time it is in place. In fact, the lifetime of the installation, if done properly, may be limited to that of the building or equipment on which it is installed rather than

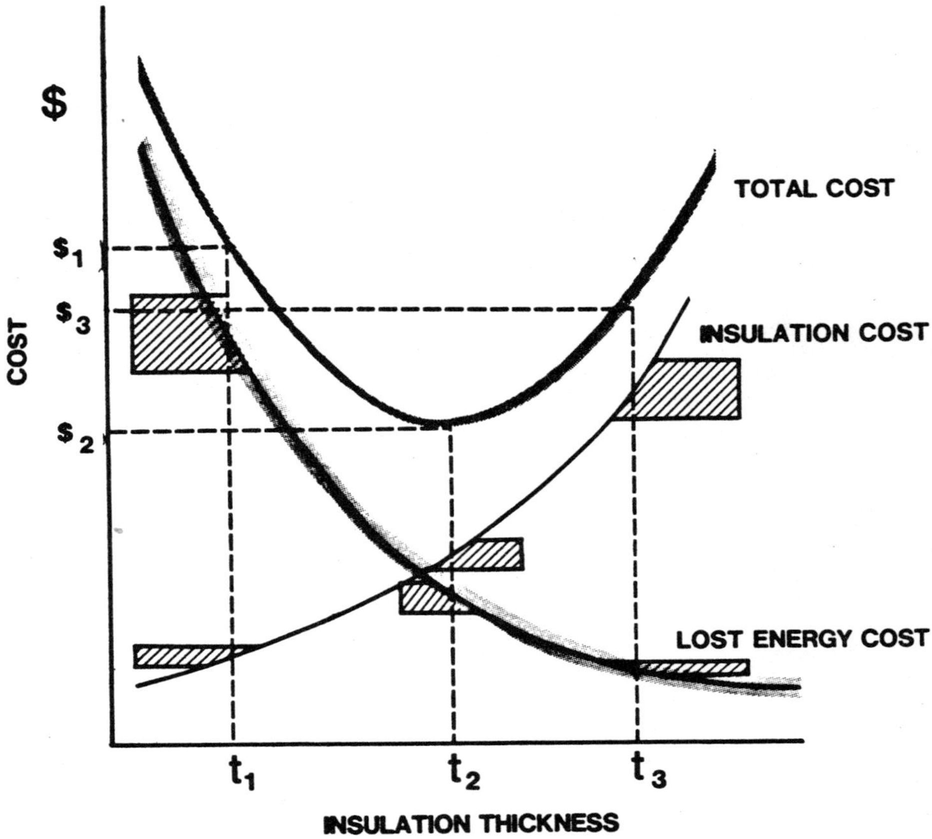

Figure 5-1 Total cost of insulation

to the life of the insulation itself. In this case, the energy cost curve should reflect the present worth of all the energy saved over the expected lifetime of the entire system. Similarly, the insulation cost curve should include the present worth of any financing costs as well as material and installation costs. As mentioned in the previous section, any other incremental costs or savings in operation, maintenance, taxes, insurance, depreciation and salvage value should also be included in the appropriate curve for a complete LCC analysis.

With thermal insulation, a negative total net present worth is rare. Most insulation jobs will not only have positive net present worths, but they will improve cash flow immediately because the total cost for energy plus a loan payment (when the insulation job is financed) will be less than the cost of energy alone without the insulation [8].

An LCC analysis can be accomplished manually using a handheld scientific calculator to aid in determining present worth factors. A number of step-by-step techniques are also commercially available. One of the most widely known techniques for the evaluation of the economics of insulation is called "Economical Thickness of Industrial Insulation", or ETI [9-10]. Developed under a contract from the Federal Energy Administration, ETI is a method which uses a series of nomographs to determine the most economic insulation thickness for any particular case. It has also been developed into a computer program; blank data input sheets may be obtained from the Thermal Insulation Manufacturers Association (TIMA), filled out and mailed back to TIMA. The computer printouts will be returned to the requestor. Both new and retrofit situations are handled by ETI.

The ETI method differs slightly from the net-present-worth method described earlier. Instead of determining present worths, ETI compares the average annual cost of the insulation job to the average annual cost of lost energy. These annual average costs can, however, be converted to present worth [9]. ETI requires the following information:

- Pipe size.
- Pipe length.
- Pipe operating temperature.
- Pipe jacket emissivity.
- Ambient temperature.
- Number of operating hours per year.
- Fuel type used to produce heat.
- Heating value of the fuel.
- Fuel conversion efficiency.

- First year fuel price.
- Annual fuel price inflation factor.
- Heating plant cost per million Btu.
- Heating plant life.
- Insulation life.
- Installed insulation cost for each thickness.
- Cost of money.
- Annual maintenance cost.

The ETI manual [9] presents a method of estimating installed insulation costs based on a number of parameters, including the estimated material cost, the local labor rate, a regional labor productivity factor, and a piping complexity factor. However, the manual also recommends that a qualified insulation contractor be consulted whenever possible for a more accurate determination of installed prices.

The outputs of the ETI computer program are:

- The most economic insulation thickness.
- The insulation surface temperature.
- The average annual heat saved over the next smaller insulation thickness.
- The present worth of the energy saved.

Several major insulation manufacturers offer telephone access to the ETI computer program. They could be contacted for further information on these services. The National Insulation Contractors Association (NICA) also has an ETI-based program available for the HP-97 desktop programmable calculator. Included in NICA's packet of tapes and programs are:

- Thermal conductivity data for fiberglass, mineral wool, calcium silicate, cellular glass, and polyurethane foam.
- Economic data and annual cost calculations.
- Heat loss and annual cost calculations for piping.
- Heat loss and annual cost calculations for flat surfaces.
- Simple and discounted payback calculations.

For additional information on investment criteria and life cycle cost analysis, the reader may consult Reference [11] or any of a number of textbooks in engineering economics. Other references deal with various aspects of the economics of insulation and other energy conserving techniques relative to residential buildings [12-15], commercial buildings [16-19], commercial building roofs [20], mechanical systems in general [14, 21], piping [22-25], piping and equipment [26-28], and HVAC ductwork [29-31]. A technique for finding present worth factors with a programmable calculator is given in Reference [32]. References [16, 17, 19, 22, 26, and 30] also discuss computer programs in their respective subjects.

Before using any economic analysis method, the reader is urged to completely understand the assumptions made in the method and its applicability. Criticisms of various methods are not often documented, and the user of any method is ultimately responsible for the results of any decisions reached.

ECONOMIC EQUATIONS

(1) Simple Payback

$$N_s = I/S \qquad \text{(Eq. 5-1)}$$

where N_s = simple payback period (years)
I = initial investment ($)
S = savings in first year ($)

(2) Discounted Payback

$$N_d = \frac{\log\left[\frac{I}{S}\left(\frac{f-d}{1+f}\right)+1\right]}{\log(1+f) - \log(1+d)} \qquad (f \neq d) \qquad \text{(Eq. 5-2)}$$

$$N_d = N_s \qquad (f = d) \qquad \text{(Eq. 5-3)}$$

where N_d = discounted payback period (years)
d = the available discount rate (%/100)
f = the estimated fuel cost inflation rate (%/100)

(3) Present Worth

$$\frac{PW}{S} = \frac{(1+d)^n - (1+f)^n}{\left(\frac{d-f}{1+f}\right)(1+d)^n} \qquad (f \neq d) \qquad \text{(Eq. 5-4)}$$

$$\frac{PW}{S} = n \qquad (f = d) \qquad \text{(Eq. 5-5)}$$

where PW = present worth ($)
PW/S = PWF, the present worth factor
n = the period of the analysis (years)

This equation is actually a rearrangement of Equation 5-2 with n substituted for N_d and PW substituted for I. Alternative formulas for PW are:

$$\frac{PW}{S} = \frac{r(r^n - 1)}{r - 1} \qquad (f \neq d) \qquad \text{(Eq. 5-6)}$$

$$\frac{PW}{S} = n \qquad (f = d) \qquad \text{(Eq. 5-7)}$$

where $r = (1 + f)/(1 + d)$

or

$$\frac{PW}{S} = \frac{1 - (1 - y)^{-n}}{y} \qquad (f \neq d) \qquad \text{(Eq. 5-8)}$$

$$\frac{PW}{S} = n \qquad (f = d) \qquad \text{(Eq. 5-9)}$$

where $y = (d-f)/(1+f)$

If f = 0, then y = d, and Equation 5-8 is seen to be the reciprocal of the capital recovery factor (CRF), which is the factor that determines what loan payments will be when a loan is made at an interest rate d (CRF = payment/principal).

(4) Total Net Present Worth

= (PW of all savings)-(PW of all costs) (Eq. 5-10)

where = net present worth ($)

(5) Benefit/Cost Ratio

B/C = (PW of all savings)/(PW of all costs) (Eq. 5-11)

where B/C = benefit/cost ratio (%/100)

(6) Rate of Return

First, find

$$CRF = S/I \qquad \text{(Eq. 5-12)}$$

Then, if S is constant for n years,

$$CRF = \frac{i}{1 - (1 + i)^{-n}} \qquad \text{(Eq. 5-13)}$$

where n = the period of the analysis (years), and may be equal to the expected life of the project, or some number of years in which the investment is required to be paid off (not necessarily the same as N_s or N_d).

i = interest rate (%/100)

Finally, ROR = rate of return = the value of i that solves Equation 5-13 (%/100).

If S increases each year,

$$CRF = \frac{y}{1 - (1 + y)^{-n}} \quad (f \neq i) \qquad \text{(Eq. 5-14)}$$

$$CRF = 1/n \quad (f = i) \qquad \text{(Eq. 5-15)}$$

where y = (i-f)/(1+f)

f = annual rate of increase in S (%/100)

and n, i, and ROR are as defined after Equation 5-13.

(7) Rate of Return on Investment, or Real Rate of Return

$$ROI = ROR - d \qquad \text{(Eq. 5-16)}$$

where ROI = rate of return on investment (%/100).

REFERENCES FOR SECTION 5.0

[1] Cohn, Lisa. "Fuel Cost Uncertainty Clouds Boiler Payback Calculations." Energy User News, Dec. 7, 1981.

[2] Montag, Geraldine M. "A Commercial Building Ownership Energy Conservation Cost Analysis Model." ASHRAE Journal. pp. 49-52, June 1979.

[3] Manian, V.S. "Toward an Accurate View of Payback." ASHRAE Journal. pp. 28-30, Feb. 1979, and letter by Chapman, W.P., pg. 6, April 1979.

[4] Department of Defense. "Economic Analysis and Program Evaluation for Resource Management." DODINST 7041.3, 18 October 1972.

[5] Department of Defense. "DOD Construction Criteria Manual." DOD 4270.1M (Advance Edition), 1 June 1978.

[6] Department of the Navy. "Economic Analysis Handbook." NAVFAC P-442. Naval Facilities Engineering Command. Alexandria, VA. July 1980.

[7] McMillan, L.B. "Heat Transfer Through Insulation." Transactions of the ASME, 1926.

[8] Guntermann, Alfred E. "Retrofit Now for Best Energy System ROI." Specifying Engineer, Vol. 42. pp. 83-85. Aug. 1979.

[9] Federal Energy Administration. "ETI-Economic Thickness of Industrial Insulation." Conservation Paper No. 46. Washington, D.C. August 1976.

[10] National Insulation Contracts Association. "Principles of Heat Transfer and Introduction to ETI." Washington, D.C.

[11] ASHRAE Handbook, 1981 Fundamentals. American Society of Heating Refrigerating and Air-Conditioning Engineers, Inc. Atlanta, GA. Chapter 45. 1981.

[12] NAHB Research Foundation, Inc. Insulation Manual-Homes, Apartments. Rockville, MD. 1979.

[13] Petersen, Stephen R. "Retrofitting Existing Housing for Energy Conservation: An Economic Analysis." NBS Building Science Series 64. National Bureau of Standards. Washington, D.C. Dec. 1974.

[14] ASHRAE Handbook, 1981 Fundamentals. American Society of Heating Refrigerating and Air-Conditioning Engineers, Inc. Atlanta, GA. Chapter 20. 1981.

[15] Stephenson, D.G. "Determining the Optimum Thermal Resistance for Walls and Roofs." Building Research Note No. 105. National Research Council of Canada. Ottawa, Canada. Jan. 1976.

REFERENCES FOR SECTION 5.0 (cont.)

[16] Montag, Geraldine M. "A Commercial Building Ownership Energy Conservation Cost Analysis Model." ASHRAE Journal. pp. 49-52, June 1979.

[17] Montag, Geraldine M. "Commercial Building Ownership Energy Conservation Cost Analysis Model - A Follow-Up." ASHRAE Journal. pp. 45-47. Sept. 1981.

[18] Russell, Alan D. "Economic Risks in Energy Conservation Strategies." Building and Environment, Vol. 16, No. 2, pp. 109-121. 1981.

[19] Kirk, Stephen J. "Life Cycle Costing - Controlling Return on Investment by Computer." Specifying Engineer. Vol. 42, pp. 91-93. Aug. 1979.

[20] Cash, Carl G. "Optimization of the Thermal Resistance of Roof Insulation in Reroofing." Journal of Thermal Insulation. Vol. 1, pp. 192-200. Jan. 1978.

[21] Turner, W.C. and Malloy, J.F. Thermal Insulation Handbook. Robert E. Krieger Publishing Company, Malabar, FL, and McGraw Hill Book Company, New York, NY. 1981.

[22] Reinhart, Robert D. "How Much Insulation for Retrofit?" Energy Engineering. Vol. 77, No. 3, pp. 32-52. Apr/May 1980.

[23] Curt, Robert J. "Economic Thickness for Thermal Piping Insulation Above Ground." Ch. 32 in Roose, Robert W. (ed.) Handbook of Energy Conservation in Mechanical Systems in Buildings. Van Nostrand Reinhold Company. New York, NY. 1978.

[24] Koenig, Alan R. "Choosing Economic Insulation Thickness." Chemical Engineering. (date required).

[25] Wepfer, W. J., Gaggioli, R.A. and Obert, E.F. "Economic Sizing of Steam Piping and Insulation." ASME Journal of Engineering for Industry, Vol. 101, pp. 427-433. Nov. 1979.

[26] Dixon, John G. and Dean, Bob W. "Applying a Computer Analysis Program to the Selection of Insulation Materials." ASHRAE Journal. pp. 62-66. March 1979.

[27] Harrison Michael R. and Pelanne, Charles M. "Cost-Effective Thermal Insulation." Chemical Engineering. pp. 62-76. Dec. 19, 1977.

[28] Turner, W. C. and Malloy, J. F. Handbook of Thermal Insulation Design Economics for Pipes and Equipment. Robert E. Krieger Publishing Company, Inc., Huntington, NY, and McGraw-Hill Book Company, New York, NY. 1980

[29] ASHRAE Handbook, 1981 Fundamentals. American Society of Heating, Refrigerating and Air-Conditioning Engineers, Inc. Atlanta, GA. 1981.

[30] Arkin, Hillel and Shitzer, Avraham. "Computer Aided Optimal Life-Cycle Design of Rectangular Air Supply Duct Systems." Paper No. 2523, ASHRAE Transactions. (date required).

[31] Shitzer, Avraham and Arkin, Hillel. "Study of Economic and Engineering Parameters Related to the Cost of an Optimal Air Supply Duct System." Paper No. 2551, ASHRAE Transactions (date required).

[32] Ng, Warren. "Life Cycle Costing With a Programmable Calculator." Heating/Piping/Air Conditioning. Vol. 53, No. 4, pp. 90-93. April 1981.

6.0

How to Use This Handbook

This section presents some examples of how to use the information given in this handbook to estimate energy savings attainable with thermal insulation. Two examples, one for a plane wall and one for a pipe, illustrate suggested procedures to follow. The plane wall case can be adapted to tanks and vessels over 3 feet in diameter, as the slight curvatures of such equipment do not affect heat loss calculations significantly.

Detailed methods for calculating building energy loads are not given here because such methods are beyond the scope of this handbook. The reader will, however, learn how to calculate heat loss coefficients step by step, and these coefficients are usually the basis for total load calculations.

Both examples follow similar procedures. For any particular case of interest, the existing or proposed building section or equipment configuration can be located in the Plates, immediately following page 121 in Section 4.0. The Plates list all the components to consider in a heat loss calculation. Candidate schemes to upgrade thermal insulation can also be found in the Plates and the components identified. Thermal conductivities or resistances for the components can be found in Tables 3-14 through 3-17 on pages 77 through 92. The overall heat transfer coefficient is then determined using techniques discussed in Section 2.0. Comparison of insulated case U-values to base case or uninsulated values indicates the potential for energy conservation with each alternative. Results for different insulations can be obtained by simply substituting different conductivities in the examples; different insulation thicknesses may be similarly examined.

The reader may wish to use the computer programs described in Section 7.0 as an alternative to the hand calculations presented in these examples. The programs, one in FORTRAN and one in BASIC, are listed in Appendix F and handle both plane and radial geometry.

After U-values and total loads have been estimated, local installed costs should be obtained and an economic analysis performed. A number of economic decision-making methods are discussed in Section 5.0.

6.1 Wall Example

Suppose energy conservation opportunities are investigated for a residence. One such opportunity might be adding insulation to a currently uninsulated frame wall. This action would result in reducing the heat loss through the wall. To estimate the magnitude of the reduction and also of the energy saved, it is necessary to calculate the heat loss coefficient for the original wall and for each insulation scheme considered. This task can be accomplished using the following procedure.

- o Find the existing wall in the Plates in Section 4.0.
- o Find the Plates illustrating insulation schemes of interest.
- o In a vertical column, list the components occurring in each wall and their thicknesses. This may be done separately for each wall, or in one composite list. Be certain to include all air films and air spaces. Include parallel framing members if applicable.
- o Using any of the building material property tables (Tables 3-14 through 3-17), construct a table of R-values corresponding to each component in the list for the existing case and each candidate insulation scheme. A separate column of R-values should be made for each parallel heat flow path within each insulation scheme. In each column of R-values, list a value only if the particular component actually occurs in the heat flow path.
- o At the bottom of the table, total the component R-values in each heat flow path.
- o Compute the U-value (the heat loss coefficient) for each path as the reciprocal of the path R-value.
- o Find the net U-value for the wall using the area-weighting rules given in Section 2.4 for parallel heat paths (the total of the UAs for each path divided by the total area gives the net U-value).
- o The net R-value, if desired, is the reciprocal of the net U-value.
- o The fraction indicating the heat lost through the insulated wall as compared to the existing wall is found by dividing the U-value of the insulated wall by the U-value of the existing wall.
- o The design heat loss through either wall is calculated by multiplying the U-value by the design temperature difference obtained from the ASHRAE Handbook of Fundamentals.

The above procedure can be illustrated in this example as follows. Call the existing uninsulated frame wall "Case 1". Consider two possible insulating schemes: (a) blowing glass fiber loose fill insulation into the wall cavities (Case 2), and (b) Case 2, plus the addition of 2 inches of polystyrene foam and a new stucco exterior over the existing siding (Case 3). Plate 1a shows the existing wall, Plate 2a shows the addition of loose fill insulation to the wall cavities, and Plate 7c shows one way of adding insulation to the outside of the existing siding. A cross section of a wall with all the components involved is shown in Figure 6-1 on the following page. The numbers in parentheses indicate the cases in which each component is present.

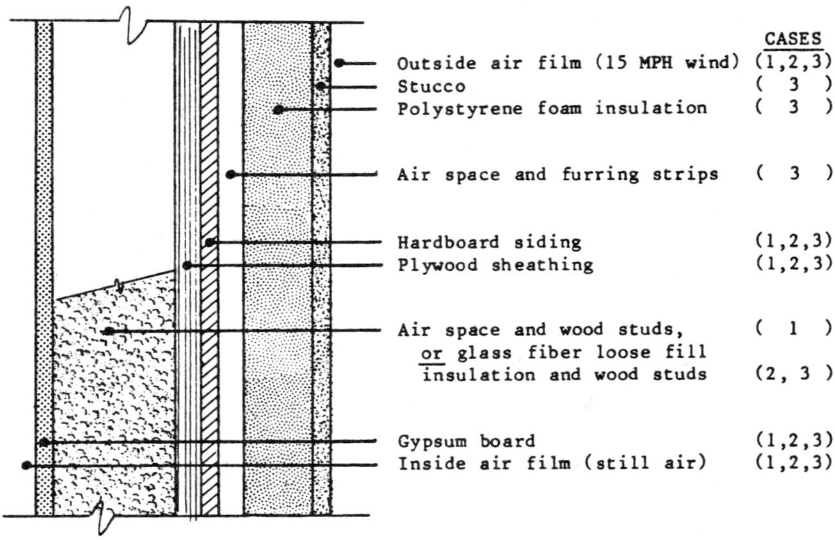

Figure 6-1. Wall Example Components

Table 6-1 on the following page shows the list of components and thicknesses, and the R-values for each component. R-values are listed by case number and heat flow path. At the bottom of the table, the path R-value, path U-value, the portion of total wall area applicable to each path, the net U-value, and the net R-value are all listed. The last item in the table is the fraction of heat lost for each case compared to the existing uninsulated wall.

The net U-value and fraction of existing heat loss can be used to estimate the energy and dollar savings gained with each insulation scheme. If the total heat loss through a building component has been previously determined, the fractional heat loss with the upgraded insulation scheme will indicate the energy savings possible for the component with the scheme. Alternatively, U-values may be used in simplified degree-day methods, bin temperature methods, or more complex computer simulations to determine seasonal energy usage and to size heating equipment (either new or replacement).

Costs and benefits may be evaluated by following the procedures outlined in Section 5.0; these procedures include various payback calculations and life-cycle cost analyses. One interesting conclusion can be drawn from Table 6-1: adding the loose fill insulation to the wall cavities saves 67% (100%-33%) of the energy lost through the wall, but also adding the exterior insulation saves only an additional 17% (33%-16%). However, adding the exterior insulation may cost as much, if not more, than installing the loose fill insulation. A proper economic analysis will determine whether or not the benefit of the exterior insulation justifies its cost.

Table 6-1

R-VALUES AND U-VALUES FOR WALL EXAMPLE

	Thickness (inches)	Case 1 Uninsulated		Case 2 Glass Fiber Loose Fill		Case 3 Glass Fiber Loose Fill and Polystyrene Foam	
		through air space	through framing	through insulation	through framing	through insulation	through framing
Outside Air Film (15 MPH wind)	-	0.17	0.17	0.17	0.17	0.17	0.17
Stucco	0.50	-	-	-	-	0.10	0.10
Polystyrene Foam	2.00	-	-	-	-	9.98	9.98
Air Space	0.75	-	-	-	-	1.01	-
Furring Strips	0.75	-	-	-	-	-	0.96
Hardboard Siding	0.50	0.34	0.34	0.34	0.34	0.34	0.34
Plywood Sheathing	0.75	0.78	0.78	0.78	0.78	0.78	0.78
Air Space	3.50	1.01	-	-	-	-	-
Glass Fiber Loose Fill	3.50	-	-	10.06	-	10.06	-
Wood Studs	3.50	-	4.49	-	4.49	-	4.49
Gypsum Board	0.50	0.45	0.45	0.45	0.45	0.45	0.45
Inside Air Film (still air)	-	0.68	0.68	0.68	0.68	0.68	0.68
Path R-Value		3.43	6.91	12.48	6.91	23.57	6.91
Path U-Value		0.292	0.145	0.080	0.145	0.042	0.145
Path portion of total area		85%	15%	85%	15%	85%	15%
Net U-Value		0.270		0.090		0.044	
Net R-Value		3.7		11.1		22.5	
Portion of Case 1 heat loss		100%		33%		16%	

6.2 Pipe Example

Consider a 4-inch pipe carrying steam at 300°F; the pipe is uninsulated and is indoors in a 75°F environment The calculation of energy saved if the pipe is insulated depends on how much insulation is added to the pipe. Typical examples of pipe insulation are shown on Plate 40.

- Find the energy lost from the bare pipe in Table C-2 (in Appendix C). The temperature difference above ambient is:

  $$\Delta T = 300 - 75 = 225°F$$

  Interpolating between ΔT = 200 and 250°F for the 4-inch pipe gives the heat loss per foot per degree temperature difference:

  $$Q/(L\ \Delta T) = 3.37 \text{ Btu/hr ft°F}$$

  Multiplying by ΔT = 225°F gives:

  $$Q/L = 758 \text{ Btu/hr ft}$$

- From Table 3-17, which begins on page 3-51, choose a pipe insulation to apply to the pipe At 300°F, the choices are limited to heavy-density glass fiber, calcium silicate, and cellular glass. For this example, choose glass fiber.

- Determine the approximate mean temperature of the insulation. In this example,

  $$T_{mean} = (300 + 75)/2 = 188°F$$

  In reality, since the exterior surface temperature of the insulation will be somewhat higher than the ambient temperature, T_{mean} will be higher also. For example, assume T_{mean} = 200°F.

- From Table 3-17, find the mean thermal conductivity of the insulation at the mean temperature just calculated. Interpolate if necessary. For the present example, choosing the glass fiber pipe insulation with a 500°F maximum service temperature results in k = 0.33 Btu in/hr ft^2°F.

- Find the formula for combined heat transfer in cylindrical geometry in Section 2.4. Assuming the outside pipe surface temperature is the same as the steam temperature, one can ignore the terms using the inside film coefficient and pipe conductivity. The heat lost from an insulated pipe is, therefore:

$$Q/L = \frac{2\pi r_o (T_i - T_o)}{\dfrac{r_o \ln(r_o/r_i)}{k} + \dfrac{1}{(h_o + h_r)}}$$

where r_o = the insulation outside radius,
and r_i = the insulation inside radius.

- Decide which insulation thicknesses to examine and determine r_o for each case. If the outside pipe diameter (from Table C-3 in Appendix C) equals r_i, and r_i plus the actual insulation thickness (from Table C-4) equals r_o.

  In this example, examine nominal thicknesses of 1, 2, 3, and 4 inches. For the 4-inch nominal pipe size, the actual outer diameter is 4.500 inches (0.3750 ft), so r_i = 2.25 inches or 0.1875 ft. The values of r_o for insualtion thicknesses of 1, 2, 3, and 4 inches are 3.29 (0.274), 4.29(0.358), 5.36(0.447), and 6.36(0.530) inches (feet), respectively.

- Determine values of h_o and h_r (or their sum, if available).

  In this example, a simplified formula for free convection of air around a horizontal pipe is used:

$$h_o = 0.27 \left(\frac{T_s - T_o}{D}\right)^{1/4}$$

where T_s is the insulation exterior surface temperature (°F) and D is the insualtion outer diamter (feet). This equation is generally valid for pipe diameters up to 2 feet. Also, from Section 2.3,

$$h_r = \sigma \varepsilon \ (T_s^2 + T_o^2)\ (T_s + T_o)$$

In this example, assume the insulation jacket emissivity (ε) is 0.85. Remember that absolute temperatures must be used to calculate h_r.

Using these two formulas, which depend on the initially unknown value of T_s, requires an initial guess for T_s. The guess can be checked after finding Q/L since the heat transfer rate through the insulation alone (i.e., between temperatures T_i and T_s) is the same as the overall rate (between temperatures T_i and T_o). From Section 2.1,

$$Q/L = \frac{2\pi r_o (T_i - T_s)}{\dfrac{r_o \ln (r_o/r_i)}{k}}$$

and when solved for T_s, this equation yields:

$$T_s = T_i - \frac{Q}{L} \cdot \frac{\ln (r_o/r_i)}{2\pi k}$$

Find the heat loss rate for the insulated pipe using the following procedure:

Step 1: Set r_i and r_o for the first insulation thickness.

Step 2: Guess T_s.

Step 3: Calculate h_o and h_r.

Step 4: Calculate Q/L.

Step 5: Calculate T_s from the equation just given.

Step 6: If the calculated value of T_s is not within a degree or two of the initial guess, replace the guess with the calculated value and go back to Step 3.

If the two values are sufficiently close, Q/L for this insulation thickness is as calculated.

Step 7: Set r_i and r_o for the next insulation thickness and go back to Step 2. Repeat this procedure until Q/L is found for every desired insulation thickness.

For the present example, with 1 inch of insulation, make an initial guess for T_s of 100°F. This value of T_s gives h_o = 0.702 Btu/ft^2 hr°F and h_r = 0.966 Btu/ft^2 hr°F, leading to Q/L = 88 Btu/hr ft. This value of Q/L gives a calculated value of T_s = 106°F. The iterations leading to the final values are shown below:

Table 6-2

PARAMETER VALUES FOR SEVERAL ITERATIONS, 1-INCH INSULATION

Trial	h_o (Btu/ft^{2}hr°F)	h_r (Btu/ft^{2}hr°F)	Q/L (Btu/hr ft)	T_s (°F)
1	0.702	0.966	88	106
2	0.739	0.981	89	105
3	0.734	0.979	89	105

Note that only a few iterations are required to arrive at the final answer, Q/L = 89 Btu/hr ft. Repeating the procedure for 2, 3, and 4 inch insulation thicknesses gives these results:

Table 6-3

FINAL TEMPERATURES AND HEAT LOSS VALUES FOR EACH INSULATION THICKNESS

Insulation thickness (inches)	T_s initial guess (°F)	T_s final value (°F)	Q/L (Btu/hr ft)
None	-	300	758
1	100	105	89
2	100	91	55
3	90	86	43
4	85	85	36

The values of Q/L can be used to determine the total amount of energy lost from uninsulated and insulated piping, as well as the amount of energy saved by adding insulation, on an annual basis. In the present example, if a plant has 1,000 ft of uninsulated 4-inch steam pipe at 300°F, and the pipe is constantly hot (8,760 hours per year), then the energy lost, energy saved, and incremental energy saved (the energy saved by each inch of insulation over the previous inch) are as follows:

Table 6-4

ENERGY VALUES FOR EACH INSULATION THICKNESS

Insulation thickness (in)	Q/L (Btu/hrft)	Energy lost (10^6Btu/yr)	Energy saved (10^6Btu/yr)	Incremental energy saved (10^6Btu/yr)
None	758	6,640	0	-
1	89	780	5,860	5,860
2	55	480	6,160	300
3	43	380	6,260	100
4	36	320	6,320	60

The above information can be used to construct a "cost of lost energy" curve, as described in Section 5.0. Cost savings occur in two places: in energy supplied to the steam plant, and also in energy supplied to the space cooling equipment, which previously may have been required to remove the energy lost from the uninsulated piping from the building. If either steam generators or cooling equipment is being replaced, lower-capacity equipment may be justified, resulting in additional savings.

7.0

Heat Transmission, Mass, and Thermal Capacity Computer Programs

7.1 General Information and Program Descriptions

Two computer programs for building and industrial thermal insulation design and analysis are described in this section. Entitled HTMCP (Heat Transmission, Mass, and Thermal Capacity Program), the programs are written in two commonly available computer languages - FORTRAN and BASIC. In their present form, these programs can be used for a variety of computer systems.

In recent years, the use of microcomputers by many architectural and engineering firms has expanded at a rapid rate. With few changes, the BASIC version of this program is compatible with microcomputers such as Apple, TRS-80, Osborne 1, Vectorgraphic, and North Star.

The HTMCP programs will calculate the overall heat transfer coefficient and total heat transfer through a composite, as well as the interfacial temperature between consecutive layers in the composite. The interfacial temperature can be used to determine whether potential moisture condensation at design conditions will take place within the composite. Condensation will occur at or below the dewpoint temperature. The dewpoint temperature can be derived from the psychrometric chart shown in Appendix E by using a given design dry-bulb temperature and relative humidity.

The programs will also calculate the composite total unit mass and unit thermal capacity. Up to 20 layers of materials for each composite can be input to this program. The unit mass can be used in structural calculations, and the unit thermal capacity can be used in passive solar applications.

Material data from the thermal physical material properties (Tables 3-14 through 3-17 on pages 77 through 92) can be used as input for the programs. Listings of the programs are contained in Appendix F.

7.1.1 Inputs:

The following inputs are required to run the program:

BASIC	FORTRAN	
T2	ITYPE	- Flag indicating composite type slab (1) or cylinder (2).
	TITLE	- Slab or cylinder description or title.
N	LYRS	- Number of layers in the composite including the inside and outside film layers--maximum of 20.

B(O), T(1), B(K) T(k) - Inside and outside temperatures (°F).

XL(I), TK(I), DEN(I), SPHT(I), RES(I), DES(I,J)
*(X(X1), T(X1), D(X1), S(X1), Z(X1), E$(X1)

Where

X(XC)* XL(I) = layer thickness (IN)

T(X1)*, TK(I) = layer thermal conductivity (Btu in/hr ft^2 °F)

D(XC)*, DEN(I) = layer density (lb/ft^3)

S(X1)*, SPHT(I) = layer specific heat (Btu/lb °F)

Z(X1)*, RES(I) = layer thermal resistance (ft^2 hr °F/Btu)

E$(X1)*,DES(I,J) = layer description

*BASIC variable

7.1.2 Outputs

The following outputs are computed by both versions of the program.

Title	- Run description
Layer Properties	- Echo check of layer properties input
CO	- Overall heat transfer coefficient in Btu/hr ft^2 °F
QO	- Unit heat transfer through composite in Btu/hr ft^2
DO	- Unit composite mass in lb/ft^2
TDO	- Unit composite thermal mass in Btu/lb °F
T(I)	- Interface temperatures

7.2 Input and Program Variables

The following variables are used as input variables and as program variables in the intermediate calculations.

7.2.1 HTMCP - FORTRAN Version - Input Variables

ITYPE - Flag indicating composite type or program completion
= 0 End program execution
= 1 Slab composite
= 2 Cylinder composite

TITLE(J) - The slab or cylinder composite description or title.

LYRS - Number of layers in the composite, including the inside and outside film layers--maximum of 20.

T(1),T(k) - The inside and outside temperatures (°F)

XL(I) - The layer thickness (slab) or the radius of the layer to the outside of the layer (cylinder) (inches)

I - The layer number (dimensionless)

TK(I) - The thermal conductivity of the layer material (Btu/hr ft^2 °F)

DEN(I) - The density of the layer material (lbm/ft^3)

SPHT(I) - The specific heat of the layer material (Btu/lb °F)

RES(I) - The thermal resistance of the layer material (ft^2 hr °F/Btu)

DES(I,J) - The description of the layer (dimensionless)

7.2.2 <u>HTMCP - FORTRAN Version - Program Variables</u>

K - The number of temperatures determined by the program. k is equal to number of layers + 1 (dimensionless).

RAD(I) - Radius to the outside of layer I (inches)

RR(I) - Resistance of layer I (Ft^2 hr °F/Btu)

DD(I) - Mass of layer I (lb/ft^2)

TDD(I) - Thermal mass of layer I (Btu/ft^2 °F)

DO - Composite system mass (lb/ft^2)

TDO - Composite system thermal mass (Btu/ft^2 lb °F)

RO - Composite system thermal resistance (ft^2 hr °F/Btu)

CO - Composite system overall heat transfer coefficient (U-value) (Btu/ft^2 hr °F)

QO - The unit heat transfer through the composite (Btu/hr ft^2)

T(I) - The interface temperature between layer (I-1) and layer (I). (°F)

7.2.3 <u>HTMCP - BASIC Version - Input Variables</u>

T2 - Flag for slab (=1) or cylinder (=2)

N - Number of layers in the composite, including the inside and outside film layers--maximum of 20.

B(0) - The inside fluid or material temperature (°F)

B(K) - The outside medium temperature (°F)

X(XI) - Layer (X1) thickness (slab) or the radius of layer (X1) to the outside of the layer (cylinder) (inches).

T(X1) - The thermal conductivity of the layer material (Btu in/hr ft °F).

D(X1) - The density of the layer material (lb/ft^3)

S(X1) - The specific heat of the layer material (Btu/lb in °F).

Z(X1) - The thermal resistance of the layer material (ft^2 hr °F/Btu).

E$ (X1) - The description of each layer (dimensionless).

7.2.4 HTMCP - BASIC Version - Program Variables

M(X1) - The mass of layer (X1) (lb/ft^2)

L(X1) - The thermal mass of layer (X1) (Btu/ft^2 °F)

M2 - Composite system mass (lb/ft^2)

L2 - Composite system thermal mass (Btu/ft^2 °F)

R2 - Composite system thermal resistance (Ft^2 hr °F/Btu)

C1 - Composite system overall heat transfer coefficient (U-value) (Btu/ft^2 hr °F)

Q1 - The unit heat transfer through the composite (Btu/hr ft^2)

B(X1) - The interface temperature between layer (X1-1) and layer (X1) (°F).

7.3 Algorithms and Analytical Relationships

7.3.1 Properties of Composite (See Figure 7-1, next page):

IX(I) - Thickness of layer I (inches)

TK(I) - Thermal conductivity of layer I (Btu in/hr ft°F)

DEN(I) - Density of layer I (lb/ft^3)

SPHT(I) - Specific heat of layer I (Btu/lb °F)

RES(I) - Thermal resistance of layer I (ft^2 hr °F/Btu)

h_i, h_o - Film or surface conductance (Btu/hr ft^2 °F); inside or outside films, composite layers

7.3.2 Equations:

Layer Resistance (RES(I)) = $1/h_i$ or $1/h_o$
= XL(I)/TK(I)

Layer Mass (DD(I)) = DEN(I)*XL(I)

Layer Thermal Mass(TDD(I)) = DEN(I)*XL(I)*SPHT(I)

Total Composite Thermal Resistance (RO):

$$RO = 1/h_i + \frac{XL(2)}{TK(2)} + \frac{XL(3)}{TK(3)} + \frac{XL(4)}{TK(4)} + 1/h_o$$

Total Composite Heat Transfer Coefficient (CO):

CO = 1/RO

Total Composite Slab Mass (DO):

DO = XL(2) * DEN(2) * XL(3) * DEN(3) * XL(4) * DEN(4)

Total Composite Slab Thermal Mass (TDO):

TDO = XL(2) * DEN(2) * SPHT(2) * XL(3) * DEN(3) * SPHT(3) * XL(4) * DEN(4) * SPHT(4)

Unit Heat Transfer Through The Slab (QO):

$$QO = (T_1 - T_6)/RO = CO * (T_1 - T_6)$$

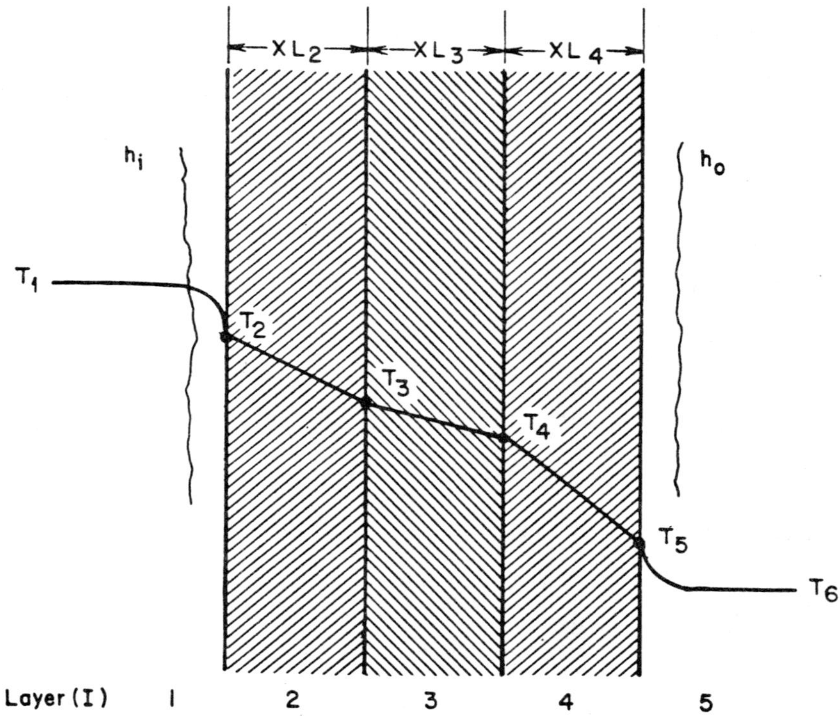

Figure 7-1 Example of slab heat transfer

7.3.3 Properties of Composite (See Figure 7-2, next page):

$r_1 - r_6$ - radii of layers (inches)

TK(I) - Thermal conductivity (Btu in/hr ft°F)

DEN(I) - Density of layer I (lb/ft^3)

RES(I) - Resistance of layer I (ft^2 hr °F/Btu)

h_i, h_o - Film or surface conductance (Btu/hr ft^2 °F)

7.3.4 Equations:

Layer Resistance (RE(I))	$= 1/h_i 2\pi r_1$	Inside film
	$= \dfrac{\ln(r(I)/r(I-1))}{2*\pi*TK(I)}$	material layers
	$= 1/h_o\ 2\pi r_4$	outside film

Total Composite Thermal Resistance (RO):

$$RO = 1/h_i\ 2\pi r_1 + \frac{\ln(r_2/r_1)}{2\pi\ TK(2)} + \frac{\ln(r_3/r_2)}{2\pi\ TK(3)} + \frac{\ln(r_4/r_3)}{2\pi\ TK(4)} + 1/h_o\ 2\pi r_4$$

Total Composite Heat Transfer Coefficient (CO):

$$CO = 1/RO$$

Total Composite Cylinder Mass (DO):

$$DO = DEN(2) * \pi * (r_2^2 - r_1^2) + DEN(6) * \pi * (r_3^2 - r_2^2) + DEN(4) * \pi * (r_4^2 - r_3^2)$$

Unit Heat Transfer Through Cylinder (QO):

$$QO = (T_1 - T_6)/RO = CO * (T_1 - T_6)$$

$$\frac{T_1 - T_2}{RES(1)} = \frac{T_2 - T_3}{RES(2)} = \frac{T_3 - T_4}{RES(3)} = \frac{T_4 - T_5}{RES(4)} = \frac{T_5 - T_6}{RES(5)}$$

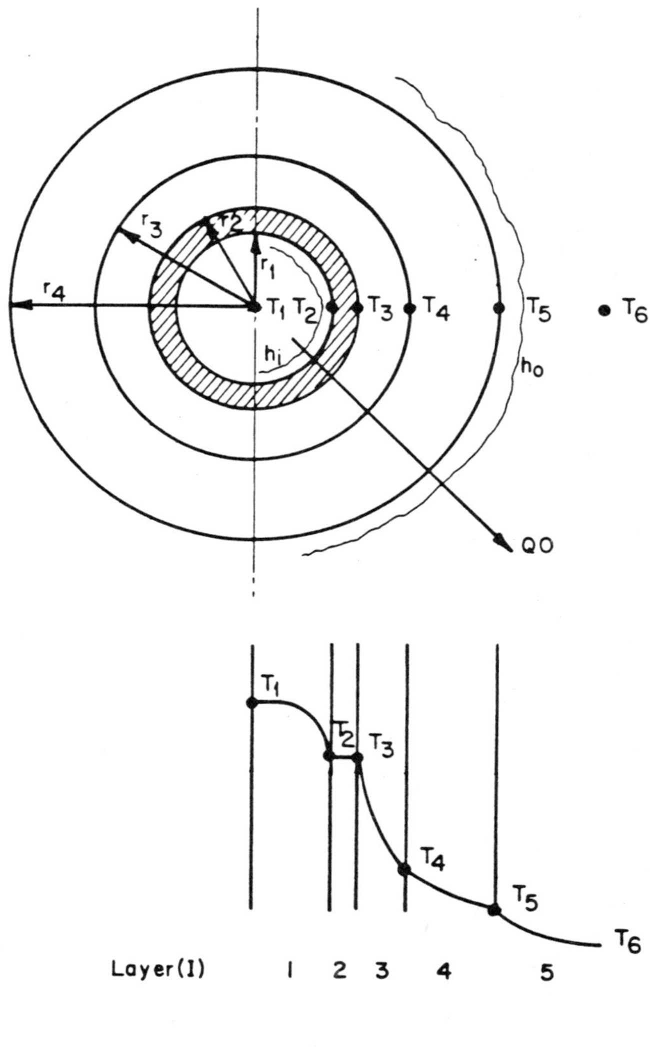

Figure 7-2 Example of radial heat transfer

7.4 Examples

7.4.1 Slab Composite

It is desired to determine the thermal and physical properties of a wall section. The following conditions apply:

Inside Temperature:		72°F
Outside Temperature:		1°F
Inside Film Coefficient:		1.5 Btu/hr ft^2 °F
Outside Film Coefficient:		5.88 Btu/hr ft^2 °F
Materials	1:	0.5 in gypsum wallboard
	2:	3.5 in fiberglass batt
	3:	0.5 in plywood sheathing
	4:	1.0 in insulated sheathing
	5:	4.0 in facebrick

The FORTRAN version input for this problem is shown in Figure 7-3. The BASIC version input would normally be in the BASIC program itself. The BASIC program listing in Appendix F currently shows the inputs for the cylindrical composite example discussed in the next section. The output of the programs is shown on pages 206 and 207.

```
7
1 'SLAB COMPOSITE'  72 1
1 0 0 0 0 .68 'INSIDE FILM'
2 .5 13.2 50. .26 0 'GYP.BOARD'
3 3.5 .318 2. .2 0 'R-11 BATT'
4 .5 9.6 34. .29 0 'PLYWOOD'
5 1. .2 1.8 .29 0 'INS.SHEATH'
6 4. 9. 130. .22 0 'FACE BRICK'
7 0. 0. 0. 0. .17 'OUTSIDE FILM'
0/
```

Figure 7-3. FORTRAN input, slab composite

```
NAVAL CIVIL ENGINEERING LABORATORY
------------------------------------------------------------------------
HTMCP-FORTRAN                                        EMC ENGINEERS,INC.

************************************************************************

****SLAB COMPOSITE                                                  ****
```

LAYER	THICKNESS (RADIUS) (IN)	THERMAL COND (K) (BTU-IN/ FT2-HR-F)	DENSITY (LB/FT3)	SPECIFIC HEAT (BTU/ LB-F)	RESISTANCE (FT2-HR-F/ BTU)	DESCRIPTION
1	0.0000	0.0000	0.0000	0.0000	0.6800	INSIDE FILM
2	0.5000	13.2000	50.0000	0.2600	0.0379	GYP.BOARD
3	3.5000	0.3180	2.0000	0.2000	11.0063	R-11 BATT
4	0.5000	9.6000	34.0000	0.2900	0.0521	PLYWOOD
5	1.0000	0.2000	1.8000	0.2900	5.0000	INS.SHEATH
6	4.0000	9.0000	130.0000	0.2200	0.4444	FACE BRICK
7	0.0000	0.0000	0.0000	0.0000	0.1700	OUTSIDE FILM

```
*** FOR SLAB COMPOSITES ***

COMPOSITE THERMAL AND PHYSICAL PROPERTIES
THE OVERALL HEAT TRANSFER COEFFICIENT IS *    0.058 BTU/HR-FT2-F
THE UNIT HEAT TRANSFER THROUGH THE SLAB IS    4.083 BTU/HR-FT2
THE COMPOSITE SYSTEM MASS IS ************   47.567 LBM/FT2
THE COMPOSITE SYSTEM THERMAL MASS IS *****  10.646 BTU/FT2-F

TEMPERATURES AT THE LAYER INTERFACES ARE

T1 IS THE INSIDE FLUID OR MATERIAL TEMPERATURE
T(LAYERS+1) IS THE OUTSIDE TEMPERATURE
T(I) IS THE TEMPERATURE BETWEEN LAYER I-1 AND LAYER I

TEMPERATURE 1 **   72.00 F

TEMPERATURE 2 **   69.22 F

TEMPERATURE 3 **   69.07 F

TEMPERATURE 4 **   24.13 F

TEMPERATURE 5 **   23.92 F

TEMPERATURE 6 **    3.51 F

TEMPERATURE 7 **    1.69 F

TEMPERATURE 8 **    1.00 F

**** STOP
```

NAVAL CIVIL ENGINEERING LABORATORY

HTMCP-BASIC EMC ENGINEERS, INC.

SLAB COMPOSITE

LAYER	THICKNESS (OR RADIUS) (INCHES)	THERMAL COND (K) (BTU-IN/ HR-FT2-F)	DENSITY (LB/FT3)	SPECIFIC HEAT (BTU/ LB-F)	THERMAL RESISTANCE (HR-FT2-F/ BTU)	DESCRIPTION
1	0	0	0	0	.68	INSIDE FILM
2	.5	13.2	50	.26	.0378788	GYP.BOARD
3	3.5	.318	2	.2	11.0063	R-11 BATT
4	.5	9.6	34	.29	.0520833	PLYWOOD
5	1	.2	1.8	.29	5	INS.SHEATH
6	4	9	130	.22	.444445	FACE BRICK
7	0	0	0	0	.17	OUTSIDE FILM

*** FOR A SLAB COMPOSITE ***

OVERALL HEAT TRANSFER COEFFICIENT IS *** 0.058 BTU/HR-FT2-F
OVERALL HEAT TEANSFER IS *************** 4.08264 BTU/HR-FT2
THE COMPOSITE SYSTEM MASS IS ******** 47.5667 LBM/FT2
THE COMPOSITE SYSTEM THERMAL MASS IS *** 10.646 BTU/FT2-F

THE TEMPERATURES AT THE SYSTEM INTERFACES ARE:

TEMPERATURE 1 ** 72 F (INSIDE TEMPERATURE)

TEMPERATURE 2 ** 69.2238 F (BETWEEN LAYER 1 AND 2)

TEMPERATURE 3 ** 69.0692 F (BETWEEN LAYER 2 AND 3)

TEMPERATURE 4 ** 24.1344 F (BETWEEN LAYER 3 AND 4)

TEMPERATURE 5 ** 23.9218 F (BETWEEN LAYER 4 AND 5)

TEMPERATURE 6 ** 3.50856 F (BETWEEN LAYER 5 AND 6)

TEMPERATURE 7 ** 1.69406 F (BETWEEN LAYER 6 AND 7)

TEMPERATURE 8 ** 1.00001 F (OUTSIDE TEMPERATURE)

7.4.2 Cylindrical Composite

It is desired to compute the thermal and physical properties of a pipe section. The following conditions apply.

Inside Temperature:	350°F
Outside Temperature:	80°F
Inside Film Coefficient:	24. Btu/hr ft^2 °F
Outside Film Coefficient:	0.95 Btu/hr ft^2 °F

Composite consists of: Nominal 2" pipe
2" calcium silicate insulation
2" glass fiber insulation

The FORTRAN code input in this problem is shown in Figure 7-4. The BASIC version input for this problem is contained in the BASIC program listing in Appendix F. The output for the cylindrical composite example is shown on pages 209 and 210.

```
4
2 'CYLINDRICAL COMPOSITE'  350 80
1 .95 0 0 0 .0833 'INSIDE FILM'
2 2 .44 12 0 0 'CA. SILICATE'
3 4 .26 6 0 0 'GLASS FIBER'
4 4 0 0 0 .5 'OUTSIDE FILM'
0/
```

Figure 7-4. FORTRAN input, cylindrical composite

```
NAVAL CIVIL ENGINEERING LABORATORY
------------------------------------------------------------------------
HTMCP-FORTRAN                                       EMC ENGINEERS,INC.

************************************************************************

****CYLINDRICAL COMPOSITE                                           ****


LAYER THICKNESS  THERMAL   DENSITY  SPECIFIC  RESISTANCE  DESCRIPTION
      (RADIUS)   COND (K)           HEAT
      (IN)       (BTU-IN/  (LB/FT3) (BTU/     (FT2-HR-F/
                 FT2-HR-F)          LB-F)       BTU)
----  -------    -------   -------  -------   -------     -----------
 1    0.9500     0.0000    0.0000   0.0000    0.0833      INSIDE FILM
 2    2.0000     0.4400   12.0000   0.0000    3.2313      CA. SILICATE
 3    4.0000     0.2600    6.0000   0.0000    5.0916      GLASS FIBER
 4    4.0000     0.0000    0.0000   0.0000    0.5000      OUTSIDE FILM


                  *** FOR CYLINDRICAL COMPOSITES ***

                COMPOSITE THERMAL AND PHYSICAL PROPERTIES
      THE OVERALL HEAT TRANSFER COEFFICIENT IS *    0.112 BTU/HR-FT-F
      THE UNIT HEAT TRANSFER THRU THE CYLINDER     30.316 BTU/HR-FT
      THE COMPOSITE SYSTEM MASS IS ***********      2.382 LBM/FT


                TEMPERATURES AT THE LAYER INTERFACES ARE

              T1 IS THE INSIDE FLUID OR MATERIAL TEMPERATURE
              T(LAYERS+1) IS THE OUTSIDE TEMPERATURE
              T(I) IS THE TEMPERATURE BETWEEN LAYER I-1 AND LAYER I

                   TEMPERATURE 1 **  350.00 F

                   TEMPERATURE 2 **  347.47 F

                   TEMPERATURE 3 **  249.51 F

                   TEMPERATURE 4 **   95.16 F

                   TEMPERATURE 5 **   80.00 F

**** STOP
```

NAVAL CIVIL ENGINEERING LABORATORY

HTMCP-BASIC EMC ENGINEERS, INC.

****

CYLINDER COMPOSITE

LAYER	THICKNESS (OR RADIUS) (INCHES)	THERMAL COND (K) (BTU-IN/ HR-FT2-F)	DENSITY (LB/FT3)	SPECIFIC HEAT (BTU/ LB-F)	THERMAL RESISTANCE (HR-FT2-F/ BTU)	DESCRIPTION
1	.95	0	0	0	.0833	INSIDE FILM
2	2	.44	12	0	3.23131	CA.SILICATE
3	4	.26	6	0	5.0916	GLASS FIBER
4	4	0	0	0	.5	OUTSIDE FILM

*** FOR A CYLINDER COMPOSITE ***

OVERALL HEAT TRANSFER COEFFICIENT IS *** 0.112 BTU/HR-FT-F
OVERALL HEAT TEANSFER IS ************** 30.3159 BTU/HR-FT
THE COMPOSITE CYLINDER MASS IS ****** 2.38172 LBM/FT.

THE TEMPERATURES AT THE CYLINDER INTERFACES ARE:

TEMPERATURE 1 ** 350 F (INSIDE TEMPERATURE)

TEMPERATURE 2 ** 347.475 F (BETWEEN LAYER 1 AND 2)

TEMPERATURE 3 ** 249.515 F (BETWEEN LAYER 2 AND 3)

TEMPERATURE 4 ** 95.158 F (BETWEEN LAYER 3 AND 4)

TEMPERATURE 5 ** 80.0001 F (OUTSIDE TEMPERATURE)

8.0

Glossary of Terms

The terms and definitions used in this section are commonly used within the insulation, mechanical, and building industry and are presented to clarify information in this field.

ABRASION RESISTANCE: The ability of a material to withstand abrasion without wearing away.

ABSOLUTE HUMIDITY: The mass of water vapor present in a unit volume of atmospheric air.

ABSOLUTE ZERO: The point at which all molecular motion ceases, with the resultant complete absence of heat. This point is -459.6° Fahrenheit, and -273.2° Celsius.

ABSORPTANCE: The ratio of the radiant flux absorbed by a body to that incident upon it.

ABSORPTION: The property of a material which allows it to take up liquids and to assimilate them.

ABUSE COVERINGS AND FINISHES: Jackets or mastics used to protect insulation from mechanical abuse.

ADHESION: The property of a material which allows it to bond to the surface to which it is applied.

ADHESIVE: A substance capable of holding materials together by surface attachment.

AIR CONDITIONED SPACE: Building area supplied directly with conditioned air.

AIR INFILTRATION BARRIER: Barrier to wind currents but allows moisture to pass.

ALKALINITY: The tendency of a material to have a basic alkaline reaction. The tendency is measured on the pH scale, with all readings above 7.0 alkaline, and below 7.0 acidic.

AMBIENT: (adj.) Surrounding. (Generally applied to temperature, humidity and atmospheric conditions.)

AMBIENT TEMPERATURE (Ta): The temperature of the environment, usually air, surrounding the object under consideration.

APPEARANCE COVERING: A material or materials used over insulation to provide the desired color or texture for aesthetic purposes.

APPLICATION TEMPERATURE LIMITS: Temperature range of a surface to which insulation materials are being applied that will not endanger the integrity of the insulation material and or finish at the point of application.

ASBESTOS (Asbestos Fiber); A group of fibrous minerals which occur as small veins in the massive body of natural hydrous silicates of serpentine or amphobole, and have heat-, fire-, and solvent-resistant properties. Used as a reinforcement in the manufacture of mastics.

ASPHALT: A dark brown to black cementitious material, solid or semisolid in consistency, in which the predominating constituents are bitumens which occur in nature as such, or are obtained as residue in refining petroleum. The principal ingredient in asphalt mastics.

ASPHALT EMULSION: Petroleum asphalt in water. (This is a breather mastic.)

ASPHALT CUT-BACK: Petroleum asphalt in mineral solvents. (This is a vapor-barrier mastic.)

ATTENUATION: The sound reduction process in which sound energy is absorbed or diminished.

AUTOIGNITION TEMPERATURE: The lowest temperature of a material which will cause it to ignite without other ignition source.

BATT: A flat, flexible form of insulation, cut to predetermined widths and thickess to fit within spaces in a stud wall.

BITUMEN: Hydrocarbon material of natural or pyrogenous origin which may be liquid, semi-solid, or solid and which is completely soluble in carbon disulfide.

BLACK BODY: The ideal, perfect emitter and absorber of thermal radiation. It emits radiant energy at each wavelength at the maximum rate possible as a consequence of its temperature, and absorbs all incident radiance.

BLANKET INSULATION: A relatively flat and flexible insulation in coherent sheet form furnished in units of substantial area.

BLISTER: Rounded elevation of the surface of a mastic somewhat resembling a blister on the human skin.

BLOCK INSULATION: Rigid insulation preformed into rectangular units.

BOARD INSULATION: Semirigid insulation preformed into rectangular units having a degree of suppleness particularly related to their geometrical dimensions.

BOND STRENGTH: The force in tension, compression, impact or cleavage required to break an adhesive assembly.

BONDING TIME: Time period after application of adhesive during which the adherents may be combined.

BREAKING LOAD: That load, concentrated in the middle of a span, which will just break a measured sample of insulation under test.

BREATHER MASTIC: A mastic which permits water vapor to pass through to the low pressure side.

BRITISH THERMAL UNIT (Btu): The amount of heat necessary to raise one pound of water from 59°F to 60°F at sea level and standard atmospheric pressure.

"C" VALUE: See Thermal Conductance.

CALCIUM SILICATE: Insulation composed principally of hydrous calcium silicate, usually contains reinforcing fibers.

CANVAS: A light, plain weave cotton fabric used for jacketing.

CAPILLARITY: The action by which the surface of a liquid, where it is in contact with a solid, is raised or lowered.

CAULKING COMPOUND: A soft, plastic material, consisting of pigment and vehicle, used for sealing joints in buildings and other structures, where normal structural movement may occur.

CELLULAR ELASTOMERIC: See Elastomeric Foam.

CELLULAR GLASS: Insulation composed of glass processed to form a rigid foam having a predominately closed-cell structure.

CELLULAR POLYSTYRENE: See Polystyrene Foam.

CELLULAR POLYURETHANE: See Polyurethane Foam.

CELLULOSIC FIBER: Insulation composed principally of cellulose fibers usually derived from paper, paperboard stock, or wood, with or without binders.

CELSIUS: A metric temperature scale in which the freezing point of water is 0° and the boiling point is 100° at sea level, atmospheric pressure.

CEMENT, FINISHING: A mixture of dry fibrous or powdery materials, or both, that when mixed with water develops a plastic consistency, and when dried in place forms a relatively hard, protective surface.

CEMENT, INSULATING: A mixture of dry granular, flaky, fibrous or powdery materials that when mixed with water develops a plastic consistency, and when dried in place forms a coherent covering that affords substantial resistance to heat transmission.

CENTIGRADE: See Celsius.

CHECKING: A defect in a coated surface characterized by the appearance of fine cracks in all directions.

CHEMICAL REACTION: The property of a material to combine or react with other materials to which it may come into contact.

CHEMICAL RESISTANCE: Capability of a material to withstand exposure to acids, alkalies, salts and their solutions.

CLOSED-CELL PLASTIC: A cellular plastic with a large predominance of non-interconnecting cells.

COATING: A liquid or semiliquid that dries or cures to form a protective finish, suitable for application to thermal insualtion or other surfaces in thickness of 30 mils (0.76 mm) or less, per coat.

COEFFICIENT OF THERMAL EXPANSION/CONTRACTION: The change in a unit length of a material corresponding to a unit change in the temperature of the material.

COMBUSTIBLE: Capable of burning.

COMBUSTIBILITY: A measure of the tendency of a material to burn.

COMPACTION RESISTANCE: That property of a fibrous or loose fill material which resists compaction under load or vibratory conditions.

COMPRESSIVE STRENGTH: That property of an insulation material which resists any change in dimensions when acted upon by a compaction force.

CONCEALED SPACES: Spaces not generally visible after the project is completed such as furred spaces, pipe spaces, pipe and duct shafts, spaces above suspended ceilings, unfinished spaces, crawl spaces, attics and tunnels.

CONDENSATE: Hot - See Steam Supply and Condensate Return. Cold - See Condensate Drain.

CONDENSATE BARRIER: A coating or laminate on the inner surface of metal jacketing the protects the jacket from condensed water.

CONDENSATE DRAIN: Piping carrying condensed water from air conditioning or refrigeration drip pans to a point of discharge.

CONDENSATION: The act of water vapor turning into liquid water upon contact with a cold surface.

CONDITIONED AIR: Air treated to control simultaneously its temperature, humidity and cleanliness to meet the requirements of a conditioned space. (May be cool and/or heated and should be clearly defined.

CONDUCTION: The transfer of heat energy within a body or between two bodies in physical contact.

CONTACT ADHESIVE: An adhesive which when dry to the touch will adhere to itself instantaneously on contact.

CONVECTION: The transfer of heat by movement of fluids.

CORROSION EFFECT: The wearing away, or destruction, of a substrate caused by acid or alkaline reactions between materials contained in the insulation and the substrate.

COVER: (v.) To place insulation and/or finish materials on, over or around a surface so as to insulate, protect or seal.

COVERAGE: The rate in square feet per gallon (coatings), or gallons per hundred square feet (mastics), at which products must be applied to obtain satisfactory performance.

CRYOGENIC INSULATION: See Insulation.

CURE: To change the properties of a plastic or resin by chemical reaction, usually accomplished by the action of either heat or a catalyst.

CURING AGENT: An additive incorporated in a coating or adhesive resulting in an increase or decrease in the rate of cure.

DELAMINATION: The physical separation of the layers of material in a laminate.

DENSITY-APPARENT: The weight of a unit volume of a material in its manufactured state, including all voids. Usually expressed in pounds per cubic foot.

DENSITY-REAL: The weight of a unit volume of a material, excluding all voids. usually expressed in pounds per cubic foot.

DEW POINT: The temperature at which the quantity of water vapor within a material or in the air surrounding a material reaches saturation, with resultant condensation of the vapor into liquid by any further reduction of temperature.

DIFFUSIVITY, THERMAL: Thermal conductivity per unit heat capacity.

DIMENSIONAL STABILITY: That property of a material which enables it to hold its original size, shape and dimensions.

DRY: (v.) To change the physical state of a substance by the loss of solvent constituents by evaporation, absorption, oxidation, or a combination of these factors.

ELASTOMER: A material which at room temperature can be stretched repeatedly to at least twice its original length and, immediately upon release of the stress, will return with force to its approximate original length.

ELASTOMERIC FOAM: Insulation composed principally of natural or synthetic elastomers, or both, processed to form a flexible, semirigid, or rigid foam which has a predominately closed-cell structure.

EMITTANCE: The ratio of the radiant flux emitted by a specimen to that emitted by a blackbody at the same temperature and under the same conditions.

EMULSION: A colloidal suspension of ore liquid in another, usually a water based material.

ENERGY: the measure of the amount of work a body (or system of bodies) can do, by virtue of its motion, or position, against forces applied to it. It is also a measure of the work it can do by virtue of its chemical composition, or as a result of having been heated.

EXPANDED METAL: A lattice type of material of various gauges and sizes, used to provide reinforcement for insulation materials.

EXPOSED SPACES: Those spaces not referred to as concealed or as defined by the specifier.

FABRIC: A material used for reinforcing or finishing surfaces of insualtion materials. (See Glass Cloth, Glass Fabric, etc.).

FACING: A thin layer or laminate, usually factory applied, on the surface of an insulating material.

FAHRENHEIT (F): A temperature scale with the freezing point at 32° and the boiling point at 212°, sea level, atmospheric pressure.

FELT: An insulation material composed of fibers of one or more kinds which are interlocked and have been compacted under pressure.

FIBER GLASS: See Glass Fiber.

FILL INSULATION: See Insulation.

FILM (Wet): The applied layer of mastic or coating before curing or drying.

FINISHING AND INSULATING CEMENT: See Cements.

FIRE POINT TEMPERATURE: The lowest temperature of a material at which it gives off vapor, which, when combined with air near its surface, forms an ignitable mixture at a rate sufficient to support combustion continuously after the external ignition source is removed.

FIRE RESISTANCE: That property of a material which enables it to resist fire.

FIRE RETARDANCE: That property of a material which retards the spread of fire.

FISH MOUTH: A gap between layers of sheet materials caused by warping or bunching of one or both layers.

FITTING COVER: The insulation for a pipe fitting composed of the specified thickness of insulation material and preformed into its proper shape before application.

FLAME SPREAD: The rate, expressed in distance-time, at which a material will propagate flame on its surface. As this is a difficult property to measure in time and distance, the measure is now by flame spread index to enable the compairson of materials by test methods.

FLAMMABILITY: That property of a material which allows continuous burning, as compared to a standard material.

FLANGE COVER: The insulation for a pipe flange composed of the specified thickness of insulation material and preformed into its proper shape before application.

FLASH-IGNITION TEMPERATURE (Flash Point): The lowest temperature of a material at which it gives off vapor, which, when combined with air near its surface forms an ignitable mixture. The rate of vapor formation at this tempeature is not sufficient for burning to contine if the ignition source is removed.

FLASHING: A strip of material installed at the junction of two planes to divert water or any substance.

FLEXIBILITY: That property of a material which allows it to be bent (flexed) without loss of strength.

FREEZE-THAW RESISTANCE: Resistance to cycles of freezing and thawing that could affect application, appearance, or performance.

FRESH AIR: Air taken from outdoors.

FUEL CONTRIBUTION: Flammable by-products of fire generated by and emitted from a burning object.

GLASS-CELLULAR: See Cellular Glass.

GLASS CLOTH: Closed weave glass fiber used as a finish jacket.

GLASS FABRIC: Open weave glass fiber used for reinforcing a mastic or coating finish on insulating materials.

GLASS FIBER: Insulation composed of thin strands of glass, sometimes mixed with a resin binder. Available as batts, blankets, or boards.

GLOSS: A term used to express the shine, sheen, or luster or a dried film.

GRAY BODY: A body having the same spectral emittance (less than unity) at all wave lengths.

HANGAR (Insulation): A device such as a welded pin, stud or adhesive secured fastener which carries the weight of insulation.

HEAT: The form of energy that is transferred by virtue of a temperature difference or a change of state.

HEAT CAPACITY: See Specific Heat Capacity.

HEATED SPACE: Building area supplied directly with heat.

HEATING DEGREE-DAYS: A degree-day is a measure of "coldness" develcped as an aid in determining fuel consumption. The number of degree-days in a calendar day is the average of the high and the low temperatures subtracted from 65. For example, if the high was 60°F and the low was 30°F, there were 20 degree-days.

HEXAGONAL WIRE MESH: Poultry netting, chicken wire, etc. (See Netting.).

HIGH RIB LATH: A metal lath with a built-in rib used to provide air space under insulation applications.

HUMIDITY, ABSOLUTE: The mass of water vapor per unit volume.

HUMIDITY, RELATIVE: The ratio of the mol fraction of water vapor present in the air to the mol fraction of water vapor present in saturated air at the same temperature and barometric pressure. Approximately, it equals the ratio of the partial pressure or density of the water vapor in the air to the saturation pressure or density, respectively, at the same temperature.

IGNITION: The initiation of combustion.

IGNITION TEMPERATURE: The minimum temperature required to initiate combustion.

IMPACT RESISTANCE: Capability of an insulation material and/or finish to withstand mechanical or physical abuse.

INSULATE: To cover with a material of low conductivity in order to reduce the passage or leakage of heat.

INSULATING CEMENT: See Cement.

INSULATION: Those materials or combination of materials which retard the flow of heat.

INSULATION, CELLULAR: Insulation composed of small individual cells separated from each other. The cellular material may be glass or plastic such as polystyrene (closed cell), polyurethane and elastomeric.

INSULATION, CRYOGENIC: Insulation for extremely low temperature surfaces from (-150°F to absolute zero (-459°F)).

INSULATION, FIBROUS: Insulation composed of small diameter fibers which finely divide the air space. Fibers used are silica, rock wool, slag wool or alumina silica.

INSULATION, GRANULAR: Insulation composed of small nodules which contain voids or hollow spaces. The material may be calcium silicate, diatomaceous earth, expanded vermiculite, perlite or cellulose.

INSULATION, REFRACTORY: Insulation of extremely high temperatures above 1500°F.

INSULATION, SPRAYED-ON: Insulation of the fibrous or foam type which is applied to a surface by means of power spray devices.

INSULATION, THERMAL: Insulation applicable within the general temperature range of -150°F to 1500°F.

JACKET: A covering placed over insulation for various functions.

LAG: (V.) To apply lagging. (n.) A single piece of lagging material.

LAGGING - INSULATION: A block material for insulating tanks and boilers, usually curved or tapered and can be made from any of several insulation materials.

LAGGING - METAL: Metal covering installed over insulation. (See Metal Jacketing.)

LAMINATE: (n.) A product made by bonding together two or more layers of material or materials.

LAP ADHESIVE: The adhesive used to seal the sides and laps of insulation jackets.

LATH - PLASTER: Plasterer's lath. (See also High Rib Lath and Expanded Metal.)

LINEAR EXPANSION OR CONTRACTION: See COEFFICIENT OF THERMAL EXPANSION/CONTRACTION.

LOG MEAN (Radius): The equivalent value of insulation thickness for pipe (curved surfaces) to produce the same resistance to heat flow as per flat areas.

LOOSE FILL INSULATION: Insulation in granular, nodular, fibrous, powdery, or similar form designed to be installed by pouring, blowing, or hand placement.

MASTIC: A protective coating, usually a petroleum or base product, applied by spray or trowel to weather proof or otherwise prevent deterioration of the insulation to which it is applied.

MAT: A piece of insulation of the semi-flexible type, cut into easily handled sizes, usually square or rectangular in shape, composed of fibers of one or more kinds in which the fibers are in random arrangement.

MEAN TEMPERATURE: The average °F operating temperature and ambient temperature. (Thermal Conductivity charts are calculated to use mean temperatures.)

MEMBRANE REINFORCEMENT: See Glass Fabric.

METAL JACKETING: See Jacketing.

METAL LATH: See Expanded Metal, High Rib Lath and/or Lath-Plaster.

MIL: A unit used in measuring thickness (0.001 in.)

MINERAL FIBER: Insulation composed principally of fibers manufactured from rock, slag, or glass, with or without binders.

MOISTURE BARRIER: See Vapor - Barrier and Weather - Barrier.

MOLD AND MILDEW RESISTANCE: That property of a material which enables it to resist the formation of fungus growths.

NETTING: Interwoven wires of metal used as a reinforcement. (See Hexagonal Mesh.)

NONCOMBUSTIBLE: A material which will not contribute fuel or heat to a fire to which it is exposed.

NONFLAMMABLE: A material which will not release heat when exposed to fire or flame.

PANEL: A prefabricated unit of insulation and lagging.

PERLITE: Insulation composed of natural perlite ore expanded to form a cellular structure.

PERMEABILITY: See Water Vapor Permeability.

PERMEANCE (Perms): See Water Vapor Permeance.

PERSONNEL PROTECTION: Insulation installed for the purpose of protecting personnel from high temperature surfaces.

PINHOLE: Very small hole through a mastic or coating.

PHENOLIC FOAM: A foamed insulation made from resins of phenols condensed with aldehydes.

PIPE INSULATION: Insulation in a form suitable for application to cylindrical surfaces.

PLASTIC: A material that contains, as an essential ingredient, an organic substance of large molecular weight, is solid in its finished state, and at some state in its manaufacture or in its processing into finished articles, can be shaped by flow.

POLYMER: A compound formed by the reaction of simple molecules having functional groups that permit their combination to proceed to high molecular weights under suitable conditions. Polymers may be formed by polymerization (addition polymer) or polycondensation (condensation polymer). When two or more monomers are involved, the product is called a copolymer.

POLYISOCYANURATE FOAM: A second-generation polyurethane foam having lower fire hazard ratings.

POLYSTYRENE: A resin made by polymerization of styrene as the sole monomer.

POLYSTYRENE FOAM: Insulation composed principally of polymerized styrene resin processed to form a rigid foam having a predominately closed-cell structure.

POLYURETHANE: A resin made by the condensation of organic isocynates with compounds or resins that contain hydroxol groups.

POLYVINYL CHLORIDE (PVC): A polymerized vinyl compound.

PRESSURE SENSITIVE TAPE: A tape with adhesive preapplied.

PUNCTURE RESISTANCE: That property of a material which enables it to resist punctures or perforations under blows or pressure from sharp objects.

R-VALUE (Resistance): A measure of the insulating ability of a given thickness of a material, equal to the temperature difference at equilibrium that the material can sustain when subjected to a unit heat flow rate per unit area.

RADIANT HEAT: That heat transmitted through space by wave motion.

RADIATION: The passage of heat from one object to another without warming the space between.

REFLECTANCE: The fraction of the incident radiation upon a surface that is reflected from the surface.

REFLECTIVE INSULATION: Insulation depending for its performance upon reduction of radiant heat transfer across air spaces by use of one or more surfaces of high reflectance and low emittance.

REFRACTORY MATERIALS: Materials, usually fibers, which do not significantly deform or change chemically at high temperatures. Manufactured in blanket, block, brick or cement form.

REINFORCING CLOTH OR FABRIC: A woven cloth or fabric of glass or resilient fibers used as reinforcement to a mastic vapor/weather barrier.

REINSULATE: To repair insulation to its former condition. (If insulation is to be removed and replaced, it should be so stated.)

RELATIVE HUMIDITY: See Humidity, Relative.

RESILIENCY: That property of a material which enables it to recover its original thickness after compression.

RESIN: A solid, semisolid, or pseudo-solid organic material which has an indefinite and often high molecular weight, exhibits a tendency to flow when subjected to stress, usually has a softening or melting range, and fractures conchoidally.

RESISTANCE (R): See Thermal Resistance.

RESISTANCE TO ACIDS, CAUSTICS, AND SOLVENTS: The property of a material to resist decomposition by various acids, caustics and solvents to which it may be subjected.

RESISTANCE TO AIR EROSION: The property which indicates the ability of an insulation material to resist erosion by air currents over its surface.

RETROFIT: The application of additional insulation over existing insulation, new insulation after old insulation has been removed, or new insulation over existing, previously uninsulated surfaces.

RIGID WRAP-AROUND INSULATION: Segments of insulation material which have been adhered to a facing giving rigid insulation materials flexibility of application.

RIGIDITY: That property of a material which opposes any tendency for it to bend (flex) under load.

"S" CLIP: A support device for banding or jacketing.

SELF-EXTINGUISHING: That property of a material which enables it to stop its own ignition after external ignition sources are removed.

SEAL: (v.) To make water tight.

SEALANT: A putty-like substance, composed of various materials, used as a barrier to the passage of water vapor or liquid water into the joint formed by the mating surfaces of jackets and water- and vapor-barriers over insulation.

SEALER: A liquid coating or mastic used to prevent excessive absorption of finish coats into porous sufaces.

SECUREMENTS (Insulation): Any device, wire, strap or adhesive used to fasten insulation into its service position and hold it there.

SHEAR STRENGTH: The property of a material which indicates its ability to resist cleavage.

SHELF LIFE: The period of time during which a packaged adhesive, coating, or sealant can be stored under specified temperature conditions and remain suitable for use.

SHRINKAGE: The property of a material which indicates its proportionate loss in dimensions or volume when its temperature is changed.

SMOKE DENSITY: The amount of smoke given off by the burning material compared to the amount of smoke given off by the burning of a standard material.

SMOLDERING: The combustion of solid materials without the accompaniment of flame.

SOLAR RESISTANCE: The property of a material to resist decomposition by the ultra-violet rays from the sun or the passage of radiant heat from the sun.

SOLIDS CONTENT: The percentage of the non-volatile matter in adhesives, coatings or sealants.

SOLVENT: Any substance, usually a liquid, which dissolves another substance.

SPECIFIC HEAT CAPACITY: The amount of heat required to raise a unit mass of a material 1 degree in temperature.

SPRAYED-IN-PLACE INSULATION: See Insulation, Sprayed-On.

STUD: Used to hold heavy insulation and/or panels in place. Applied with arc welder, studs differ from pins in that studs are generally 1/4" or greater in diameter.

SUMMER COOLING HOURS: The number of equivalent full-load operating hours per year that air-conditioning equipment operates. The number of hours in the entire cooling season is greater than the number of summer cooling hours because the equipment operates intermittently and under partial load.

SUPPORT (Insulation): A device which carries the weight of insulation.

SURFACE TEMPERATURE (TS): The surface temperature of finished insulation.

TACK: The property of an adhesive that enables it to form a measurable bond immediately after adhesive and adherent are brought into contact under low pressure.

TEAR STRENGTH: That property of a material which enables it to resist being pulled apart by opposing forces.

TEMPERATURE LIMITS: See Thermal Temperature Limits.

THERMAL CONDUCTANCE, C: The time rate of heat flow through a unit area of a body, induced by a unit temperature difference between the body surfaces. Also, the reciprocal of thermal resistance.

THERMAL CONDUCTANCE, FILM, h: The time rate of heat flow from a unit area of a surface to its surroundings, induced by a unit temperature difference between the surface and the environment.

THERMAL CONDUCTIVITY, k: The time rate of heat flow per unit area through unit thickness of an infinite slab of a homogeneous material in a direction perpendicular to the surface, induced by unit temperature difference.

THERMAL INSULATION: A material or assembly of materials used to provide resistance to heat flow.

THERMAL INSULATION SYSTEM: Applied or installed thermal insulation complete with any accessories, vapor retarder, and facing required.

THERMAL RESISTANCE: The mean temperture difference, at equilibrium, between two defined surfaces of material or a construction that induces a unit heat flow rate through the material. Also, the reciprocal of thermal conductance.

THERMAL SHOCK RESISTANCE: The property of a material which indicates its ability to be subjected to rapid temperature changes without physical failure.

THERMAL TEMPERATURE LIMITS: The upper and lower temperatures at which a material will experience no change in its properties.

THERMAL TRANSMITTANCE (U): The combined thermal conductance value of all the materials in a building section, including air spaces and surface air films. Also known as the overall heat transfer coefficient, it is usually expressed as Btu/hr-sq ft-F.

TRANSMISSION, HEAT: The quantity of heat flowing through unit area due to all modes of heat transfer induced by the prevailing conditions.

VAPOR-BARRIER: A material or materials which when installed on the high vapor pressure side of a material retard the passage of the moisture vapor to the lower vapor pressure side.

VERMICULITE: Insulation composed of natural vermiculite ore expanded to form an exfoliated structure.

VIBRATION RESISTANCE: The property of a material which indicates its ability to resist mechanical vibration without wearing away, settling or dusting off.

WARPAGE: The change in the flatness of a material caused by differences in the temperatures and/or humidities applied to opposite surfaces of the material.

WASHER (Insulation): Used with weld pins to hold insulation in place.

WATER ABSORPTION: The increase in weight of a test specimen expressed as a percentage of its dry weight after immersion in water for a specified time.

WATERPROOF: (adj.) Impervious to prolonged exposure to water.

WATER RESISTANT: Capable of withstanding limited exposure to water.

WATER VAPOR DIFFUSION: The process by which water vapor spreads or moves through permeable materials caused by a difference in water vapor pressure.

WATER VAPOR PERMEABILITY: The property of a substance which permits passage of water vapor and is equal to the permeance of an 1 inch thickness of the substance. Permeability is measured in perm inches.

WATER VAPOR PERMEANCE: The ratio of water vapor flow to the vapor pressure difference between the two surfaces of a sheet of material (or the assembly between parallel surfaces). Permeance is measured in perms (1 perm = 1 grain/(ft^2-hr-(in-Hg)).

WATER VAPOR RESISTANCE: The steady vapor pressure difference that induces unit time rate of vapor flow through unit area of a flat material (or construction that acts like a homogeneous body) for specific conditions of temperature and relative humidity at each surface.

WATER VAPOR RETARDANT(Barrier): See Vapor - Barrier.

WATER VAPOR TRANSMISSION (WVT): The rate of water vapor transmission of a body between two specified parallel surfaces is the time rate of water vapor flow normal to the surfaces under steady condition through unit area, under the conditions of test. An accepted unit of WVT is 1 grain per square foot, hour (with test conditions stated).

WEATHER-BARRIER: A material or materials which, when installed on the outer surface of thermal insulation, protects the insulation from weather damage incurred by rain, snow, sleet, wind, solar radiation and atmospheric contamination.

WEATHER/VAPOR-BARRIER: A material which combines the properties of a weather-barrier and a vapor-barrier.

WELD PIN: Made of carbon steel, stainless steel or aluminum in various lengths for attaching insulation to metal surfaces. Applied by welding, manufactured in 10, 12 and 14 gauges.

STUD: See Stud.

WICKING: Action of absorbing by capillary action.

WOOD FIBER: Insulation composed of wood fibers, with or without binders. NOTE: This is a type of cellulosic fiber insulation.

"Z" CLIP: See "S" Clip.

9.0

Bibliography

I. INSULATION FUNDAMENTALS:

Allan, Graham G., Dutkiewicz, Jacek, and Gilmartin, Earl J. "Long-Term Stability of Urea-Formaldehyde Foam Insulation." Environmental Science and Technology Vol. 14, No. 10, pp. 1235-40. Oct. 1980.

Anderson, B.R. "The Thermal Resistance of Airspaces in Building Constructions." Building and Environment. Vol. 16, No. 1, pp. 35-39. 1981.

BASF Wyandotte Corporation. "EPS - Expanded Polystyrene Insulation." Wyandotte, MI. 1979.

Baumann, Gert F. "An Analysis of Rigid Urethane Foam Insulation Effectiveness." Journal of Cellular Plastics. pp. 219-223. July/August 1978.

Bejan, Adrian. "A General Variational Principle for Thermal Insulation System Design." International Journal Heat Mass Transfer. Pergamon Press Ltd. Great Britain. Vol. 22, pp. 219-228. 1979.

Bomberg, M. and Solvason, K.R. "How to Ensure Good Thermal Performance of Blown Mineral Fiber Insulation in Horizontal and Vertical Installations." Journal of Thermal Insulation. Technomic Publishing Co., Inc. Vol. 4, pp. 187-213. Jan. 1980.

Bomberg, M. and Solvason, K.R. "How to Ensure Good Thermal Performance of Cellulose Fiber Insulation. Part I, Horizontal Applications." Journal of Thermal Insulation. Technomic Publishing Co., Inc. Vol. 4, pp. 93-114. Oct. 1980.

Bomberg, M. and Solvason, K.R. "How to Ensure Good Thermal Performance of Cellulose Fiber Insulation. Part II, Exterior Walls." Journal of Thermal Insulation. Technomic Publishing Co., Inc. Vol. 4 pp. 119-133. Oct. 1980.

Bourne, J.G., et. al. Building Insulation Materials Compilation. Report No. CR-80.001, Naval Civil Engineering Laboratory, Port Hueneme, Ca. January 1980.

Bowen, R.P., Shirtliffe, C.J., and Chown, G.A. "Urea-Formaldehyde Foam Insulation: Problem Identification and Remedial Measures for Wood-Frame Construction." Building Practice Note No. 23. Division of Building Research, National Research Council of Canada. Ottawa, Canada. Aug. 1981.

Burch, D.M., and Hunt, C.M. "Retrofitting an Existing Wood-Frame Residence for Energy Conservation - An Experimental Study." NBS Building Science Series 105. National Bureau of Standards. Washington, D.C. July 1978.

Chown, G.A., Bowen, R.P., and Shirtliffe, C.J. "Urea-Formaldehyde Foam Insulation." Building Practice Note No. 19. Division of Building Research National Research Council of Canada. Ottawa, Canada. April 1981.

Department of Energy. "Insulation." DOE/CS-0017 Oak Ridge, TN. May 1979.

Doherty, D.J., Hurd, R., and Lester, G.R. "The Physical Properties of Rigid Polyurethane Foams." Chemistry and Industry. pp. 1340-1356. July 28, 1962.

Dow Chemical U.S.A. How To Insulate bulletin series. Midland, MI. 1980, 1981.

E.I. duPont de Nemours & Company. "Effect of Aluminum Cladding on the Thermal Conductivity of Aged Urethane Foam Panels." Freon Blowing Agents for Polymeric Foams. Bulletin BA-4. "Freon" Products Division, Wilmington, DE. Jan. 1961.

Elastomerics. "No Formaldehydes in New Foam Retrofit Insulation." Vol. 113, No. 3. March 1981.

Fairweather, E. W. "Thermal Performance Characteristics of EPS Insulation." Journal of Thermal Insulation. Vol. 3, pp. 142-156. Jan. 1980.

Gier, J.T. and Dunkle, R.V. "Using the Heat Flow Meter to Study Heat Transfer - A Description of Its Application to Laboratory Tests of Low Temperature Insulating Materials." Refrigerating Engineering. pp. 63-69, 90, 92, 94, 96. Oct. 1954.

Gorring, Robert L. and Churchill, Stuart W. "Thermal Conductivity of Heterogeneous Materials." Chemical Engineering Progress. Vol. 57. No. 7. pp. 53-59. July 1961.

Harding, R.H. "Heat Transfer through Low-Density Cellular Materials." I&EC Process Design and Developement. Vol. 3, No. 2. pp. 117-125. April 1964.

Jansen, Paul E. "Glass Fiber Insulating Exterior Sheathings." Journal of Thermal Insulation. Vol. 3, pp. 235-259. April 1980.

Johns-Manville Corporation. "Fiber Glass Building Insulation Installation Manual." Denver, CO. April 1981.

Knickle, Harold N. and Kalthod, Vikram. "Measurement of the Thermal Conductance of Shredded Urea-Formaldehyde in Vertical Panels." Journal of Thermal Insulation. Vol. 4, pp. 265-274. April 1981.

Knox, R.E. "Insulation Properties of Fluorocarbon Expanded Rigid Urethane Foam." ASHRAE Journal. pp. 43-49, 53. Oct. 1962.

Kowalski, G. J., and Mitchell, J.W. "An Analytical and Experimental Investigation of the Heat Transfer Mechanisms within Fibrous Media Exposed to Solar Radiation." Paper no. 81-HT-41, presented at the 20th Joint ASME/AICHE National Heat Transfer Conference, Milwaukee, WI. Aug. 1981.

Larkin, Bert K. and Churchill, Stuart W. "Heat Transfer by Radiation through Porous Insulations." AIChE Journal. Vol. 5. No. 4. pp. 467-474. Dec. 1959.

Loser, J.B., Moeller, C.E., and Thompson, M.B. Thermophysical Properties of Thermal Insulating Materials. Midwest Research Institute, Kansas City, MO. 1964.

Marciano, Jorge H., Rojas, Alfredo J., and Williams, Roberto J.J. "A Theoretical Model for the Thermal Conductivity of Plastic Foams." European Journal of Cellular Plastics. Technomic Publishing Co., Inc. Vol. 3, No. 3. 1981.

Metal Building Review. "Tests Show Laminated Fiberglass Insulation Surpasses Published Values." pp. 23-24. Dec. 1980.

Mineral Insulation Manufacturers Association, Inc. "Home Insulation Checklist." Insulation Facts #1 (Revised). Summit N.J. Dec. 1978.

Mineral Insulation Manufacturers Association Inc. "Three Reports Analyze UF Foam, Set Rules for its Acceptance". Insulation Facts #3. March 1979.

Modern Plastics. "Big New Action in Polyolefin Foams." pp. 52-53. June 1979.

Modern Plastics. "New Candidate for Insulation: Phenolic Foam." pp. 48-49. July 1980.

Modern Plastics. "The Biggest Problem With Foam Insulation: Which One to Choose?" pp. 52-55. March 1981.

Nelson, E. R. "Thermal Insulation Corrosion Problems in the Petrochemical Industry." Journal of Thermal Insulation. Vol. 3, pp. 71-79. Oct. 1979.

Norman, E. G. "Rigid Phenolic Foam in Building Insulation." European Journal of Cellular Plastics. Technomic Publishing Co., Inc. Vol. 2, No. 2, pp. 68-73. April 1979.

Palfey, Albert J. "Thermal Performance of Low Emittance Building Sheathing." Journal of Thermal Insulation. Vol. 3, pp. 129-141. Jan. 1980.

Pelanne, Charles M. "Thermal Insulation. What Is It? - How Does It Work?" Johns Manville Corp., Research and Development Center, Denver, CO. Sept. 1976.

Pelanne, Charles M. "Heat Flow Principles in Thermal Insulations." Journal of Thermal Insulation. Vol. 1, pp. 49-80. July 1977.

Pelanne, Charles M. "Thermal Insulation: What It Is and How It Works." Journal of Thermal Insulation. Vol. 1, pp. 223-235. April 1978.

Pelanne, Charles M. "Thermal Insulation Heat Flow Measurements: Requirements for Implementation." ASHRAE Journal. pp. 51-58. March 1979.

Plastics World. "Energy Awareness Heats Up Rigid Foam Insulation." Vol. 37, No. 7. pp. 42-46. June 1981.

Plastics World. "An Alternative to Urea-Formaldhyde: Polyurea Foam Insulation." Vol. 38, pp. 80-82. Sept. 1980.

Plastics World. "Plastics to Share in Insulation Boom." Vol. 37, No. 2. Feb. 1981.

Rennex, Brian G. "Thermal Parameters as a Function of Thickness for Combined Radiation and Conduction Heat Transfer in Low-Density Insulation." Journal of Thermal Insulation. Technomic Publishing Co., Inc. Vol. 1, pp. 37-61. July 1979.

Rockwool Industries, Inc. "Blowing Wool" Technical Bulletins #6, #8, #9, and #10. Englewood, CO. Aug. 1981.

Rockwool Industries, Inc. "Facts About Home Insulation." No. 78120. Denver, CO. 1981.

Rockwool Industries, Inc. "Multi-layer Foil." Technical Bulletin #5. Englewood, CO. Aug. 1981.

Roofing, Siding and Insulation. "Cellulose and Fiberglass: Dynamic Duo KO's Heat Loss." Vol. 55, p. 61. Dec. 1978.

Roofing, Siding and Insulation. "EPA Threatens CFC Use for Urethane." Vol. 58, No. 1, pp. 54, 58. Jan. 1981.

Roofing, Siding and Insulation. "Foil: An Insulation to Reflect On." Vol. 57, p. 58. Sept. 1980.

Roofing, Siding and Insulation. "Polyurea: Yours For the Retrofit." Vol. 58, No. 1. p. 52. Jan. 1981.

Rossiter, Walter J., Jr., Mathey, Robert G., Burch, Douglas M., and Pierce, E. Thomas. "Urea-Formaldehyde Based Foam Insulations: An Assessment of Their Properties and Performance." NBS Technical Note 946. National Bureau of Standards. Washington, D.C. July 1977.

Shen, Kelvin K. "Corrosion Tests on Cellulosic Insulation Containing Borates." Journal of Thermal Insulation. Vol. 3, pp. 190-202. Jan. 1980.

Skochdopole, Richard E., "The Thermal Conductivity of Foamed Plastics." Chemical Engineering Progress. Vol. 57. No. 10, pp. 55-59. Oct. 1961.

Spinney, Stewart C., Tye, Ronald P., and Weidt, John L. "A Study of Some In-Situ properties of Cellulose and Urea-Formaldehyde Retrofit Insulation". ASHRAE Transactions. DE-79-3. No. 2. 1979.

Stone, W. J. D. and Nye J. D. "Development of the Americ-Baker Cellular Mineral Thermal Insulating Materials for Buildings." Canadian Mining and Metallurgical Bulletin. Vol. 72, No. 806. June 1979.

Timm, William and Smith, Patti M. "Cause and Effect of Shrinkage in Urea Based Foams." Journal of Thermal Insulation. Vol. 3, pp. 212-234. April 1980.

Ternbah, M. and Hill, M. J. "Sprayed Polyurethane Foam - A Canadian Experience." European Journal of Cellular Plastics. Vol. 3, No. 3, pp. 82-86. July 1980.

Tong, T.W., and Tien, C.L. "Radiative Heat Transfer in Fibrous Insulations - Part I: Analytical Study." Paper No. 81-HT-42, and "-Part II: Experimental Study." Paper No. 81-HT-43, presented at the 20th Joint ASME/AIChE National Heat Transfer Conference, Milwaukee, WI. Aug. 1981.

Topper, Leonard. "Analysis of Porous Thermal Insulating Materials." Industrial and Engineering Chemistry. Vol. 47. No. 7. pp. 1377-1379. July 1955.

Walker, Lloyd. "Insulation - An Energy-Saving Home Improvement." No. 4.652. Colorado State University Extension Service. Fort Collins, CO. August 1980.

Yarbrough, David W. "Physical Properties of Residential Insulations." Ceramic Energy and Science Proc. Oak Ridge, TN. Vol. 1, No. 11-12. Nov./Dec. 1980.

II. INSULATION AND ENERGY USE:

ASHRAE Journal. "Energy Use In Buildings Could Be Cut In Half, Report Says." p. 15. June 1981.

ASHRAE Journal. "GAO Sees Possible Insulation Shortage This Winter." p. 10. Feb. 1980.

Goutte, Rene. "Time is Right For Recognition of Insulation's Role." Heating/Piping/Air Conditioning. p. 156. Jan. 1981.

McDuff, S.J. "Energy Conservation Increases Insulation Demand." Heating/Piping/Air Conditioning. p. 128. Jan. 1980.

McGrew, Jay L. "Energy Conservation, Friend or Foe?" Brick and Clay Record. Vol. 176, No. 6. pp. 42-43. June 1980.

Roofing, Siding and Insulation. "Contractors See Sharp Upswing for Urethane in '78". p. 18. Oct. 1978.

Spielvogel, Lawrence G. "More Insulation Can Increase Energy Consumption." ASHRAE Journal. pp. 61-63, Jan. 1974.

Teitsma, Gerard J. "Advancing the State-of-the-Arts, Thermal Insulations and Moisture Retarders." ASHRAE Journal. p. 32. March 1981.

Trias, Priscilla F. "Industry News." ASHRAE Journal. p. 13. March 1981.

Tye, R. P. "Thermal Insulation Evaluation: Present Status and Future Requirements." Journal of Thermal Insulation. Technomic Publishing Co., Inc. Vol. 2, pp. 109-130. Jan. 1979.

III. CODES, STANDARDS, AND NATIONAL PROGRAMS:

Achenback, P.R., et. al. The National Program Plan for Building Thermal Envelope Systems and Insulating Materials: Technology and Implementation for Energy Conservation. DOE/CS-0059, Department of Energy. Jan. 1979.

Achenback, P.R. and Freeman, E.C., Jr. "The National Program for the Thermal Performance of Building Envelope Systems and Materials." Presented at the DOE-ORNL/ASTM Conference on Thermal Insulation, Materials, and Systems for Energy Conservation in the '80s. Clearwater Beach, FL. Dec. 1981.

ASHRAE Journal. "The National Program Plan for Building Thermal Envelope Systems and Insulating Materials." pp. 39-43. March 1979.

Clyde, Gordon F. "Model Codes and Thermal Insulation Requirements." ASHRAE Journal. p. 55. Oct. 1980.

Hilado, Carlos J. "Standards for Cellulose Insulation." Journal of Thermal Insulation. Vol. 1. pp. 255-282. April 1978.

Hilado, Carlos J. and Cumming, Heather J. "Standards for Thermal Insulation." Journal of Thermal Insulation. Vol. 1, pp. 129-148. Oct. 1977.

Hildebrand, Floyd C. "Development of New Federal Criteria for Underground Heat Distribution Systems." ASHRAE Journal. pp. 49-53. Oct. 1980.

Hillier, Ray. "California Insulation Quality Standards." Journal of Thermal Insulation. Vol. 3. pp. 67-70. Oct. 1979.

Mineral Insulation Manufacturers Association, Inc. "Federal Trade Commissions's Trade Regulation Rule - Labeling and Advertising of Home Insulation." Insulation Facts #6. Summit, NJ. Nov. 1980.

Mineral Insulation Manufacturers Association, Inc. "Mineral Fiber Batts and Blankets - Fiber Glass and Rock Wool - Standard for Installation in Residential and Other Light-Frame Construction." Summit, NJ. June 1981.

Powell, Frank J. "Aspects of a National Program Plan for Industrial/Commercial Insulation for Mechanical Systems Applications." ASHRAE Journal. pp. 58-59. Oct. 1980.

Roofing, Siding and Insulation. "Insulation: Government Report." pp. 52-55. March, 1979.

"Standards and Tentatives Relating to Thermal and Cryogenic Insulating Materials, Building Seals and Sealants, Fire Standards, Building Constructions, Environmental Acoustics," Part 18, 1980 Annual Book of ASTM Standards. American Society for Testing and Materials, Philadelphia, PA. 1980.

IV. BUILDING ENVELOPE ISSUES:

Harrje, David T., Dutt, Gautam, S. and Beyea, Jan E. "Locating and Eliminating Obscure But Major Energy Losses in Residential Housing." ASHRAE Transactions. Vol. 85. Part 2. 1979.

Harrje, David T. "Building Envelope Performance Testing." ASHRAE Journal. pp. 39-41. March 1981.

Johns-Manville Sales Corporation. "A Study of the Effects of Insulation Gaps on Building Heat Losses." NTIS Report No. AD-A085222. Johns-Manville Research and Development Center. Denver, CO. Prepared for the U.S. Navy Civil Engineering Laboratory. Nov. 1979.

Johns-Manville Sales Corporation. "Effectiveness of Building Insulation Applications." NTIS Report No. AD-A053452. Johns-Manville Research and Development Center. Denver, CO. Prepared for the U.S. Navy Civil Engineering Laboratory. Nov. 1977.

Lehtinen, Merja Helen. "Building Envelope: The Issues." ASHRAE Journal. p. 23. March 1981.

Nelson, Lorne W. and Tobias, James R. "Energy Savings in Residential Buildings." ASHRAE Journal. pp. 38-45. Feb. 1974.

Strelka, C.S. and Burn, K.N. "Insulation Retrofitting - A Case History." Building Practice Note No. 21. Division of Building Research, National Research Council of Canada. Ottawa, Canada. Sept. 1981.

Weidt, John L. and Saxler, Robert J. "Field Investigation of the Performance of Residential Retrofit Insulation." NBS Technical Note 1131. National Bureau of Standards. Washington, D.C. Sept. 1980.

V. WALL INSULATION:

Aho, Arnold, J. "The Role of Brick Masonry in Energy Conservation Design." Brick Institute of America. McLean, VA. 1974.

Anderson, B.R. "On the Calculation of the U-value of Walls Containing Slotted Bricks or Blocks." Building and Environment, Vol. 16, No. 1, pp. 41-50. 1981.

Berkeley Solar Group. "Passive Design Saves Energy and Money." Concrete Masonry Association of California and Nevada. Sacramento, CA. 1979.

Berlad, A. L., Jaung, R., Joshi, N., and Westerinen, W. J. "Free and Forced Convective Effects on Air-Permeable Insulation Systems." Paper No. 81-HT-47, presented at the 20th Joint ASME/AIChE National Heat Transfer Conference, Milwaukee, WI. Aug. 1981.

Brick Institute of America. "Heat Transmission Coefficients of Brick Masonry Walls." Technical Notes on Brick Construction. No. 4 Revised. Aug/Sept. 1974.

Brick Institute of America. "Thermal Transmission Corrections for Dynamic Conditions - M Factor." Technical Note 4B. Technical Notes on Brick Construction. McLean, VA. Mar/Apr. 1977.

Chandra, Prakash. "Rating of Wall and Roof Sections - Thermal Considerations." Building and Environment. Pergamon Press Ltd. Great Britian. Vol. 15, pp. 245-251. 1980.

Childs, K.W. "An Appraisal of the M Factor and the Role of Building Thermal Mass in Energy Conservation." ORNL/CON-46. Oak Ridge National Laboratory. Oak Ridge, TN. July 1980.

Contreras, Antonio G. and Palfey, Albert J. "Thermal Resistances of Insulated Brick Veneer Walls with Reflective and Nonreflective Air Spaces." Presented at the DOE-ORNL/ASTM Conference on Thermal Insulation, Materials, and Systems for Energy Conservation in the '80s. Clearwater Beach, FL. Dec. 1981.

Dexter, Michael E. "Energy Conservation Design Guidelines for Including Mass and Insulation in Buildings Walls." ASHRAE SP28. Proceedings of the ASHRAE/DOE-ORNL Conference. Thermal Performance of the Exterior Envelopes Of Buildings. Kissimmee, FL. Dec. 3-5, 1979.

Dexter, Michael E. "Energy Conservation Design Guidelines: Including Mass and Insulation in Building Walls." ASHRAE Journal. pp. 35-38. March 1980.

Fiorato, A.E. and Cruz, C.R. "Thermal Performance of Masonry Walls." Portland Cement Association. Research and Development Bulletin RD071.01M. Skokie, IL. 1980.

Godfrey, R.D., Wilkes, K.E. and Lavine, A.G. "A Technical Review of the "M" Factor Concept." ASHRAE Journal. pp. 47-50. March 1980.

Godfrey, R.D., Wilkes, Kenneth E., and Lavine, Adrienne G. "A Technical Review of the 'M' Factor Concept." ASHRAE SP28. Proceedings of the ASHRAE/DOE-ORNL Conference, Thermal Performance of the External Envelopes of Buildings. Kissimmee, FL. Dec. 3-5, 1979.

Gujral, Parambir S., Clark, Raymond J. and Burch, Douglas M. "Transient Thermal Behavior of Externally Insulated Massive Building." ASHRAE Transactions. DV-80-4. No. 2. 1980.

Hoffman, Milo E. and Feldman, Moshe. "Calculation of the Thermal Response of Buildings by the Total Thermal Time Constant Method." Building and Environment, Vol. 16, No. 2, pp. 71-85. 1981.

Holland, Elizabeth. "Insulation Moves Outside." Solar Age. Vol. 11, No. 11, pp. 22-27. Nov. 1981.

Kusuda, Tamami, Hill, James E., Liu, Stanley T., Barnett, James P., and Bean, John W. "Pre-Design Analysis of Energy Conservation Options for a Multi-Story Demonstration Office Building." NBS Building Science Series 78. National Bureau of Standards. Washington, D.C. Nov. 1975.

National Concrete Masonry Association. "A Comparison of Very Lightweight Walls of Wood, Metal and Glass Versus Concrete Masonry in Energy Conservation." CM-230. McLean, VA. 1976.

National Concrete Masonry Association. "Estimating U-Factors for Concrete Masonry Construction." NCMA TEK 12. Herndon, VA 1969.

National Concrete Masonry Association. "Thermal Insulation of Concrete Masonry Walls." NCMA TEK 38-A. Herndon, VA. 1980.

National Concrete Masonry Association. "Energy Conservation with Concrete Masonry." NCMA TEK 58. Herndon, VA. 1974.

National Concrete Masonry Association. "Tables of U-Values for Concrete Masonry Walls." NCMA TEK 67. Herndon, VA. 1975.

National Concrete Masonry Association. "New Findings on Energy Conservation with Concrete Masonry." NCMA TEK 68. Herndon, VA. 1975.

National Concrete Masonry Association. "Energy Conscious Design for Buildings." NCMA TEK 82. Herndon, VA. 1976.

National Concrete Masonry Association. "Design of Solar Energy Walls with Concrete Masonry." NCMA TEK 97. Herndon, VA. 1978.

National Concrete Masonry Association. "'U' Values for Reinforced Concrete Masonry Walls." NCMA TEK 101. Herndon, VA. 1978.

National Concrete Masonry Association. "Testing the Thermal Performance of Concrete Masonry Walls." NCMA-TEK 112. Herndon, VA. 1979.

National Concrete Masonry Association. "Second Generation Passive Solar Systems Using Concrete Masonry." NCMA TEK 120. Herndon, VA. 1981.

Peavy, Bradley, A., Powell, Frank J., and Burch, Douglas M. "Dynamic Thermal Performance of an Experimental Masonry Building." NBS Building Science Series 45. National Bureau of Standards. Washington, D.C. July 1973.

Portland Cement Association. "Special Considerations for the Selection of Tilt-Up Concrete Sandwich Panels." Skokie, IL. 1975.

Portland Cement Association. "The Concrete Approach to Saving Energy." Skokie, IL. 1974.

Scoggin, Harry L. "Cast-in-Place Concrete Residences with Insulated Walls." Journal of the PCA Research and Development Laboratories, Vol. 8, No. 2, pp. 21-29. Portland Cement Association. Skokie, IL. May 1966.

Shipp, Paul H. and Broderick, Thomas B. "Comparison of Annual Heating Loads for Various Basement Wall Insulation Strategies Using Transient and Steady State Models." Presented at the DOE-ORNL/ASTM Conference on Thermal Insulation, Materials, and Systems For Energy Conservation in the '80s. Clearwater Beach, FL. Dec. 1981.

Stewart, David B. "Time - Domain Transient Thermal Response of Structural Elements." Building and Environment, Vol. 16, No. 2, pp. 87-91. 1981.

Structural Clay Products Institute. "SCR Insulated Cavity Walls." Technical Note No. 21. McLean, VA. 1968.

Wilkes, K.E., Godfrey, R.D., and Lavine, A.G. "Wall Massiveness - What is Known Today About Annual Energy Requirements?" ASHRAE Journal, Vol. 24, No. 2, pp. 21-23. Feb. 1982.

VI. ROOF INSULATION:

ASHRAE Journal "Improved Methods: Installing Insulation." Vol. 23, No. 3. pp. 48-49. March 1981.

Brady, Sam A. "Protecting Polyurethane Foam Roofing Insulation." Metal Building Review. p. 54-58. July 1977.

Busching, Herbert W., Mathey, Robert G., Rossiter, Walter J., Jr., and Cullen, William C. "Effects of Moisture in Built-Up Roofing - A State-of-the-Art Literature Survey." NBS Technical Note 965. National Bureau of Standards. Washington, D.C. July 1978.

Cash, Carl G. "Optimization of the Thermal Resistance of Roof Insulation In Reroofing." Journal of Thermal Insulation. Vol. 1, pp. 192-200. Jan. 1978.

Chandra, Prakash. "Rating of Wall and Roof Sections - Thermal Considerations." Building and Environment. Pergamon Press Ltd. Great Britian. Vol. 15, pp. 245-251. 1980.

Cullen, William C. "Solar Heating, Radiative Cooling, and Thermal Movement - Their Effects on Built-Up Roofing." NBS Technical Note 231. National Bureau of Standards. Washington, D.C. April 1964.

Department of the Air Force. "Built-Up Roof Management Program." ARM 91-36. Washington, D.C. Sept. 1980.

Epstein, K.A. and Putnam, L.E. "Performance Criteria for the Protected Membrane Roof System." Journal of Thermal Insulation. Vol. 1, pp. 149-167.

Griffin, C.W. "Plug the Energy Leaks in Your Roof." Journal of Thermal Insulation. Vol. 1, pp. 206-214. Jan. 1978.

Gumpertz, Werner. "Thermally Efficient Roofs Put a Lid on Heat Loss." Specifying Engineer. Vol. 42, pp. 107-109. Aug. 1979.

Knab, Lawrence I., Jenkins, David R. and Mathey, Robert G. "The Effect of Moisture on the Thermal Conductance of Roofing Systems." Roof, Insulation and Siding. Special report, pp. 61-66, Part I. July 1980, and pp. 62-67 Part II. Sept 1980.

Knab, Lawrence I., Jenkins, David R., and Mathey, Robert G. "The Effect of Moisture on the Thermal Conductance of Roofing Systems." NBS Building Science Series 123. National Bureau of Standards. Washington, D.C. April 1980.

Lewis, James E. "Thermal Evaluation of the Effects of Gaps Between Adjacent Roof Insulation Panels." Journal of Thermal Insulation. Technomic Publishing Co., Inc. Vol. 3. pp. 80-103. Oct. 1979.

Milanese, Robert. "The Super Insulating Perlite Roof Deck." Roof, Insulation and Siding. pp. 94-98. Oct. 1980.

Miller, R.G., and Sherman, M. "Thermal Performance of Insulated Metal Building Roof Deck Constructions." Presented at the DOE-ORNL/ASTM Conference on Thermal Insulation, Materials, and Systems for Energy Conservation in the '80s. Clearwater Beach, FL. Dec. 1981.

Morgenroth, Dan E. "A Cost Effective Solution To Achieve Thermal Performance of Roofing Systems." ASHRAE Journal. p. 54. March 1981.

Richards, David E. and Mirra, Edward J. "Does More Roof Insulation Cause Premature Roofing Membrane Failure or Are Roofing Membranes Adequate?" Journal of Thermal Insulation. Vol. 1, pp. 171-180. Jan. 1978.

Roof, Insulation and Siding. "Hidden Market: Insulating Metal Roofs." pp. 74-76. Oct. 1980.

Roof, Insulation and Siding. "Urethane Crusade?" pp. 34-36. Oct. 1980.

Roof, Insulation and Siding. "Blistering: Still Looking For Answers." p. 68. Feb. 1980.

Roof, Insulation and Siding. "Attaching Insulation to Steel Decks." pp. 65-67. Feb. 1980.

Roof, Insulation and Siding. "Hot Built-Up to Falter in the 80's." p. 58. Feb. 1980.

Rossiter, Walter J., Jr. and Mathey, Robert G. "Effect of Insulation on the Surface Temperature of Roof Membranes." NBSIR 76-987. National Bureau of Standards. Washington, D.C. Feb. 1976.

Sodha, M.S., Kaushik, S.C., Tiwari, G.N., Goyal, I.C., Malik, M.A.S. and Khatry, A.K. "Optimum Distribution of Insulation Inside and Outside the Roof." Building and Environment. Vol. 14, pp. 47-52. Feb. 1979.

The Roofing Industry Educational Institute. "Roof Maintenance." Englewood, CO. Feb. 1981.

Tobiasson, Wayne and Ricard, John. "Moisture Gain and Its Thermal Consequence for Common Roof Insulations." U.S. Army Corps of Engineers, Hanover, N.H. Reprinted From: Proceedings 5th Conference on Roofing Technology. April, 1979.

VII. WINDOWS:

Alereza, T. and Hossli, R.I. "A Simplified Method of Calculating Heat Loss and Solar Gain Through Residential Windows During the Heating Season." Transactions. PH-79-6. No. 1. 1979.

Bishop, Ronald R. "Performance Evaluation of Solar Films and Screens." Johns-Manville Sales Corp., Denver, CO. Prepared for the U.S. Army Facilities Engineering Support Agency, Fort Belvoir, VA. May 1979.

Chapman, William F. "Less Refrigeration: More Glass, a Compatible Idea?" ASHRAE Journal. pp. 31-33. Feb. 1979.

Grasso, Maureen M. and Buchanan, David R. "Roller Shade System Effectiveness in Space Heating Energy Conservation". ASHRAE Transactions. No. 2520. GIA-39

Hastings, S. Robert and Crenshaw, Richard W. "Window Design Strategies to Conserve Energy." NBS Building Science Series 104. National Bureau of Standards. Washington, D.C. June 1977.

McGrew, J.L., McGrew, David P. and Yeagle, George P. "Integrated Heat Flow in Windows." Applied Science and Engineering. Littleton, CO. 1978.

Quirouette, R.L. "Insulated Window Shutters." Building Practice Note No. 17. National Research Council of Canada. Ottawa, Canada. June 1980.

Rennekamp, Stephen J. "U-Value Testing of Windows Using a Modified Guarded Hot Box Technique". ASHRAE Transactions. PH-79-6. No. 1. 1979.

Shephard, P.B. "Performance Evaluation of Vinyl Replacement Windows." FESA-TS-70-78-D-0002, Task Order 10. Johns-Manville Sales Corp., Denver, CO. Prepared for the U.S. Army Facilities Engineering Support Agency, Fort Belvoir, VA. Jan. 1980.

VIII. VAPOR BARRIERS:

Adams, Ludwig. "Thermal Conductivity of Wet Insulations." ASHRAE Journal. pp. 61-62. Oct. 1974.

Building Safety. "AIA Committee Warns Against Wall Insulation of Frame Buildings." and "Thermal Insulation Testing Criteria Announced." Building Standards. Special Issue, p. 5 Mar.-Apr. 1979.

Burch, Douglas M., Contreras, Antonio, G. and Treado, Stephen J. "The Use of Low-Moisture-Permeability Insulation as an Exterior Retrofit System - A Condensation Study." National Bureau of Standards, Center for Building Technology, National Engineering Laboratory. Washington D.C. DE-79-3. No. 3, pp. 547-562.

Hedlin, C.P. "Moisture Gains by Foam Plastic Roof Insulations Under Controlled Temperature Gradients." Journal of Cellular Plastics. pp. 313-319. Sept./Oct., 1977.

Latta, J.K. "Vapor Barriors: What Are They? Are They Effective?" Canadian Building Digest. p. 175. March 1976.

Mineral Insulation Manufacturers Association, Inc. "Covering Vapor Barriers." Insulation Facts #4. Summit, N.J. Nov. 1978.

Misselhorn, Donald J. "Some Problems With Insulation Over Suspended Ceilings." ASHRAE Journal. pp. 46-49. March 1979.

Munawwar, S.M. "Concept of Vapor Barriors in Thermal Insulation." Journal Of Thermal Insulation. Technomic Publishing Co., Inc. Vol. 4, pp. 288-306. April 1981.

Ogniewicz, Y. and Tien, C.L. "Analysis of Condensation in Porous Insulation." Int. Journal Heat Mass Transfer. Vol. 24, pp. 421-429. March 1981.

Rockwool Industries. "Blowing Wool: Moisture Vapor." Englewood, CO. Technical Bulletin #7. August 1981.

Turner, William C. and Johnson, John W. "Methods For Keeping Thermal Insulation Dry to Preserve Its Function For Energy Conservation". Journal Of Thermal Insulation. Technomic Publishing Co., Inc. Vol. 2, pp. 67-86. Oct. 1978.

Tye, Ronald P. and Spinney, S.C. "A Study of the Effects of Moisture Vapor On The Thermal Transmittance Characteristics Of Cellulose Fiber Thermal Insulation." Journal Of Thermal Insulation. Vol. 2, pp. 175-196. April 1979.

Webber, Joseph F. "Cold Storage Insulation: The Vapor Barrior, Can We Get It Straight?" ASHRAE Journal. pp. 36-38. March 1979.

IX. CAULKING AND WEATHERSTRIPPING

American Concrete Institute. "Use of Epoxy Compounds with Concrete." ACI 503R-73. Detroit, MI. Sept. 1973.

American Concrete Institute. "Guide to Joint Sealants for Concrete Structures." ACI 504R-77. Detroit, MI. June, 1977.

"Exterior Caulking Compounds." Consumer Reports. Vol. 46. No. 10. pp. 579-581. Oct. 1981.

Federal Energy Administration. "Project Retro-Tech Home Weatherization Manual." Conservation paper No. 28C. Washington, D.C. May 1976.

Pemko Manufacturing Company. "What You Should Know About Interior and Exterior Weatherstrip and Thresholds." Emeryville, CA. 1970.

"Weather Stripping." Consumer Reports. Vol. 46. No. 10. pp. 576-578. Oct. 1981.

X. MECHANICAL EQUIPMENT:

Heating/Piping/Air Conditioning. "Energy Conservation and HVAC Contractors." pp. 25-26. Nov. 1980.

Heating/Piping/Air Conditioning. "Glass Fiber Ductwork." pp. 57-67. July 1980.

Heating/Piping/Air Conditioning. "Steam Pipe Insulation Could Save 305,000 Barrels of Oil a Day." p. 27. Oct. 1980.

Midwest Insulation Contractors Association. "Commercial and Industrial Standards." Omaha, NE. 1979.

Roose, Robert W. "Thermal Insulation Guide for Ducts, Equipment, and Piping." Ch. 31 in Roose, Robert W. (ed.) Handbook of Energy Conservation in Mechanical Systems in Buildings. Van Nostrand Reinhold Company. New York, NY. 1978.

Roose, Robert W. and Pannkoke, T.E. "Thermal Insulation for Buried Piping." Ch. 33 in Roose, Robert W. (ed.) Handbook of Energy Conservation in Mechanical Systems in Buildings. Van Nostrand Reinhold Company. New York, NY. 1978.

Sheetmetal and Air Conditioning Contractors' National Association, Inc. "Fibrous Glass Duct Construction Standards." 5th edition. Vienna, VA. 1979.

Thermal Insulation Manufacturers Association. "The Thermal Performance of Operating Duct Systems." Mt. Kisco, NY. Nov. 1980.

XI. HEALTH EFFECTS:

ASHRAE Journal. "UF Foam Industry Gets Some Good News and Some Bad News." p. 11. Feb. 1980.

ASHRAE Journal. "UF Foam Industry Prepares for Battle Over Proposed CPSC Ban." p. 21. March 1981.

Carney, I.F. "Toxicology of Isocyanates." European Journal of Cellular Plastics. Technomic Publishing Co., Inc. Vol. 3, No. 3. July 1980.

Federal Register. "Cellulose Insulation; Proposed Amendment to Labeling Requirement." Vol. 44, No. 201, pp. 59557-59559. Oct. 16, 1979.

Federal Register. "Cellulose Insulations." Vol. 43. No. 153. p. 35238. Aug. 8, 1978.

Federal Register. "Urea-Formaldehyde Foam Insulation; Proposed Ban; Denial of Petition." Vol. 46, No. 24, pp. 11188-11211. Feb. 5, 1981.

Haggerty, Brian. "CPSC Ban of UF Foam Insulation Won't Stand Up In Court, Declare Industry Critics." Engineering Times. Vol. 4, No. 4, p.13. April 1982.

Hilado, Carlos J. and Huttlinger, Patricia A. "Toxicity of Off-Gasses From Thermal Insulation." Journal of Thermal Insulation. Vol. 4, pp. 276-287. April 1981.

Hilado, Carlos J. and Huttlinger, Patricia A. "Toxicity Aspects of Thermal Insulation: A Bibliography." Journal of Thermal Insulation. Technomic Publishing Co., Inc. Vol. 4. pp. 214-216. Jan. 1981.

Johns Manville. "Asbestos, Health and Johns Manville." Denver, CO. Sept. 1981.

Leineweber, J.P. "Vitreous Insulations Are They Hazardous?" ASHRAE Journal. pp. 51-54. March 1980.

Martino, Robert. "For Fluorocarbon Blowing Agents, A Time for Decision." Modern Plastics. pp. 46-48. Oct. 1980.

Morgan, Robert W., Kaplan, Samuel D. and Bratsberg, Julie A. "Mortality Study of Fibrous Glass Production Workers." Archives of Environmental Health. Vol. 36, No. 4. pp. 179-183. July/Aug. 1981.

Plastics Engineering. "Polyurethane Insulation: No Harmful Effects." Vol. 37, No. 4. April 1981.

Rybicky, Jaroslav and Kambanis, Stamatis M. "Determination of Rate of Formaldehyde Release From Urea-Formaldehyde and Dhenol-Formaldehyde Foams." Journal of Thermal Insulation. Technomic Publishing Co., Inc. Vol. 4, pp. 17-181. Jan. 1981.

Timm, William and Smith, Patti M. "Formaldehyde Odor and Health Problems Within Residences." Journal of Thermal Insulation. Technomic Publishing Co., Inc. Vol. 3, pp. 104-125. Oct. 1979.

Timm, William and Smith, Patti M. "Test For Formaldehyde Off-Gassing Rates From Insulation and Other Building Materials." Journal of Thermal Insulation. Technomic Publishing Co., Inc. Vol. 4, pp. 137-153. Oct. 1980.

XII. AIR INFILTRATION AND INDOOR AIR QUALITY:

ASHRAE Journal. "Ozone Chlorofluorocarbon Issue." p. 54. August 1981.

"Can You Make a House Too Tight?" Consumer Reports. Vol. 46, No. 10, pg. 582. Oct. 1981.

Department of Energy. "Find and Fix the Leaks-A Guide to Air Infiltration Reduction and Indoor Air Quality." DOE/CS-0006. Washington, D.C. May 1981.

Hadley, John. "Energy Conservation and Indoor Air Quality." ASHRAE Journal. p. 35 March 1981.

Henning, G.N. "Energy Conservation With Air Infiltration Barriers." Presented at the DOE-ORNL/ASTM Conference on Thermal Insulation, Materials, and Systems for Energy Conservation in the '80s. Clearwater Beach, FL. Dec. 1981.

Hollowell, Craig D., Berk, James V. and Traynor, Gregory W. "Impact of Reduced Infiltration and Ventilation on Indoor Air Quality in Residential Buildings." ASHRAE Transactions. PH-79-10, No. 2. 1979.

Hollowell, Craig D., Berk, James V. and Traynor, Gregory W. "Impact of Reduced Infiltration and Ventilation on Indoor Air Quality." ASHRAE Journal. pp. 49-53. July 1979.

Kusuda, T., Hunt, C.M. and McNall, P.E. "Radioactivity (Radon and Daughter Products) As a Potential Factor in Building Ventilation." ASHRAE Journal. pp. 30-34. July 1979.

Lee, B.E., Hussain M. and Soliman, B. "Predicting Natural Ventilation Forces Upon Low-Rise Buildings." ASHRAE Journal. pp. 35-39. Feb. 1980.

Liptak, Bela G. "Savings Through CO_2 Based Ventilation." ASHRAE Journal. pp. 38-41. July 1979.

Peterson, Joel E. "Estimating Air Infiltration into Houses." ASHRAE Journal. pp. 60-62. Jan. 1979.

Sandberg, Mats. "What is Ventilation Efficiency?" Building and Environment. Great Britain. Vol. 16, No. 2. pp. 123-135. 1981.

Sepsy, Charles F. "The HVAC Industry: State-of-the-Art." Consulting Engineer. Vol. 58. No. 1. pp. 67-71. Jan. 1982.

Shair, F. H., Wolbrink, D.W., Bowen, L.O., Neelley, C.E. and Sampsel, K.E. "Influence of Mechanical Ventilation." ASHRAE Journal. pp. 54-60. July 1979.

Shepherd, P.B. and Gerharter, J.E. "Techniques for Control of Air Infiltration in Buildings." Johns-Manville Sales Corp., Denver, CO. Prepared for the U.S. Army Facilities Engineering Support Agency, Fort Belvoir, VA. April 1979.

Tucker, Henry W. "Simplified Determination of Air Infiltration." ASHRAE Journal. pp. 44-47. July 1979.

Woods, James E. "Ventilation, Health and Energy Consumption: A Status Report." ASHRAE Journal. pp. 23-27. July 1979.

Woods, James E., Maldonado, Edwardo A.B. and Reynolds Gary L. "How Ventilation Influences Energy Consumption and Indoor Air Quality." ASHRAE Journal. pp. 40-43. Sept. 1981.

XIII. INSULATION FLAMMABILITY:

Anderson, Robert W. and Freischel, William. "Evaluation of Proposed Smoldering Testing Methodology for Cellulosic Insulation." Journal of Thermal Insulation. Vol. 2. p. 7-23. July 1978.

Day, M., Suprunchuk, T. and Wiles, D.M. "A Combustibility Study of Cellulose Insulation." Journal of Thermal Insulation. Technomic Publications Co., Inc. Vol. 3. pp. 260-271. April 1980.

Day, M., Suprunchuk, T., and Wiles, D.M. "The Fire Properties of Cellulose Insulation." Journal of Thermal Insulation. Technomic Publishing Co., Inc. Vol. 4, pp. 157-170. Jan. 1981.

Day, M. and Wiles, D.M. "Combustibility of Loose Fiber Fill Cellulose Insulation: The Role of Borax and Boric Acid." Journal of Thermal Insulation. Vol. 2. pp. 30-39. July 1978.

Hilado, Carlos J. and Cumming, Heather J. "Fire Safety Aspects of Thermal Insulation." Journal of Thermal Insulation. Vol. 1, pp. 116-128. Oct. 1977.

Hilado, Carlos J., Cumming, Heather J., and Murphy, Regina M. "Flash-fire Propensity of Some Insulation and Building Materials." Journal of Thermal Insulation. pp. 201-205. Jan. 1978.

Hilado, Carlos J., Cumming, Heather J., and Schneider, Jane E. "Relative Toxicity of Pyrolysis Gases From Some Rigid Foams Uses As Thermal Insulation." Journal of Thermal Insulation. Vol. 1, pp. 215-220. Jan. 1980.

Hilado, Carlos J. and Murphy Regina M. "Flash-Fire Propensity of Rigid Foam Insulation." Journal of Thermal Insulation. Vol. 1, pp. 283-286. April 1978.

Hilado, Carlos J. and Murphy, Regina M. "Ignitability of Some Insulation and Building Materials." Journal of Thermal Insulation. Vol. 1, pp. 237-240. April 1978.

Hilado, Carlos J., Murphy, Regina M., and Sprague, Robert W. "Flash-Fire Propensity of Cellulose Insulation Samples Containing Various Borate Fire Retardants." Journal of Thermal Insulation. Vol. 2, pp. 3-6. July 1978.

Hilado, Carlos J. and Kosola, Kay L. "Toxicity of Pyrolysis Gases from Cellulose Insulation." Journal of Thermal Insulation. Vol. 2, pp. 24-29. July 1978.

Hilado, Carlos J. and Huttlinger, Nancy V. "Toxicity of Pyrolysis Gases From Phenolic, Isocyanurate and Polystyrene Rigid Foam Insulation." Journal of Thermal Insulation. Vol. 2, pp. 40-47. July 1978.

Hilado, Carlos J., Murphy, Regina M. and Spraque, Robert W. "Flash-Fire Propensity of Cellulose Insulation Samples Containing Various Borate Fire Retardants." Journal of Fire Retardant Chemistry. Vol. 5, pp. 138-143. August 1978.

Hilado, Carlos J. and Brauer, Diana P. "Low-Cost Fire Response Tests For Cellulose Insulation." Journal of Thermal Insulation. Vol. 2, pp. 51-56. Oct. 1978.

Hilado, Carlos J. and Olcomendy, Elaine M. "Ignitability of Rigid Foam Insulation." Journal of Thermal Insulation. Vol. 2, pp. 87-89. Oct. 1978.

Hilado, Carlos J. and Brauer, Diana P. "Toxicity of Pyrolysis Gases From Thermal Insulation Materials." Journal of Thermal Insulation. Technomic Publishing Co., Inc. Vol. 2, pp. 90-101. October 1978.

Hilado, Carlos J. and Huttlinger, Patricia A. "Smoldering Combustion: A Bibliography of Published Information." Journal of Thermal Insulation. Technomic Publishing Co., Inc. Vol. 2, pp. 102-106. Oct. 1978.

Hilado, Carlos J., Olcomendy, Elaine M. and Brauer, Diana P. "Toxicity of Gases Generated From Smoldering Cellulose Insulation." Journal of Thermal Insulation. Technomic Publishing Co., Inc. Vol. 2, pp. 131-134. Jan. 1979.

Hilado, Carlos J. and Huttlinger, Patricia A. "Fire Tests of Pipe and Vessel Insulation: A Bibliography of Published Information." Journal of Thermal Insulation. Technomic Publishing Co., Inc. Vol. 2, p. 205. April 1979.

Hilado, Carlos J. "A Review of Some Full-Scale Fire Tests on Polyurethane Foam Insulated Pipes and Vessels." Journal of Thermal Insulation. Technomic Publishing Co., Inc. Vol. 3, pp. 22-31. July 1979.

Hilado, Carlos J. and Murphy, Regina M. "The Measurement of Smoke From Thermal Insulation." Journal of Thermal Insulation. Vol. 3, pp. 272-280. April 1980.

Hilado, Carlos J. and Huttlinger, Patricia A. "Additive Content and Residue Weight in Cellulose Insulation." Journal of Thermal Insulation. Technomic Publishing Co., Inc. Vol. 4, pp. 134-136. Oct. 1980.

Hilado, Carlos J. and Huttlinger, Patricia A. "Carbon Monoxide Production From Overheated Thermal Insulation Materials." Journal of Thermal Insulation. Technomic Publishing Co., Inc. Vol. 4, pp. 87-92. Oct. 1980.

Hilado, Carlos J. and Huttlinger, Patricia A. "Smoldering Combustion: A Bibliography of Published Information: Part 2." Journal of Thermal Insulation. Technomic Publishing Co., Inc. Vol. 4, pp. 115-118. Oct. 1980.

Hilado, Carlos J. and Huttlinger, Patricia A. "Relative Flammability and Toxicity of Thermal Insulation." Journal of Thermal Insulation. Technomic Publishing Co., Inc. Vol. 4, pp. 217-222. Jan. 1981.

Hilado, Carlos J. and Huttlinger, Patricia A. "Integration of Toxic Effects With Foam Insulation." Journal of Thermal Insulation. Technomic Publishing Co., Inc. Vol. 4, pp. 307-314. April 1981.

Hilado, Carlos J, and Huttlinger, Patricia A. "A Study of Boron Additives in Cellulose Insulation." Journal of Thermal Insulation. Technomic Publishing Co., Inc. Vol. 4, pp. 255-264. April 1981.

Kanakia, M., Herrera, William R., and Hutto, Francis B. "Fire Resistance Tests For Thermal Insulation." Journal of Thermal Insulation. Vol. 1, pp. 241-254. Arpil 1978.

Lie, T.T. "Effects of Insulation on Fire Safety." Canadian Building Digest. pp. 218-218-4. August 1981.

Madacsi, John P. and Knoepfler, Nestor B. "Treatment of Home-Blown Cellulosic Insulation with a Boric Acid Vapor Process to Impart Smolder Resistance." Journal of Thermal Insulation. Technomic Publishing Co., Inc. Vol. 3, pp. 207-211. April 1980.

Mineral Insulation Manufacturers Association, Inc. "Insulation and Fire Safety." Insulation Facts. #5. Summit N.J. Feb. 1980.

Nosse, John H. "Enforcement of Foam Plastic Regulations." Building Safety. p. 14.

Roofing, Siding and Insulation. "Cellulose Proves Effective as Fire Fighter in Florida." p. 59. July 1978.

Roofing, Siding and Insulation. "Cellulose Finished Last in Fire Race." p. 8. Oct. 1978.

Roofing, Siding and Insulation. "Recessed Light Fixtures: Danger in the Attic." p. 74. March 1979.

Rockwool Industries Inc. "Blowing Wool" Technical Bulletin #3. Englewood, CO.

Schaffer, Erwin L. "Smoldering Combustion Tendency: An Introduction." U.S. Dept. of Agriculture, Forest Products Laboratory. Madison, WI. Technomic Publishing Co., Inc. June 1978.

Shafizadeh, Fred and Bradbury, Allan G.W. "Smoldering Combustion of Cellulosic Materials." Journal of Thermal Insulation. Technomic Publishing Co., Inc. Vol. 2, pp. 141-152. Jan. 1979.

Sprague, Robert W. "Present Variability in Chemical Add-on in the Cellulose Insulation Industry." Journal of Thermal Insulation. Technomic Publishing Co., Inc. Vol. 2, pp. 197-204. April 1979.

Sprague, Robert W. and Shen, Kelvin K. "The Use of Boron Products in Cellular Insulation." Journal of Thermal Insulation. Technomic Publishing Co., Inc. Vol. 2, pp. 161-174. April, 1979.

Shorter, T.W. and McGuire, J.H. "Fire and Plastic Foam Insulation Material." Canadian Building Digest. pp. 178-178-4. June 1976.

Sumi, Kikuo and Tsuchiya, Yoshio. "Evaluating the Toxic Hazard of Fires." Canadian Building Digest. pp. 197-197-4. July 1978.

Tsuchiya, Yoshio and Sumi, Kikuo. "Spontaneous Ignition." Canadian Building Digest. pp. 189-189-4. Sept. 1977.

XIV. ECONOMICS:

Arkin, Hillel and Shitzer, Avraham. "Computer Aided Optimal Life-Cycle Design of Rectangular Air Supply Duct Systems." Paper No. 2523, ASHRAE Transactions. (date required).

Cash, Carl G. "Optimization of the thermal Resistance of Roof Insulation in Reroofing." Journal of Thermal Insulation. Vol. 1, pp. 192-200. Jan. 1978.

Cohn, Lisa. "Fuel Cost Uncertainty Clouds Boiler Payback Calculations." Energy User News, Dec. 7, 1981.

Curt, Robert J. "Economic Thickness for Thermal Piping Insulation Above Ground." Ch. 32 in Roose, Robert W. (ed.) Handbook of Energy Conservation in Mechanical Systems in Buildings. Van Nostrand Reinhold Company. New York, NY. 1978.

Dixon, John G. and Dean, Bob W. "Applying a Computer Analysis Program to the Selection of Insulation Materials." ASHRAE Journal. pp. 62-66. March 1979.

Federal Energy Administration. "ETI-Economic Thickness of Industrial Insulation." Conservation paper No. 46. Washington, D.C. August 1976.

Guntermann, Alfred E. "Retrofit Now for Best Energy System ROI." Specifying Engineer, Vol. 42, pp. 83-85. Aug. 1979.

Harrison, Michael R. and Palanne, Charles M. "Cost-Effective Thermal Insulation." Chemical Engineering. pp. 62-76. Dec. 19, 1977.

Kirk, Stephen J. "Life Cycle Costing - Controlling Return on Investment by Computer." Specifying Engineer. Vol. 42. pp. 91-93. Aug. 1979.

Koenig, Alan R. "Choosing Economic Insulation Thickness." Chemical Engineering. (date required).

Manian, V.S. "Toward an Accurate View of Payback." ASHRAE Journal. pp. 28-30, Feb. 1979, and letter by Chapman, W.P., pg. 6, April 1979.

McMillan, L.B. "Heat Transfer Through Insulation." Transactions of the ASME, 1926.

Montag, Geraldine M. "A Commercial Building Ownership Energy Conservation Cost Analysis Model." ASHRAE Journal. pp. 49-52, June 1979.

Montag, Geraldine M. "Commercial Building Ownership Energy Conservation Cost Analysis Model - A Follow-Up." ASHRAE Journal. pp. 45-47. Sept. 1981.

National Insulation Contracts Association. "Principles of Heat Transfer and Introduction to ETI." Washington, D.C.

Ng, Warren. "Life Cycle Costing With a Programmable Calculator." Heating/Piping/Air Conditioning. Vol. 53. No. 4. pp. 90-93. April 1981.

Petersen, Stephen R. "Retrofitting Existing Housing for Energy Conservation: An Economic Analysis." NBS Building Science Series 64. National Bureau of Standards. Washington, D.C. Dec. 1974.

Reinhart, Robert D. "How Much Insulation for Retrofit?" Energy Engineering. Vol. 77. No. 3, pp. 32-52. Apr/May 1980.

Russell, Alan D. "Economic Risks in Energy Conservation Strategies." Building and Environment, Vol. 16, No. 2. pp. 109-121. 1981.

Shitzer, Avraham and Arkin, Hillel. "Study of Ecnomic and Engineering Parameters Related to the Cost of an Optimal Air Supply Duct System." Paper No. 2551, ASHRAE Transactions (date required.)

Stephenson, D.G. "Determining the Optimum Thermal Resistance for Walls and Roofs." Building Research Note No. 105. National Research Council of Canada. Ottawa, Canada, Jan. 1976.

Wepfer, W. J., Gaggioli, R.A. and Obert, E.F. "Economic Sizing of Steam Piping and Insulation." ASME Journal of Engineering for Industry. Vol. 101, pp. 427-433. Nov. 1979.

XV. MISCELLANEOUS:

ASHRAE Journal. "Occupant Acceptance of Temperature Drifts." and "Comfort During Cyclical Temperature Fluctuations." p. 51. March 1981.

ASHRAE Handbook, 1981 Fundamentals. American Society of Heating, Refrigerating, and Air-Conditioning Engineers, Inc. Atlanta, GA 1981.

Brick Institute of America. "Brick Passive Solar Heating Systems Materials Properties - Part IV." Tech. Note 43D. Technical Notes on Brick Construction. Sept/Oct. 1980.

Brick Institute of America. "Heat Gain." Tech. Note 4A Rev. Apr/May 1974; "Energy Conservation-Estimating Energy Use - Part 1." Tech. Note 4C Rev. May/Jun 1979; "-Part 2." Tech. Note 4D Rev. Nov/Dec 1979; "-Part 3." Tech. Note 4E Rev. Jan. 1980; "-Part 4.' Tech. Note 4F. July 1980; "-Part 5." Tech. Note 4G. Aug. 1980; Technical Notes on Brick Construction. McLean, VA.

Department of Energy. Minimum Energy Dwelling (MED) Workbook - An Investigation of Techniques and Materials for Energy Conscious Design. SAN/1198-1. Dec. 1977.

Hager, N.E., Jr. "Energy Conservation and Floor Covering Materials." ASHRAE Journal. pp. 34-39. Sept. 1977.

Hickman, William C. "Successful Insulation Hanger Bonding." Heating/Piping/Air Conditioning. p. 92. Jan. 1981.

Hilado, Carlos J. "Some Laboratory Studies on the Use of Polyurethane Rigid Foam as a Thermal Insulation Material in Low Temperature Service." Journal of Thermal Insulation. Vol. 1, pp. 15-46. July 1977.

Hilado, Carlos J. "A Laboratory Method in Determining the Effectiveness of Thermal Insulation Systems at Cryogenic Temperatures." Journal of Thermal Insulation. Technomic Publishing Co., Inc. Vol. 3, pp. 32-36. July 1979.

Horak, H.L., et. al. DOE-2 Reference Manual. Report Nos. LA-7689-M, Ver. 211 and LBL-8706 Rev. 1. Lawrence Berkeley Laboratory and Los Alamos Scientific Laboratory. May, 1980.

Munawwar, S.M. "Insulation Facing Pressure Sensitive Tapes: General Review: Technical Application Requirements." Journal of Thermal Insulation. Technomic Publishing Co., Inc. Vol. 4, pp. 45-60. July 1980.

Musgrave, Dwight S. "Thermal Performance of Urethane Foam Pipe Insulation at Cryogenic Temperatures." Journal of Thermal Insulation. Technomic Publishing Co., Inc. Vol. 3, pp. 3-20. July 1979.

NAHB Research Foundation, Inc. Insulation Manual-Homes, Apartments. Rockville, MD 1979.

National Concrete Masonry Association. "Thermal Comfort in Housing with Concrete Masonry." NCMA-TEK 26-A. Herndon, VA. 1979.

National Concrete Masonry Association. "Concrete Masonry in Passive Solar Buildings." NCMA-TEK 90, Herndon, VA. 1977.

National Concrete Masonry Association. "Concrete Masonry Passive Solar Design - Rules of Thumb." NCMA-TEK 116. Herndon, VA. 1980.

National Concrete Masonry Association. "Estimating Temperature Swings in Direct Gain Passive Solar Buildings." NCMA-TEK 118. Herndon, VA. 1981.

National Concrete Masonry Association. "Earth-Sheltered Housing with Concrete Masonry." Herndon, VA.

Portland Cement Association. Simplified Thermal Design of Building Envelopes for Use with ANSI/ASHRAE/IES Standard 90A-80. 1981.

Rossiter, Walter J., Jr. and Mathey, Robert G. (eds.) "Criteria for Retrofit Materials and Products for Weatherization of Residences." NBS Technical Note 982. National Bureau of Standards. Washington, D.C. Sept. 1978.

Sabine, Hale J. and Lacher, Myron B. "Acoustical and Thermal Performance of Exterior Residential Walls, Doors and Windows." NBS Building Science Series 77. National Bureau of Standards. Washington, D.C. Nov. 1975.

Saxby, Lewis W. and Denny, James W. "Engineered Systems Approach: Insulating the Alaska Pipeline". ASHRAE Journal. pp. 30-34. March 1979.

Shipp, P.H., Pfender, E., and Bligh, T.P. "Thermal Characteristics of a Large Earth - Sheltered Building (Parts I and II)." Underground Space. Vol. 6, pp. 53-64. July/Aug. 1981.

Steven Winter Associates, Inc. Passive Solar Construction Handbook. SSEC/SP-41187. Southern Solar Energy Center (U.S. Dept. of Energy). Atlanta, GA. Feb. 1981.

Sundberg, D.G. "Acoustical Differences of Various Building Insulations In Residential Construction." Sound and Vibration. pp. 6-7. June 1981.

Szoke, Stephen S. "Brick Properties and Performance in Thermal Storage." Solar Age. Vol. 6, No. 7, pp. 37-40. July 1981.

Turner, W.C. and Malloy, J.F. Thermal Insulation Handbook. Robert E. Krieger Publishing Company, Malabau, FL, and McGraw Hill Book Company, New York, NY 1981.

Wallis, F.A. "Evolution of Cold Storage Building Techniques." Insulation Outlook, Vo. 26, No. 5. pp. 24-27. May 1981.

Watson, Don A. Construction Materials and Processes. McGraw-Hill Book Company. New York, NY. 1978.

The following papers are contained in the ASTM publication STP 781, Thermal Insulation Performance, D. L. McElroy and R. P. Tye, Eds., American Society for Testing and Materials, 1980.

Session I. Building Insulation Materials

Tye, R. P., Ashare, E., Guyer, E. C., and Sharon, A. C.

"An Assessment of Thermal Insulation Materials for Building Applications"

Derbyshire, Brian

"Settled Density of Wood Fiber (Cellulose Based) Loose-Fill Thermal Insulation Using the CS-204 Method"

Shirtlife, C. J.

"Effect of Thickness on the Thermal Properties of Thick Specimens of Low-Density Thermal Insulation"

Shen, K. K.

"Some Problems Concerning the Corrosion Tests of Cellulose Insulating Material"

Tye, R. P. and Spinney, S. C.

"Effects of Moisture on Thermal Performance of Cellulose Insulation"

Session II. Applications

Tao, S. S., Bomberg, M., and Hamilton, J. J.

"Glass Fiber as Insulation and Drainage Layer on Exterior of Basement Walls"

Bomberg, Mark

"Some Performance Aspects of Glass Fiber Insulation on the Outside of Basement Walls"

Johnson, R. J. and Lee, O. G.

"The Energy-Efficient Residence: Research Results"

Carroll, W. L.

"Thermal Performance of Buildings and Building Envelope Systems: An Annotated Bibliography"

Vinieratos, E. R. and Verschoor, J. D.

"Influence of Insulation Deficiencies on Heat Loss in Walls and Ceilings"

Session III. Testing/Evaluation

Stainton, W. D.

"Interpretation of Building Thermograms"

Berlad, A. L., Tutu, N., Yeh, Y-J., Jaung, R., Krajewski, R., Hoppe, R., and Salzano, F. J.

"Air Intrusion Effects on the Performance of Permeable Insulation Systems"

Mumaw, J. R.

"Thermal Research Facility - A Large Calibrated Hot Box for Horizontal Building Elements"

Chaille, C. E. and Wahle, H. W.

"Evaluation of ASTM C 201 Apparatus for Measuring the Thermal Conductance of Ceramic Fiber Insulation"

Rohles, F. H., Jr., and Munson, D. M.

"Quantifying the Thermal Protection and Comfort Characteristics of Sleeping Bags"

De Ponte, F. and Maccato, W.

"The Calibration of Heat Flow Meters"

Hollingsworth, M., Jr.

"Experimental Determination of the Thickness Effect in Glass Fiber Building Insulation"

Bhattacharyya, R. K.

"Heat-Transfer Model for Fibrous Insulations"

McCaa, D. J.

"Implementation of a New Quality-Control Instrument for Thermal Measurements of a Low-Density Building Insulation"

Sherman, Morton

"Sampling Faced Foam Insulation Board for Heat Flow Meter Thermal Performance Testing"

Hedlin, C. P., Orr, H. W., and Tao, S. S.

"A Method for Determining the Thermal Resistances of Experimental Flat Roof Systems Using Heat Flow Meters"

Pelanne, C. M.

"Discussion on Experiments to Separate the "Effect of Thickness" from Systematic Equipment Errors in Thermal Transmission Measurements"

Klarsfeld, S. and Hyrien, M.

"Measurement of the Thermal Resistance of Thick Low-Density Mineral Fiber Insulation"

Session IV. Implementation

Siu, M. C. I.

"Fibrous Glass Board as a Standard Reference Material for Thermal Resistance Measurement Systems"

Bowles, A. M. and Shirtliffe, C. J.

"Development of a Canadian Standard for Urea Formaldehyde Thermal Wall Insulation"

Chinneck, J. W., Chandrashekar, M. and Hahn, C. K. G.

"Life-Cycle Energy Costs of Thermal Insulation"

Bomberg, Mark

"Problems in Predicting the Thermal Properties of Faced Polyurethane Foam"

Session V. Industrial Insulations

Sparks, L. L.

"Low-Temperature Properties of Expanded Polyurethane and Polystyrene"

Cunningham, C. R., Roni, J. P., and Oliver, L. T.

"Insulation for Cryogenic Service"

Barton, R. N.

"Thermal Transfer Calculation Methods for Insulated Underground Piping Systems and Comparisons of Results"

Eilhardt, B., Stoeckl, G., and Ziemek, G.

"Thermal Insulation for a Novel Flexible District Heating Pipe System"

Simmons, L. D. and Beckman, J. W.

"Optimum Economic Thickness of Pipe Insulation Accounting for Variable Heat-Transfer Parameters"

Shuman, E. C.

"Field Measurements of Heat Flow Through a Roof with Saturated Thermal Insulation and Covered with Black and White Granules"

Session VI. Research Program and Needs

Stenger, L. A.

"Research Needs in the Insulation Industry"

The following papers are contained in ASHRAE publication SP 28, Proceedings of the ASHRAE/DOE-ORNL Conference, Thermal Performance of the Exterior Envelopes of Buildings. Kissimmee, FL. Dec. 3-5, 1979.

Session I. Thermal Analysis of Building Envelopes

Twenty, John E. and McQueen, Lee V.

"A Calculator Programming Technique for Estimating Thermally Optimum Wall Construction"

Hans, Gunard E.

"Proposed Streamlined Residential Heating Energy Budget Analysis by a Variable Temperature Design Method"

Agnoletto, L., Brunello, P., and Zecchin, R.

"Simple Methods for Predicting Thermal Behaviour and Energy Consumption in Buildings"

Dexter, Michael E.

"Energy Conservation Design Guidelines for Including Mass and Insulation in Building Walls"

Cuplinskas, E. L.

"A Rational Manual Method for Determination of Space Temperature Swing Due to Solar Gains"

Mueller, Gerhard R. and Pedreyra, Donald C.

"Computer-Aided Energy Savings Redesign of the Exterior Envelope of an Apartment Building"

Session II. Air Infiltration

Keast, David N. and Pei, Hsien-Sheng

"The Use of Sound to Locate Infiltration Openings in Buildings"

Sherman, M. H., Grimsrud, D. T., and Sonderegger, R. C.

"The Low Pressure Leakage Function of a Building"

Blomsterberg, Ake K., Sherman, M. H., and Grimsrud, D. T.

"A Model Correlating Air Tightness and Air Infiltration in Houses"

Stewart, Michael B., Jacob, Thomas R., and Winston, Jeffrey G.

"Analysis of Infiltration by Tracer Gas Technique, Pressurization Tests, and Infrared Scans"

Weidt, J. L., Weidt, J., and Selkowitz, S.

"Field Air Leakage of Newly Installed Residential Windows"

Hunt, C. M. and Treado, S. J.

"Air Exchange Measurements in a High-Rise Office Building"

Grot, Richard A. and Clark, Roy E.

"Air Leakage Characteristics and Weatherization Techniques for Low-Income Housing"

Collins, John O., Jr.

"Air Leakage Measurement and Reduction Techniques on Electrically Heated Homes"

Session III. Test Procedure Development

Fiorato, A. E.

"Laboratory Tests of Thermal Performance of Exterior Walls"

Rucker, James L. and Mumaw, John R.

"Calibration Procedures and Results for a Large Calibrated Hot Box"

Janssen, J. E. and Pearman, A. N.

"Comparison of BTR Measurements"

Brown, W. C. and Schuyler, G. D.

"A Calorimeter for Measuring Heat Flow Through Walls"

Bryson, J. O.

"Thermal Insulation Laboratory Accreditation Program"

Rennex, Brian G.

"Thermal Parameters as a Function of Thickness for Combined Radiation and Conduction Heat Transfer in Low-Density Insulation"

Perrine, E. L., Linehan, P. W., Howanski, J. W., and Shu, L. S.

"The Design and Construction of a Calibrated/Guarded Hot Box Facility"

Achenbach, Paul R.

"Design of a Calibrated Hot-Box for Measuring the Heat, Air, and Moisture Transfer of Composite Building Walls"

Session IV. Windows and Skylights

Burkhardt, William C.

"Effective U-Values for Skylights"

Klems, J. H.

"A Calibrated Hotbox for Testing Window Systems -- Construction, Calibration, and Measurements on Prototype High-Performance Windows"

Berlad, A. L.

"Dynamic Interpane Systems for Multipane Solar Collector Windows"

Brakefield, Charles and Hein, Dale E.

"Thermal Performance of Vinyl Covered Fiberglass Solar Screens"

Gillette, Gary

"The Energy Potential of Daylighting in a Classroom"

Cumali, Zulfikar O., Sezgen, A. Osman, Sullivan, Robert, and Hossli, Robert [illegible].

"Dynamic Response of Conventional Residential Construction to Solar Radiation Through Windows During the Heating Season"

Selkowitz, S. E. and Bazjanac, V.

"Thermal Performance of Managed Window Systems"

Session V. Modeling Building Envelope Systems

Cumali, Zulfikar O., Sezgen, A. Osman, Sullivan, Robert and Kammerud, Ronald C.

"Extension of Methods Used in Analyzing Building Thermal Loads"

Shu, L. S., Fiorato, A. E., and Howanski, J. W.

"Heat Transmission Coefficients of Concrete Block Walls with Core Insulation"

Wilkes, Kenneth E.

"Modeling of Residential Attics"

Wang, F. S.

"Mathematical Modeling and Computer Simulation of Insulation Systems in Below Grade Applications"

Godfrey, Richard D., Wilkes, Kenneth E., and Lavine, Adrienne G.

"A Technical Review of the "M" Factor Concept"

Session VIA. Economic Optimization

Blancett, Robert S.

"A Method of Rating the Thermal Efficiency of a Home Envelope Relative to an Energy Efficient Standard"

Ercules, Anna Maria

"A Statistical Decision Method for Economic Evaluation of the Exterior Envelopes of Buildings"

Petersen, Stephen R.

"Economic Considerations in Insulating Masonry and Wood-Frame Walls of Single-Family Housing"

Yuill, Grenville K. and Proskiw, Gary

"Economic Comparison and Thermal Performance of Several Different Insulated Wall Constructions"

Session VIB. Standards For Energy Conservation

Levine, Mark D., Goldstein, David B., Lokmanhekim, Metin, and Rosenfeld, Arthur H.

"Evaluation of Residential Building Energy Performance Standards"

Hart, Michael N., Jones, Jerold W., and Bowen, William

"Building Envelope Thermal Performance Standard, The Texas Approach"

Miller, David R.

"Energy Conservation Opportunities in the Building Envelope Through the Enforcement of a Mandatory Energy Code"

Albrecht, Raymond J. and Jones, J. Delaine

"Implementation of Building Energy Standards in New York State"

Session VII. Innovative Construction Practices

Roseme, G. D., Berk, J. V., Boegel, M. L., Halsey, H. I., Hollowell, C. D., Rosenfeld, A. H., and Turiel, I.

"Residential Ventilation with Heat Recovery: Improving Indoor Air Quality and Saving Energy"

Sodergren, David and Dahl, Carl E.

"Ventilated Windows as Solar Collectors, Air as Thermal Carrying Media and Thermal Storage in Hollow Core Concrete Floor Slabs"

Waite, Harold J. and Snyder, Marvin K.

"Thermal Spacers Improve Roof Insulation Performance of Metal Buildings"

Larsson, Lars-Erik

"Sandwich Panels with Foamed Polyurethane Insulation"

Johnson, Ralph J.

"The Performance of an Energy Efficient Residence vs a Conventional Comparison Home"

Wilkes, Kenneth E.

"Thermophysical Properties Data Base Activities at Owens-Corning Fiberglas"

Meixel, George D. Jr., Shipp, Paul H. and Ramsey, James W.

"Analysis and Measurement of the Thermal Behavior of the Walls and Surrounding Soil for a Large Underground Building"

Session VIII. Field Measurement of Energy Use

Carroll, William L.

"Annual Heating and Cooling Requirements and Design Day Performance for a Residential Model in Six Climates: A Comparison of NBSLD, BLAST 2, and DOE-2.1"

Condon, P. E., Carroll, W. L., and Sonderegger, R. C.

"A New Measurement Strategy for In-Situ Testing of Wall Thermal Performance"

Condon, P. E. and Carroll, William L.

"Measurement and Analysis of In-Situ Dynamic Thermal Performance of Building Envelopes Using Heat Flow Meter Arrays"

Gujral, Parambir S., Clark, Raymond J., and Burch, Douglas M.

"Transient Thermal Response of an Intermittently Cooled Massive Building"

Mill, Peter A. D.

"The Principles of Building Science and Thermography Needed to Diagnose the Performance of Building Enclosures"

Richtmyer, T. E., May, W. B., Hunt, C. M., and Hill, J. E.

"Thermal Performance of the Norris Cotton Federal Building in Manchester, New Hampshire"

Session IX. Moisture and Condensation

Tsongas, George A., Odell, F. Glen, and Thompson, James C.

"A Field Study of Moisture Damage in Walls Insulated Without a Vapor Barrier"

Knab, Lawrence I., Jenkins, David R., and Mathey, Robert G.

"The Effect of Moisture on the Thermal Conductance of Roofing Systems"

Wang, F. S.

"Comparative Studies of Vapor Condensation Potentials in Wood Framed Walls"

Hart, G. H. and Carlson, Tage C. G.

"Moisture Condensation Above Insulated Suspended Ceilings - Experimental Results"

Homma, Hiroshi and Guy, Richard W.

"Ventilation of Back Space of Building Enclosure Siding for Solar Heat Gain Reduction"

Carlson, Tage C. G. and Luce, Ross G.

"An Experimental Investigation of the Effect of Added Insulation on Water Vapor Condensation in Mobile Home Roof Cavities"

Stewart, Michael B.

"Annual Cycle Moisture Analysis"

Session X. Retrofitting For Energy Conservation

Rossiter, Walter J. Jr., Weidt, John L., and Saxler, Robert J.

"A Field Survey of the Performance Properties of Insulation Used to Retrofit Cavity Walls of Residences"

Carlson, Tage C. G. and Bemis, Richard S.

"Recommended Insulation Retrofit Techniques for Mobile Homes with a Simple Payback Analysis"

Klems, J. H. and Selkowitz, S. E.

"The Mobile Window Thermal Test Facility (MoWitt)"

Fogleman, Sam R.

"Energy Conservation Study for Puerto Rico Aqueduct and Sewer Authority Building"

Sherman, Morton

"Aged Thermal Resistance (R Value) of Foil-Faced Polyisocyanurate Foam Thermal Insulation Board"

Tye, R. P. Desjarlais, A. O., Bourne, J. G. and Spinney, S. C.

"The Effective Thermal Performance of a.. Insulated Standard Stud Wall Containing Air Gaps"

Deal, Peter N.

"The Direct Approach to Retrofitting - Protected Membrane Roofing

Meixel, George D., Shipp, Paul H., and Bligh, Thomas P.

"The Impact of Insulation Placement on the Seasonal Heat Loss through Basement and Earth-Sheltered Walls"

The following papers were presented at the DOE-ORNL/ASTM Conference on Thermal Insulation, Materials, and Systems for Energy Conservation in the '80s, Clearwater Beach, Florida, December 8-11, 1981.

Session I. Federal and State Energy Programs

Achenback, P. R. and Freeman, E. C., Jr.

"The National Program for the Thermal Performance of Building Envelope Systems and Materials"

Tartaglione, L. C. and Macartney, D. C.

"Innovative Building Insulation Systems - Code Acceptance, Standards, and Laboratory Accreditation Practices In Massachusetts"

Feinbaum, R. and Ruby, E.

"Implementation and Enforcement of Residential Energy Conservation Standards - The California Experience"

Kempton, W., Gladhart, P. and Keefe, D.

"Home Insulation: The User's View"

Session II. Insulation Safety and Health Issues

Lancer, P. M.

"Tackling Safety and Health Issues on the Use of Thermal Insulation"

Wegner, T. H. and Holmes, C. A.

"Efficient Application of Borbon Fire Retardant to Cellulosic Loose - Fill Insulation"

Dudney, C. S., Hinkle, N. F. and Becker, J. M.

"Fungi In Loose-Fill Attic Insulation"

Sheppard, K. G. and Weil, R.

"Corrosion Testing of Thermal Insulation and Its Relationship To Some Field Conditions"

Sheckler, C. L.

"Health Aspects Of Man-made Vitreous Fiber Insulations"

Lord, D.

"Indoor Air Pollution, Energy Conservation, and The Use Of Building Materials"

Session III. Economic Evaluation

Wong, Y. C. and Sauer, H. J. Jr.

"Total Energy Costs of Building Construction and Operation"

Albrecht, R. J.

"Study Periods and Energy Price Escalation Rates - Important Factors In The Economic Evaluation Of Insulation Systems"

Robinson, D. A.

"Life-Cycle Cost Economic Optimization Of Insulation, Infiltration, and Solar Aperture in Energy Efficient Houses"

Genter, R. E.

"Economic Performance Factors In Adding Insulation To Existing Facilities"

Woller, B. E.

"Engineering and Economic Evaluation Of A Commercial Roof/Insulation Retrofit"

Session IV. Thermal Testing Apparatus

Howanski, J. W., Derderian, G. D., Shu, L. S., and Orlandi, R. D.

"A Start-Up Procedure and Its Application For The Grace Hot-Box Facility"

Goss, W. P. and Olpak, A.

"Design and Calibration Of A Rotatable Thermal Test Facility"

Lavine, A. G., Rucker, J. L. and Wilkes, K. E.

"Flanking Loss Calibration For A Calibrated Hot Box"

Powell, F. J. and Bales, E. L.

"Design Of Round-Robin Tests Using Guarded/Calibrated Hot Boxes - Guarded Hot Plates/Heat Flow Meters"

Messmer, D. A.

"The Design and Construction Of A Full-Thickness Guarded Hot-Plate Test System"

Bomberg, M. and Solvason, K. R.

"New Designs Of Heat Flow Meter Apparatus For Testing Low-Density Thermal Insulations"

Wright, R. E., Jr., Kantsios, A. G., and Henley, W. C.

"The Effect Of Mounting On The Performance Of Surface Heat Flowmeters Used To Evaluate Building Heat Losses"

Buckley, R. E.

"The R Meter - What It Is And How It Is Used"

Session V. Field and Laboratory Testing Of Building Components

Grot, R. A.

"The Thermographic Inspection Of Exterior Wall Insulation Retrofits"

Flanders, S. N. and Marshall, S. J.

"In Situ Building R-Value Measurement"

Infante, L. J., Aller, P. A., and Fay, R. E.

"Insulation Case Histories, Including Experiences And Problems In The Field Application of Loose Fill"

Mill, P. A. D. and Kaplan, A. G.

"Thermographic And Thermologic Diagnosis Of Conventional North American Timber Frame Construction"

Adams, J. W., Sherman, M. H., and Sonderegger, R. C.

"Dynamic Measurement of Wall Thermal Performance"

Fiorato, A. E. and Julien, J. T.

"On Testing Thermal Performance Of Walls By The Calibrated Hot-Box Method"

Contreras, A. G. and Palfey, A. J.

"Measured Thermal Resistances of Reflective and Nonreflective Air Spaces In Masonry Walls"

Miller, R. G. and Sherman, M.

"Thermal Performance Of Insulated Metal Building Roof Deck Constructions"

Larson, D. C. and Corneliussen, R. D.

"Thermal Testing Of Roof Systems"

Deacon, P. C.

"Glass Fibre As A Draining Insulation System For The Exterior Of Basement Walls"

Ovstaas, G., Smith, S., Strzepek, W., and Titley, G.

"Thermal Performance Of Various Insulations In Below-Earth-Grade Perimeter Application"

Shipp, P. H. and Broderick, T. B.

"Comparison Of Annual Heating Loads For Various Basement Wall Insulation Strategies Using Transient And Steady State Models"

Session VI. Convection and Air Infiltration Effects

Taylor, B. and Phillips, A. J.

"The Thermal Transmittance and Conductance Of Roof Constructions Incorporating Fibrous Insulation"

Scanlan, T. F., Bayne, C. K. and Johnson, D. R.

"Investigation Of Attic Insulation Effectiveness Using Actual Energy Consumption Data"

de Marne, H.

"New and Retrofit Insulation Of Single-Member Cathedral Ceiling, A-Frame, and Flat Residential Roofs"

Yarbrough, D. W. and Toor, I. A.

"The Effect Of Air Movement On The Thermal Resistance Of Loose-Fill Thermal Insulations"

Schuyler, G. D. and Solvason, K. R.

"The Effectiveness Of Insulation In Wall Systems"

Henning, G. N.

"Energy Conservation With Air Infiltration Barriers"

Session VII. Moisture Effects

Langlais, C., Klarsfeld, S. and Eyrien, M.

"Influence Of The Moisture Distribution Inside A Fibrous Insulating Material On Its Thermal Resistance As A Function Of The Applied Thermal Gradient"

Thomas, W. C., Bal, G. P. and Onega, R. J.

"Heat and Moisture Transfer In A Glass Fiber Roof Insulating Material"

Hedlin, C. P.

"Effect Of Moisture On Thermal Resistance Of Some Insulations"

Tobiasson, W., Korhonen, C., Coutermarsh, B., and Greatorex, S.

"Can Wet Roof Insulation Be Dried Out?"

Stewart, M. B.

"Moisture Control In Retrofit Commercial Roof Insulations"

Kelso, R. M.

"Water Vapor Flow And High Thermal Resistance Insulation Systems For Metal Buildings"

Session VIII. Materials Behavior

Rossiter, W. J., Jr., Ballard, D. B. and Sleater, G. A.

"Elevated Temperature And Humidity Effects On Urea-Formaldehyde Foam Insulations Observed By Scanning Electron Microscopy"

Glicksman, L. R. and Valenzuela, J.

"Thermal Resistance And Aging Of Rigid Urethane Foam Insulation"

Yarbrough, D. W., Wright, J. H., McElroy, D. L. and Scanlan, T. F.

"Settling Of Loose-Fill Insulations Due To Vibration"

Singleton, E. F.

"Test Results As Related To Reflective Insulation Systems"

Low, N. M. P.

"Glass-Mica Composite: A New Thermal Insulating Material For Building Applications"

Session IX. Mechanical, Power, and Process Systems Insulation

Tye, R. P. and Desjarlais, A. O.

"Factors Influencing The Thermal Performance Of Thermal Insulations For Industrial Applications"

Marks, J. B.

"The Protection Of Thermal And Cryogenic Insulating Materials By The Use Of Metal Jacketing And Mastic Coatings"

Saatdjian, E., Demars, Y., Klarsfeld, S., and Buck, Y.

"Effects Of The Binder Decomposition On High-Temperature Mineral Fiber Thermal Insulation"

Sullivan, J. M., Jr.

"The Thermal Performance Of Insulated Pipe Systems"

McAllister, J. D. and Biggers, R. K.

"A Lump Sum, Unit-Price Bid Proposal Evaluation Method"

Kusuda, T. and Ellis, W. M.

"Performance Of Thermal Insulation In Conduit-Type Underground Heat Distribution Systems"

Rogus, B. J.

"Reduction Of Heat Stress In Naval Ships Through Improved Insulation Installations"

Allmon, B. A., Rausch, D. A., and Wahle, H. W.

"Total System Heat Loss Measurements"

Charter, K. F.

"Finite Difference Thermal Analysis Of An Insulation System On A Precipitator Building In A Power Plant"

Appendix A

Material Specifications—ASTM Insulation Standards

ASTM STANDARDS
(SORTED BY ASTM CODE AND CATEGORY)

ASTM NO.	TITLE

ADHESION OF THERMAL INSULATING CEMENTS

C 353	TEST FOR ADHESION OF DRIED THERMAL INSULATING OR FINISHING CEMENT
C 383	TEST FOR WET ADHESION OF THERMAL INSULATING CEMENTS TO METAL

AIR LEAKAGE

E 283	TEST FOR RATE OF AIR LEAKAGE THROUGH EXTERIOR WINDOWS, CURTAIN WALLS, AND DOORS
E 741	PRACTICE FOR MEASURING AIR LEAKAGE RATE BY THE TRACER DILUTION METHOD

BREAKING LOAD

C 203	TEST FOR BREAKING LOAD AND CALCULATED FLEXURAL STRENGTH OF PREFORMED BLOCK-TYPE THERMAL INSULATION
C 446	TEST FOR BREAKING LOAD AND CALCULATED MODULUS OF RUPTURE OF PREFORMED INSULATION FOR PIPES
D 781	TEST FOR PUNCTURE AND STIFFNESS OF PAPERBOARD,AND CORRUGATED AND SOLID FIBERBOARD
D 790	TEST FOR FLEXURAL PROPERTIES OF PLASTICS AND ELECTRICAL INSULATING MATERIALS
D 828	TEST FOR TENSILE BREAKING STRENGTH OF PAPER AND PAPERBOARD

COMPRESSIVE STRENGTH

C 165	RECOMMENDED PRACTICE FOR MEASURING COMPRESSIVE PROPERTIES OF THERMAL INSULATION
C 354	TEST FOR COMPRESSIVE STRENGTH OR THERMAL INSULATING OR FINISHING CEMENT
C 495	TEST FOR COMPRESSIVE STRENGTH OF LIGHTWEIGHT INSULATING CONCRETE
D 1621	TEST FOR COMPRESSIVE PROPERTIES OF RIGID CELLULAR PLASTICS

CORROSION

C 464	TEST FOR CORROSION EFFECT OF THERMAL INSULATING CEMENTS ON BASE METAL
C 590	TEST FOR ACTION ON SUBSTRATES BY COATINGS, ADHESIVES, AND JOINT SEALANTS USED ON OR WITH THERMAL INSULATION
C 692	EVUALATING THE INFLUENCE OF WICKING-TYPE THERMAL INSULATIONS ON THE STRESS CORROSION CRACKING TENDENCY OF AUSTENITIC STAINLESS STEEL
D 1654	EVALUATION OF PAINTED OR COATED SPECIMENS SUBJECTED TO CORROSIVE ENVIRONMENTS

DENSITY

C 167	TESTS FOR THICKNESS AND DENSITY OF BLANKET OR BATT TYPE THERMAL INSULATING MATERIALS

ASTM STANDARDS
(SORTED BY ASTM CODE AND CATEGORY)

ASTM NO.	TITLE

DENSITY (Cont.)

C	302	TEST FOR DENSITY OF PREFORMED PIPE-COVERING-TYPE INSULATION
C	303	TEST FOR DENSITY OF PREFORMED BLOCK-TYPE THERMAL INSULATION
C	519	TEST FOR DENSITY OF FIBROUS LOOSE FILL BUILDING INSULATIONS
C	520	TEST FOR DENSITY OF GRANULAR LOOSE FILL INSULATIONS
D	70	TEST FOR SPECIFIC GRAVITY OF SEMI SOLID BITUMINOUS MATERIALS
D	71	TEST FOR SPECIFIC GRAVITY OF SOLID PITCH AND ASPHALT-DISPLACEMENT METHOD
D	1622	TEST FOR APPARENT DENSITY OF RIGID CELLULAR PLASTICS
E	605	TESTS FOR THICKNESS AND DENSITY OF SPRAYED FIRE-RESISTIVE MATERIAL APPLIED TO STRUCTURAL MEMBERS

DIMENSIONAL STABILITY

C	167	TESTS FOR THICKNESS AND DENSITY OF BLANKET OR BATT TYPE THERMAL INSULATING MATERIALS
C	548	TEST FOR DIMENSIONAL STABILITY OF LOW TEMPERATURE THERMAL BLOCK AND PIPE INSULATION
C	550	TEST FOR TRUENESS AND SQUARENESS OF BLOCK THERMAL INSULATION
C	733	TEST FOR VOLUME SHRINKAGE OF LATEX SEALING COMPOUNDS
D	1204	TEST FOR LINEAR DIMENSIONAL CHANGES OF NONRIGID THERMOPLASTIC SHEETING OR FILM AT ELEVATED TEMPERATURES
D	2126	TEST FOR RESPONSE OF RIGID CELLULAR PLASTICS TO THERMAL AND HUMID AGING
D	2453	TEST FOR SHRINKAGE AND TENACITY OF OIL- AND RESIN-BASE CAULKING COMPOUNDS
E	605	TESTS FOR THICKNESS AND DENSITY OF SPRAYED FIRE-RESISTIVE MATERIAL APPLIED TO STRUCTURAL MEMBERS

DROPPING (Resistance to)

C	487	TEST FOR RESISTANCE TO DROPPING OF PREFORMED BLOCK-TYPE THERMAL INSULATION

EMMITANCE

C	835	TEST FOR TOTAL HEMISPHERICAL EMITTANCE OF SURFACES FROM 20 TO 1400 C (68 TO 2550 F)

EXTERNAL LOADS (Resistance to)

C	589	TEST FOR APPARENT IMPACT STRENGTH OF PREFORMED BLOCK-TYPE INSULATING MATERIALS
C	686	TEST FOR PARTING STRENGTH OF MINERAL FIBER BATT- AND BLANKET-TYPE INSULATION
C	854	TEST FOR RESISTANCE TO EXTERNAL LOADS ON METAL REFLECTIVE PIPE INSULATION
E	72	CONDUCTING STRENGTH TESTS OF PANELS FOR BUILDING CONSTRUCTION
E	518	TEST FOR FLEXURAL BOND STRENGTH OF MASONRY
E	519	TEST FOR DIAGONAL TENSION (SHEAR) IN MASONRY ASSEMBLAGES

ASTM STANDARDS
(SORTED BY ASTM CODE AND CATEGORY)

ASTM NO.	TITLE

EXTERNAL LOADS (Cont.)

E	529	FLEXURAL TESTS ON BEAMS AND GIRDERS FOR BUILDING CONSTRUCTION
E	546	STATIC LOAD TEST FOR SHEAR RESISTANCE OF FRAMED WALLS FOR BUILDINGS
E	575	REC. PRACTICE FOR REPORTING DATA FROM STRUCTURAL TESTS OF BUILDING QCONSTRUCTIONS ELEMENTS, CONNECTIONS, AND ASSEMBLIES

FLAMMABILITY

D	56	TEST FOR FLASH POINT BY TAG CLOSED TESTER
D	92	TEST FOR FLASH AND FIRE POINTS BY CLEVELAND OPEN CUP TESTER
D	93	TEST FOR FLASH POINT BY PENSKY-MARTENS CLOSED TESTER
D	568	TEST FOR RATE OF BURNING AND OR EXTENT AND TIME OF BURNING OF FLEXIBLE PLASTICS IN A VERTICAL POSITION
D	635	TEST FOR RATE OF BURNING AND OR EXTENT AND TIME OF BURNING OF SELF-SUPPORTING PLASTICS IN A HORIZONTAL POSITION
D	777	TEST FOR FLAMMABILITY OF TREATED PAPER AND CARDBOARD
D	1310	TEST FOR FLASH POINT OF LIQUIDS BY TAG OPEN CUP APPARATUS
D	1525	TEST FOR VICAT SOFTENING TEMPERATURE OF PLASTICS
D	1929	TEST FOR IGNITION PROPERTIES OF PLASTICS
D	2015	TEST FOR GROSS CALORIFIC VALUE OF SOLID FUEL BY THE ADIABATIC BOMB CALORIMETER
D	2020	TEST FOR MILDEW (FUNGUS) RESISTANCE OF PAPER AND PAPERBOARD
D	2582	TEST FOR PUNCTURE-PROPAGATION TEAR RESISTANCE OF PLASTIC FILMS AND THIN SHEETING
D	2843	TEST FOR DENSITY OF SMOKE FROM THE BURNING OR DECOMPOSITION OF PLASTICS
D	2863	TEST FOR FLAMMABILITY OF PLASTICS USING OXYGEN INDEX METHOD
D	3014	TEST FOR FLAME HEIGHT, TIME OF BURNING, AND LOSS OF WEIGHT OF RIGID CELLULAR PLASTICS IN A VERTICAL POSITION
D	3211	TEST FOR RELATIVE DENSITY OF BLACK SMOKE (RINGELMANN METHOD)
E	69	TEST FOR COMBUSTION PROPERTIES OF TREATED WOOD BY THE FIRE-TUBE APPARATUS
E	84	TEST FOR SURFACE BURNING CHARACTERISTICS OF BUILDING MATERIALS
E	136	TEST FOR BEHAVIOR OF MATERIALS IN A VERTICAL TUBE FURNACE AT 750 C
E	162	TEST FOR SURFACE FLAMMABILITY OF MATERIALS USING A RADIANT HEAT ENERGY SOURCE
E	176	DEF. OF TERMS RELATING TO FIRE TESTS OF BUILDING CONSTRUCTION AND MATERIALS
E	286	TEST FOR SURFACE FLAMMABILITY OF BUILDING MATERIALS USING AN 8-FT(2.44 M) TUNNEL FURNACE
E	535	REC. PRACTICE FOR PREPARATION OF FIRE TEST STANDARDS
E	662	TEST FOR SPECIFIC OPTICAL DENSITY OF SMOKE GENERATED BY SOLID MATERIALS

GLAZINGS

C	669	SPEC. FOR GLAZING COMPOUNDS FOR BACK BEDDING AND FACE GLAZING OF METAL SASH
C	681	TEST FOR VOLATILITY OF OIL- AND RESIN-BASED, KNIFE-GRADE, CHANNEL GLAZING COMPOUNDS
C	713	TEST FOR SLUMP OF AN OIL-BASE KNIFE-GRADE CHANNEL GLAZING COMPOUND
C	741	TEST FOR ACCELERATED AGING OF WOOD SASH FACE GLAZING COMPOUND
C	742	TEST FOR DEGREE OF SET FOR WOOD SASH GLAZING COMPOUND

ASTM STANDARDS
(SORTED BY ASTM CODE AND CATEGORY)

ASTM NO.	TITLE

GLAZINGS (Cont.)

C 797	REC. PRACTICES AND TERMINOLOGY FOR USE OF OIL- AND RESIN-BASED PUTTY AND GLAZING COMPOUNDS
D 2249	PREDICTING THE EFFECT OF WEATHERING OF FACE GLAZING AND BEDDING COMPOUNDS ON METAL SASH
D 2376	TEST FOR SLUMP OF FACE GLAZING AND BEDDING COMPOUNDS ON METAL SASH
D 2451	TEST FOR DEGREE OF SET FOR GLAZING COMPOUNDS ON METAL SASH

HEAT FLUX

C 745	TEST FOR HEAT FLUX THROUGH EVACUATED INSULATIONS USING A GUARDED FLAT PLATE BOILOFF CALORIMETER

HANDLING

C 929	PRACTICE FOR HANDLING, TRANSPORTING, SHIPPING, STORAGE, RECEIVING, AND APPLICATION OF THERMAL INSULATION MATERIALS TO BE USED OVER AUSTENITIC STAINLESS STEEL

HARDNESS

C 569	TEST FOR INDENTATION HARDNESS OF PREFORMED THERMAL INSULATIONS
C 661	TEST FOR INDENTATION HARDNESS OF ELASTOMERIC-TYPE SEALANTS BY MEANS OF A DUROMETER

HOT SURFACE PERFORMANCE

C 411	TEST FOR HOT-SURFACE PERFORMANCE OF HIGH-TEMPERATURE THERMAL INSULATION

INSULATING CEMENT

C 163	MIXING THERMAL INSULATING CEMENT SAMPLES
C 195	SPEC. FOR MINERAL FIBER THERMAL INSULATING CEMENT
C 166	TEST FOR COVERING CAPACITY AND VOLUME CHANGE UPON DRYING OF THERMAL INSULATING CEMENT
C 353	TEST FOR ADHESION OF DRIED THERMAL INSULATING OR FINISHING CEMENT
C 354	TEST FOR COMPRESSIVE STRENGTH OR THERMAL INSULATING OR FINISHING CEMENT
C 383	TEST FOR WET ADHESION OF THERMAL INSULATING CEMENTS TO METAL
C 405	TESTS FOR CONSISTENCY OF WET-MIXED THERMAL INSULATING CEMENT
C 449	SPEC. FOR MINERAL FIBER HYDRAULIC-SETTING AND FINISHING CEMENT
C 464	TEST FOR CORROSION EFFECT OF THERMAL INSULATING CEMENTS ON BASE METAL

ASTM STANDARDS
(SORTED BY ASTM CODE AND CATEGORY)

ASTM NO.		TITLE

INSULATING MATERIALS

C	168	DEF. OF TERMS RELATING TO THERMAL INSULATING MATERIALS
C	195	SPEC. FOR MINERAL FIBER THERMAL INSULATING CEMENT
C	196	SPEC. FOR EXPANDED OR EXFOLIATED VERMICULITE THERMAL INSULATING CEMENT
C	208	SPEC. FOR INSULATING BOARD (CELLULOSIC FIBER),STRUCTURAL AND DECORATIVE
C	240	TESTING CELLULAR GLASS INSULATING BLOCK
C	390	CRITERIA FOR PREFORMED THERMAL INSULATION LOTS
C	450	REC. PRACTICE FOR PREFABRICATION AND FIELD FABRICATION OF THERMAL INSULATING FITTING COVERS FOR NPS PIPING, VESSEL LAGGING, AND DISHED HEAD SEGMENTS
C	516	SPEC. FOR VERMICULITE LOOSE FILL INSULATION
C	517	SPEC. FOR DIATOMACEOUS EARTH BLOCK AND PIPE THERMAL INSULATION
C	532	SPEC. FOR STRUCTURAL INSULATING FORMBOARD (CELLULOSIC FIBER)
C	533	SPEC. FOR CALCIUM SILICATE BLOCK AND PIPE THERMAL INSULATION
C	534	SPEC. FOR PREFORMED FLEXIBLE ELASTOMERIC CELLULAR THERMAL INSULATION IN SHEET AND TUBULAR FORM
C	547	SPEC. FOR MINERAL FIBER PREFORMED PIPE INSULATION
C	549	SPEC. FOR PERLITE LOOSE FILL INSULATION
C	552	SPEC. FOR CELLULAR GLASS BLOCK AND PIPE THERMAL INSULATION
C	553	SPEC. FOR MINERAL FIBER BLANKET AND FELT INSULATION (INDUSTRIAL TYPE)
C	578	SPEC. FOR PREFORMED,BLOCK-TYPE CELLULAR POLYSTYRENE THERMAL INSULATION
C	585	REC. PRACTICE FOR INNER AND OUTER DIAMETERS OF RIGID THERMAL INSULATION FOR NOMINAL SIZES OF PIPE AND TUBING (NPS SYSTEM)
C	591	SPEC. FOR RIGID PREFORMED CELLULAR URETHANE THERMAL INSULATION
C	592	SPEC. FOR MINERAL FIBER BLANKET INSULATION AND BLANKET-TYPE PIPE INSULATION (METAL MESH COVERED)(INDUSTRIAL TYPE)
C	610	SPEC. FOR EXPANDED PERLITE BLOCK AND PIPE THERMAL INSULATION
C	612	SPEC. FOR MINERAL FIBER BLOCK AND BOARD THERMAL INSULATION
C	640	SPEC. FOR CORKBOARD AND CORK PIPE INSULATION FOR LOW TEMP THERMAL INSULATION
C	656	SPEC. FOR STRUCTURAL INSULATING BOARD, CALCIUM SILICATE
C	665	SPEC. FOR MINERAL FIBER BLANKET THERMAL INSULATION FOR WOOD FRAME AND LIGHT CONSTRUCTION
C	667	REC. PRACTICE FOR PREFABRICATED REFLECTIVE INSULATION SYSTEMS FOR EQUIPMENT AND PIPE OPERATING AT TEMPERATURES ABOVE AMBIENT AIR
C	720	SPEC. FOR SPRAY-APPLIED FIBROUS THERMAL INSULATION FOR ELEVATED TEMPERATURE
C	726	SPEC. FOR MINERAL FIBER ROOF INSULATION BOARD
C	727	REC. PRACTICE FOR USE OF REFLECTIVE INSULATION IN BUILDING CONSTRUCTIONS
C	728	SPEC. FOR PERLITE THERMAL INSULATION BOARD
C	739	SPEC. FOR CELLULOSIC FIBER (WOOD BASE) LOOSE FILL THERMAL INSULATION
C	740	REC. PRACTICE FOR EVACUATED REFLECTIVE INSULATION IN CRYOGENIC SERVICE
C	762	REC. PRACTICE FOR APPLICATION OF SPRAY-APPLIED FIBROUS THERMAL INSULATION
C	764	SPEC. FOR MINERAL FIBER LOOSE FILL THERMAL INSULATION
C	795	SPEC. FOR WICKING-TYPE THERMAL INSULATION FOR USE OVER AUSTENITIC STAINLESS STEEL
C	800	SPEC. FOR GLASS FIBER BLANKET INSULATION (AIRCRAFT TYPE)
C	846	SPEC. FOR APPLICATION OF STRUCTURAL INSULATING BOARD (FIBERBOARD) SHEATHING
C	870	REC. PRACTICE FOR CONDITIONING OF THERMAL INSULATING MATERIALS
C	871	CHEMICAL ANALYSIS OF THERMAL INSULATION MATERIALS FOR LEACHABLE CHLORIDE, SILICATE AND SODIUM IONS
C	892	SPEC. FOR HIGH-TEMPERATURE FIBER BLANKET THERMAL INSULATION

ASTM STANDARDS
(SORTED BY ASTM CODE AND CATEGORY)

ASTM NO.		TITLE

INSULATING MATERIALS (Cont.)

C	916	SPEC. FOR ADHESIVES FOR DUCT THERMAL INSULATION
E	546	TEST FOR FROST POINT OF SEALED INSULATING GLASS UNITS
E	576	TEST FOR DEW/FROST POINT OF SEALED INSULATING GLASS UNITS IN VERTICAL POSITION
E	737	PRACTICE FOR INSTALLATION OF STORM WINDOWS, REPLACEMENT WINDOWS, MULTI-GLAZING STORM DOORS, AND REPLACEMENT DOORS

JACKETING

C	921	PRACTICE FOR DETERMINING THE PROPERTIES OF JACKETING MATERIALS FOR THERMAL INSULATION

MAXIMUM USE TEMPERATURE

C	447	RECOMMENDED PRACTICE FOR ESTIMATING THE MAXIMUM USE TEMPERATURE OF PREFORMED HOMOGENOUS THERMAL INSULATIONS

MECHANICAL STABILITY

C	421	TEST FOR TUMBLING FRIABILITY OF PREFORMED BLOCK-TYPE THERMAL INSULATION

MOISTURE

C	240	TESTING CELLULAR GLASS INSULATING BLOCK
D	529	REC. PRACTICE FOR ACCELERATED WEATHERING TEST OF BITUMINOUS MATERIALS
D	2842	TEST FOR WATER ABSORPTION OF RIGID CELLULAR PLASTICS
E	96	TESTS FOR WATER VAPOR TRANSMISSION OF MATERIALS IN SHEET FORM

SAFETY PROPERTIES

C	930	CLASSIFICATION FOR IDENTIFYING AND CATEGORIZING POTENTIAL HEALTH AND SAFETY HAZARDS OF THERMAL INSULATION MATERIALS AND ACCESSORIES DURING INSTALLATION
D	56	TEST FOR FLASH POINT BY TAG CLOSED TESTER
D	92	TEST FOR FLASH AND FIRE POINTS BY CLEVELAND OPEN CUP
D	93	TEST FOR FLASH POINT BY PENSKY-MARTENS CLOSED TESTER
D	127	TEST FOR DROP MELTING POINT OF PETROLEUM WAX INCLUDING PETROLATUM
D	568	TEST FOR RATE OF BURNING AND OR EXTENT AND TIME OF BURNING OF FLEXIBLE PLASTICS IN A VERTICLE POSITION
D	635	TEST FOR RATE OF BURNING AND OR EXTENT AND TIME OF BURNING OF SELF-SUPPORTING PLASTICS IN A HORIZONTAL POSITION

ASTM STANDARDS
(SORTED BY ASTM CODE AND CATEGORY)

ASTM NO.		TITLE

SAFETY PROPERTIES (Cont.)

D	777	TEST FOR FLAMMABILITY OF TREATED PAPER AND CARDBOARD
D	781	TESTS FOR PUNCTURE AND STIFFNESS OF PAPERBOARD, AND CORRUGATED AND SOLID FIBERBOARD
D	828	TEST FOR TENSILE BREAKING STRENGTH OF PAPER AND PAPERBOARD
D	1310	TEST FOR FLASH POINT OF LIQUIDS BY TAG OPEN CUP APPARATUS
D	1525	TEST FOR VICAT SOFTENING TEMPERATURE OF PLASTICS
D	1929	TEST FOR IGNITION PROPERTIES OF PLASTICS
D	2015	GROSS CALORIFIC VALUE OF SOLID FUEL BY ADIABATIC BOMB CALORIMETER
D	2020	TEST FOR MILDEW (FUNGUS) RESISTANCE OF PAPER AND PAPERBOARD
D	2117	TEST FOR MELTING POINT OF SEMICRYSTALLINE POLYMERS
D	2382	TEST FOR HEAT OF COMBUSTION OF HYDROCARBON FUELS BY BOMB CALORIMETER (HIGH-PRECISION METHOD)
D	2843	TEST FOR DENSITY OF SMOKE FROM THE BURNING OR DECOMPOSITION OF PLASTICS
D	2863	TEST FOR FLAMMABILITY OF PLASTICS USING OXYGEN INDEX METHOD
D	3014	TEST FOR FLAME HEIGHT, TIME OF BURNING, AND LOSS OF WEIGHT OF RIGID CELLULAR PLASTICS PLASTICS IN A VERTICAL POSITION
D	3211	TEST FOR RELATIVE DENSITY OF BLACK SMOKE (RINGELMANN METHOD)
E	69	TEST FOR COMBUSTION PROPERTIES OF TREATED WOOD BY THE FIRE-TUBE APPARATUS
E	84	TEST FOR SURFACE BURNING CHARACTERISTICS OF BUILDING MATERIALS
E	136	TEST FOR BEHAVIOR OF MATERIALS IN A VERTICAL TUBE FURNACE AT 750 C
E	154	TESTING MATERIALS FOR USE AS VAPOR BARRIERS UNDER CONCRETE SLABS AND AS GROUND COVERS IN CRAWL SPACES
E	162	TEST FOR SURFACE FLAMMABILITY OF MATERIALS USING A RADIANT HEAT ENERGY SOURCE
E	176	DEF. OF TERMS RELATING TO FIRE TESTS OF BUILDING CONSTRUCTION AND MATERIALS
E	286	TEST FOR SURFACE FLAMMABILITY OF BUILDING MATERIALS USING AN 8-FT(2.44 M)TUNNEL FURNACE
E	324	TEST FOR RELATIVE INITIAL AND FINAL MELTING POINTS AND THE MELTING RANGE OF ORGANIC CHEMICALS
E	535	REC. PRACTICE FOR PREPARATION OF FIRE TEST STANDARDS
E	605	TESTS FOR THICKNESS AND DENSITY OF SPRAYED FIRE-RESISTIVE MATERIAL APPLIED TO STRUCTURAL MEMBERS
E	662	TEST FOR SPECIFIC OPTICAL DENSITY OF SMOKE GENERATED BY SOLID MATERIALS

SEALANTS

C	509	SPEC. FOR CELLULAR ELASTOMERIC PREFORMED GASKET AND SEALING MATERIAL
C	510	TEST FOR STAINING AND COLOR CHANGE OF SINGLE- OR MULTI-COMPONENT JOINT SEALANTS
C	542	SPEC. FOR LOCK-STRIP GASKETS
C	570	SPEC. FOR OIL- AND RESIN-BASE CAULKING COMPOUND FOR BUILDING CONSTRUCTION
C	603	TEST FOR EXTRUSION RATE AND APPLICATION LIFE OF ELASTOMERIC SEALANTS
C	639	TEST FOR RHEOLOGICAL (FLOW) PROPERTIES OF ELASTOMERIC SEALANTS
C	661	TEST FOR INDENTATION HARDNESS OF ELASTOMERIC-TYPE SEALANTS BY MEANS OF A DUROMETER
C	679	TEST FOR TACK-FREE TIME OF ELASTOMERIC-TYPE JOINT SEALANTS
C	711	TEST FOR LOW-TEMPERATURE FLEXIBILITY AND TENACITY OF ONE-PART ELASTOMERIC, SOLVENT-RELEASE TYPE SEALANTS
C	712	TEST FOR BUBBLING OF ONE-PART, ELASTOMERIC SOLVENT-RELEASE TYPE SEALANTS
C	716	REC. PRACTICE FOR USE OF LOCK-STRIP GASKETS
C	717	DEF. OF TERMS RELATING TO BUILDING SEALS

ASTM STANDARDS
(SORTED BY ASTM CODE AND CATEGORY)

ASTM NO.		TITLE

SEALANTS (Cont.)

C	718	TEST FOR UV-COLD BOX EXPOSURE OF ONE-PART, ELASTOMERIC, SOLVENT-RELEASE TYPE SEALANTS
C	719	TEST FOR ADHESION AND COHESION OF ELASTOMERIC JOINT SEALANTS UNDER CYCLIC MOVEMENT
C	731	TEST FOR EXTRUDABILITY, AFTER PACKAGE AGING, OF LATEX SEALING COMPOUNDS
C	732	TEST FOR AGING EFFECTS OF ARTIFICIAL WEATHERING ON LATEX SEALING COMPOUNDS
C	733	TEST FOR VOLUME SHRINKAGE OF LATEX SEALING COMPOUNDS
C	734	TEST FOR LOW-TEMPERATURE FLEXIBILITY OF LATEX SEALING COMPOUNDS AFTER ARTIFICIAL WEATHERING
C	736	TEST FOR EXTENSION-RECOVERY AND ADHESION OF LATEX SEALING COMPOUNDS
C	765	TEST FOR LOW-TEMPERATURE FLEXIBILITY OF PREFORMED SEALING TAPES
C	766	TEST FOR ADHESION AFTER IMPACT OF PREFORMED SEALING TAPES
C	771	TEST FOR WEIGHT LOSS AFTER HEAT AGING OF PREFORMED SEALING TAPES
C	772	TEST FOR OIL MIGRATION OR PLASTICIZER BLEED-OUT OF PREFORMED SEALING TAPES
C	782	TEST FOR SOFTNESS OF PREFORMED SEALING TAPES
C	790	REC. PRACTICES FOR USE OF LATEX SEALING COMPOUNDS
C	792	TEST FOR EFFECTS OF HEAT AGING ON WEIGHT LOSS, CRACKING, AND CHALKING OF ELASTOMERIC SEALANTS
C	793	TEST FOR EFFECTS OF ACCELERATED WEATHERING ON ELASTOMERIC JOINT SEALANTS
C	794	TEST FOR ADHESION-IN-PEEL OF ELASTOMERIC JOINT SEALANTS
C	804	REC. PRACTICES FOR USE OF SOLVENT-RELEASE TYPE SEALANTS
C	834	SPEC. FOR LATEX SEALING COMPOUNDS
C	864	SPEC. FOR DENSE ELASTOMERIC COMPRESSION SEAL GASKETS, SETTING BLOCKS, AND SPACERS
C	879	TESTING RELEASE PAPERS USED WITH PREFORMED SEALING TAPES
C	907	TEST FOR TENSILE ADHESIVE STRENGTH OF PREFORMED SEALING TAPES BY DISK METHOD
C	908	TEST FOR YIELD STRENGTH OF PREFORMED SEALING TAPES
C	910	TEST FOR BOND AND COHESION OF ONE-PART ELASTOMERIC SOLVENT RELEASE-TYPE SEALANTS
C	919	PRACTICE FOR USE OF SEALANTS IN ACOUSTICAL APPLICATIONS
C	920	SPEC. FOR ELASTOMERIC JOINT SEALANTS
D	2202	TEST FOR SLUMP OF CAULKING COMPOUNDS AND SEALANTS
D	2203	TEST FOR STAINING OF CAULKING COMPOUNDS AND SEALANTS
D	2377	TEST FOR TACK-FREE TIME OF CAULKING COMPOUNDS AND SEALANTS
D	2450	TEST FOR BOND OF OIL- AND RESIN-BASE CAULKING COMPOUNDS
D	2452	TEST FOR EXTRUDABILITY OF OIL- AND RESIN-BASE CAULKING COMPOUNDS

SHRINKAGE

C	356	TEST FOR LINEAR SHRINKAGE OF PREFORMED HIGH-TEMPERATURE THERMAL INSULATION SUBJECTED
D	2453	TEST FOR SHRINKAGE AND TENACITY OF OIL- AND RESIN-BASE CAULKING COMPOUNDS TO SOAKING HEAT

SOUND ABSORPTION

C	367	TESTS FOR STRENGTH PROPERTIES OF PREFABRICATED ARCHITECTURAL ACOUSTICAL TILE OR LAY-IN CEILING PANELS

ASTM STANDARDS
(SORTED BY ASTM CODE AND CATEGORY)

ASTM NO.	TITLE

SOUND ABSORPTION (Cont.)

C	384	TEST FOR IMPEDANCE AND ABSORPTION OF ACOUSTICAL MATERIALS BY THE IMPEDENCE TUBE METHOD
C	423	TEST FOR SOUND ABSORPTION AND SOUND ABSORPTION COEFFICIENTS BY THE REVERBERATION ROOM METHOD
C	634	DEF. OF TERMS RELATING TO ENVIRONMENTAL ACOUSTICS
C	635	SPEC. FOR METAL SUSPENSION SYSTEMS FOR ACOUSTICAL TILE AND LAY-IN PANEL CEILINGS
C	636	REC. PRACTICE FOR INSTALLATION OF METAL CEILING SUSPENSION SYSTEMS FOR ACOUSTICAL TILE AND LAY-IN PANELS
C	643	TEST FOR CHANGE IN ACOUSTICAL ABSORPTION OF CEILING MATERIALS DUE TO REPAINTING
E	90	LABORATORY MEASUREMENT OF AIRBORNE-SOUND TRANSMISSION LOSS OF BUILDING PARTITIONS
E	336	TEST FOR MEASUREMENT OF AIRBORNE SOUND INSULATION IN BUILDINGS
E	413	CLASSIFICATION FOR DETERMINATION OF SOUND TRANSMISSION CLASS
E	447	TESTING DUCT LINER MATERIALS AND PREFABRICATED SILENCERS FOR ACOUSTICAL AND AIRFLOW PERFORMANCE
E	492	LABORATORY MEASUREMENT OF IMPACT SOUND TRANSMISSION THROUGH FLOOR-CEILING ASSEMBLIES USING THE TAPPING MACHINE
E	596	LABORATORY MEASUREMENT OF THE NOISE REDUCTION OF SOUND-ISOLATING ENCLOSURES

SPECIFIC HEAT

C	351	TEST FOR MEAN SPECIFIC HEAT OF THERMAL INSULATION

TENSILE PROPERTIES

D	638	TEST FOR TENSILE PROPERTIES OF PLASTICS
D	882	TEST FOR TENSILE PROPERTIES OF THIN PLASTIC SHEETING

THERMAL HEAT TRANSFER

C	177	TEST FOR STEADY-STATE THERMAL TRANSMISSION PROPERTIES BY MEANS OF THE GUARDED HOT PLATE
C	236	TEST FOR THERMAL CONDUCTANCE AND TRANSMITTASNCE OF BUILT-UP SECTIONS BY MEANS OF THE GUARDED HOT BOX
C	335	TEST FOR STEADY-STATE HEAT TRANSFER PROPERTIES OF HORIZONTAL PIPE INSULATIONS
C	518	TEST FOR STEADY-STATE THERMAL TRANSMISSION PROPERTIES BY MEANS OF THE HEAT FLOW METER
C	653	RECOMMENDED PRACTICE FOR DETERMINATION OF THE THERMAL RESISTANCE OF LOW-DENSITY MINERAL FIBER BLANKET-TYPE BUILDING INSULATION
C	680	REC. PRACTICE FOR DETERMINATION OF HEAT GAIN OR LOSS AND SURFACE TEMPERATURES OF INSULATED PIPE AND EQUIPMENT SYSTEMS BY THE USE OF A COMPUTER PROGRAM
C	687	RECOMMENDED PRACTICE FOR DETERMINATION OF THE THERMAL RESISTANCE OF LOW-DENSITY FIBROUS LOOSE FILL TYPE BUILDING INSULATION
C	691	TEST FOR STEADY-STATE THERMAL TRANSMISSION PROPERTIES OF NONHOMOGENOUS PIPE INSULATION INSTALLED HORIZONTALLY
C	855	THERMAL RESISTANCE FACTORS FOR PREFORMED ABOVE-DECK ROOF INSULATION

ASTM STANDARDS
(SORTED BY ASTM CODE AND CATEGORY)

ASTM NO.	TITLE

WATER ABSORPTION

C	272	TEST FOR WATER ABSORPTION OF CORE MATERIALS FOR STRUCTURAL SANDWICH CONSTRUCTIONS
D	2020	TESTS FOR MILDEW (FUNGUS) RESISTANCE OF PAPER AND PAPERBOARD
D	2842	TEST FOR WATER ABSORPTION OF RIGID CELLULAR PLASTICS
E	331	TEST FOR WATER PENETRATION OF EXTERIOR WINDOWS, CURTAIN WALLS, AND DOORS BY UNIFORM STATIC AIR PRESSURE DIFFERENCE
E	514	TEST FOR WATER PERMEANCE OF MASONRY
E	547	TEST FOR WATER PENETRATION OF EXTERIOR WINDOWS, CURTAIN WALLS, AND DOORS BY CYCLIC STATIC AIR PRESSURE DIFFERENTIAL

WATER VAPOR TRANSMISSION

C	355	TESTS FOR WATER VAPOR TRANSMISSION OF THICK MATERIALS
C	677	RECOMMENDED PRACTICE FOR USE OF A STANDARD REFERENCE SHEET FOR THE MEASUREMENT OF THE TIME-AVERAGED VAPOR PRESSURE IN A CONTROLLED HUMIDITY SPACE
C	755	RECOMMENDED PRACTICE FOR SELECTION OF VAPOR BARRIERS FOR THERMAL INSULATIONS
C	945	TEST FOR WATER VAPOR TRANSMISSION OF MATERIALS (PENDING)
E	96	TEST FOR WATER VAPOR TRANSMISSION OF MATERIALS IN SHEET FORM
E	154	TESTING MATERIALS FOR USE AS VAPOR BARRIERS UNDER CONCRETE SLABS AND AS GROUND COVERS IN CRAWL SPACES
E	398	REC. PRACTICE FOR DYNAMIC MEASUREMENT OF WATER VAPOR TRANSFER

WEATHER BARRIERS

C	355	TESTS FOR WATER VAPOR TRANSMISSION OF THICK MATERIALS
C	419	MAKING AND CURING TEST SPECIMENS OF MASTIC THERMAL INSULATION COATINGS
C	461	TESTING OF BITUMINIOUS MASTIC COATINGS USED IN CONJUNCTION WITH THERMAL INSULATION
C	488	REC. PRACTICE FOR CONDUCTING EXTERIOR EXPOSURE TESTS OF FINISHES FOR THERMAL INSULATION
C	647	GUIDE FOR PROPERTIES AND TESTS OF MASTICS AND COATINGS FOR THERMAL INSULATION
C	836	SPEC. FOR HIGH SOLIDS CONTENT, COLD LIQUID-APPLIED ELASTOMERIC WATERPROOFING MEMBRANE MEMBRANE FOR USE WITH SEPARATE WEARING COURSE
C	898	GUIDE FOR USE OF HIGH SOLIDS CONTENT, COLD LIQUID-APPLIED ELASTOMERIC WATERPROOFING MEMBRANE WITH SEPARATE WEARING COURSE
D	70	SPECIFIC GRAVITY OF SEMI SOLID BITUMINOUS MATERIALS
D	71	SPECIFIC GRAVITY OF SOLID PITCH AND ASPHALT--DISPLACEMENT METHOD
D	529	REC. PRACTICE FOR ACCELERATED WEATHERING TEST OF BITUMINOUS MATERIALS
D	543	TEST FOR RESISTANCE OF PLASTICS TO CHEMICAL REAGENTS
D	638	TEST FOR TENSILE PROPERTIES OF PLASTICS
D	658	TEST FOR ABRASION RESISTANCE OF COATINGS OF PAINT,VARNISH,LACQUER,AND RELATED PRODUCTS WITH THE AIR BLAST ABRASION TESTER
D	747	TEST FOR STIFFNESS OF PLASTICS BY MEANS OF A CANTILEVER BEAM
D	781	TEST FOR PUNCTURE AND STIFFNESS OF PAPERBOARD,AND CORRUGATED AND SOLID FIBERBOARD

ASTM STANDARDS
(SORTED BY ASTM CODE AND CATEGORY)

ASTM NO.		TITLE

WEATHER BARRIERS (Cont.)

D	790	TEST FOR FLEXURAL PROPERTIES OF PLASTICS AND ELECTRICAL INSULATING MATERIALS
D	828	TEST FOR TENSILE BREAKING STRENGTH OF PAPER AND PAPERBOARD
D	903	TEST FOR PEEL OR STRIPPING STRENGTH OF ADHESIVE BONDS
D	968	TEST FOR ABRASION RESISTANCE OF COATINGS OF PAINT,VARNISH,LACQUER,AND RELATED PRODUCTS BY THE FALLING SAND METHOD
D	1640	TEST FOR DRYING,CURING,OR FILM FORMATION OF ORGANIC COATINGS AT ROOM TEMPERATURE
D	1654	EVALUATION OF PAINTED OR COATED SPECIMENS SUBJECTED TO CORROSIVE ENVIRONMENTS
D	1729	VISUAL EVALUATION OF COLOR DIFFERENCES OF OPAQUE MATERIALS
D	1823	TEST FOR APPARENT VISCOSITY OF PLASTISOLS AND ORGANOSOLS AT HIGH SHEAR RATES BY CASTOR-SEVERS VISCOMETER
D	1824	TEST FOR APPARENT VISCOSITY OF PLASTISOLS AND ORGANOSOLS AT LOW SHEAR RATES BY BROOKFIELD VISCOMETER
E	96	TEST FOR WATER VAPOR TRANSMISSION OF MATERIALS IN SHEET FORM
E	154	TESTING MATERIALS FOR USE AS VAPOR BARRIERS UNDER CONCRETE SLABS AND AS GROUND COVERS IN CRAWL SPACES
G	20	TEST FOR CHEMICAL RESISTANCE OF PIPELINE COATINGS
G	21	REC. PRACTICE FOR DETERMINING RESISTANCE OF SYNTHETIC POLYMERIC MATERIALS TO FUNGI
G	22	REC. PRACTICE FOR DETERMINING RESISTANCE OF PLASTICS TO BACTERIA

ASTM INSULATION STANDARDS
(SORTED BY ASTM CODE)

ASTM NO.		TITLE
C	163	MIXING THERMAL INSULATING CEMENT SAMPLES
C	165	RECOMMENDED PRACTICE FOR MEASURING COMPRESSIVE PROPERTIES OF THERMAL INSULATION
C	166	TEST FOR COVERING CAPACITY AND VOLUME CHANGE UPON DRYING THERMAL INSULATING CEMENT
C	167	TESTS FOR THICKNESS AND DENSITY OF BLANKET OR BATT TYPE THERMAL INSULATING MATERIALS
C	168	DEF. OF TERMS RELATING TO THERMAL INSULATING MATERIALS
C	177	TEST FOR STEADY-STATE THERMAL TRANSMISSION PROPERTIES BY MEANS OF THE GUARDED HOT PLATE
C	195	SPEC. FOR MINERAL FIBER THERMAL INSULATING CEMENT
C	196	SPEC. FOR EXPANDED OR EXFOLIATED VERMICULITE THERMAL INSULATING CEMENT
C	203	TEST FOR BREAKING LOAD AND CALCULATED FLEXURAL STRENGTH OF PREFORMED BLOCK-TYPE THERMAL INSULATION
C	208	SPEC. FOR INSULATING BOARD (CELLULOSIC FIBER),STRUCTURAL AND DECORATIVE
C	236	TEST FOR THERMAL CONDUCTANCE AND TRANSMITTANCE OF BUILT-UP SECTIONS BY MEANS OF THE GUARDED HOT BOX
C	240	TESTING CELLULAR GLASS INSULATING BLOCK
C	272	TEST FOR WATER ABSORPTION OF CORE MATERIALS FOR STRUCTURAL SANDWICH CONSTRUCTIONS
C	302	TEST FOR DENSITY OF PREFORMED PIPE-COVERING-TYPE INSULATION
C	303	TEST FOR DENSITY OF PREFORMED BLOCK-TYPE THERMAL INSULATION
C	335	TEST FOR STEADY-STATE HEAT TRANSFER PROPERTIES OF HORIZONTAL PIPE INSULATIONS
C	351	TEST FOR MEAN SPECIFIC HEAT OF THERMAL INSULATION
C	353	TEST FOR ADHESION OF DRIED THERMAL INSULATION OR FINISHING CEMENT
C	354	TEST FOR COMPRESSIVE STRENGTH OF THERMAL INSULATING OR FINISHING CEMENT
C	355	TESTS FOR WATER VAPOR TRANSMISSION OF THICK MATERIALS
C	356	TEST FOR LINEAR SHRINKAGE OF PREFORMED HIGH-TEMPERATURE THERMAL INSULATION SUBJECTED TO SOAKING HEAT
C	367	TESTS FOR STRENGTH PROPERTIES OF PREFABRICATED ARCHITECTURAL ACOUSTICAL TILE OR LAY-IN CEILING PANELS
C	383	TEST FOR WET ADHESION OF THERMAL INSULATING CEMENTS TO METAL
C	384	TEST FOR IMPEDENCE AND ABSORPTION OF ACOUSTICAL MATERIALS BY THE IMPEDENCE TUBE METHOD
C	390	CRITERIA FOR SAMPLING AND ACCEPTANCE OF PREFORMED THERMAL INSULATION LOTS
C	405	TESTS FOR CONSISTENCY OF WET-MIXED THERMAL INSULTING CEMENTS
C	411	TEST FOR HOT-SURFACE PERFORMANCE OF HIGH-TEMPERATURE THERMAL INSULATION
C	419	MAKING AND CURING TEST SPECIMENS OF MASTIC THERMAL INSULATION COATINGS
C	421	TEST FOR TUMBLING FRIABILITY OF PREFORMED BLOCK-TYPE THERMAL INSULATION
C	423	TEST FOR SOUND ABSORPTION AND SOUND ABSORPTION COEFFICIENTS BY THE REVERBERATION ROOM METHOD
C	446	TEST FOR BREAKING LOAD AND CALCULATED MODOLUS OF RUPTURE OF PREFORMED INSULATION FOR PIPES
C	447	RECOMMENDED PRACTICE FOR ESTIMATING THE MAXIMUM USE TEMPERATURE OF PREFORMED HOMOGENOUS THERMAL INSULATIONS
C	449	SPEC. FOR MINERAL FIBER HYDRAULIC-SETTING THERMAL INSULATING AND FINISHING CEMENT
C	450	REC. PRACTICE FOR PREFABRICATION AND FIELD FABRICATION OF THERMAL INSULATING FITTING COVERS FOR NPS PIPING, VESSEL LAGGING, AND DISHED HEAD SEGMENTS
C	461	TESTING BITUMINIOUS MASTIC COATINGS USED IN CONJUNCTION WITH THERMAL INSULATION
C	464	TEST FOR CORROSION EFFECT OF THERMAL INSULATING CEMENTS ON BASE METAL
C	487	TEST FOR RESISTANCE TO DROPPING OF PREFORMED BLOCK-TYPE THERMAL INSULATION
C	488	REC. PRACTICE FOR CONDUCTING EXTER OR EXPOSURE TESTS OF FINISHES FOR THERMAL INSULATION
C	495	TEST FOR COMPRESSIVE STRENGTH OF LIGHTWEIGHT INSULATING CONCRETE
C	509	SPEC. FOR CELLULAR ELASTOMERIC PREFORMED GASKET AND SEALING MATERIAL
C	510	TEST FOR STAINING AND COLOR CHANGE OF SINGLE- OR MULTICOMPONENT JOINT SEALANTS

ASTM INSULATION STANDARDS
(SORTED BY ASTM CODE)

ASTM NO.	TITLE
C 516	SPEC. FOR VERMICULITE LOOSE FILL INSULATION
C 517	SPEC. FOR DIATOMACEOUS EARTH BLOCK AND PIPE THERMAL INSULATION
C 518	TEST FOR STEADY-STATE THERMAL TRANSMISSION PROPERTIES BY MEANS OF THE HEAT FLOW METER
C 519	TEST FOR DENSITY OF FIBROUS LOOSE FILL BUILDING INSULATIONS
C 520	TEST FOR DENSITY OF GRANULAR LOOSE FILL INSULATIONS
C 532	SPEC. FOR STRUCTURAL INSULATING FORMBOARD (CELLULOSIC FIBER)
C 533	SPEC. FOR CALCIUM SILICATE BLOCK AND PIPE THERMAL INSULATION
C 534	SPEC. FOR PREFORMED FLEXIBLE ELASTOMERIC CELLULAR THERMAL INSULATION IN SHEET AND TUBULAR FORM
C 542	SPEC. FOR LOCK-STRIP GASKETS
C 547	SPEC. FOR MINERAL FIBER PREFORMED PIPE INSULATION
C 548	TEST FOR DIMENSIONAL STABILITY OF LOW TEMPERATURE THERMAL BLOCK AND PIPE INSULATION
C 549	SPEC.FOR PERLITE LOOSE FILL INSULATION
C 550	TESTS FOR TRUENESS AND SQUARENESS OF BLOCK THERMAL INSULATION
C 552	SPEC.FOR CELLULAR GLASS BLOCK AND PIPE THERMAL INSULATION
C 553	SPEC. FOR MINERAL FIBER BLANKET AND FELT INSULATION (INDUSTRIAL TYPE)
C 569	TEST FOR INDENTATION HARDNESS OF PREFORMED THERMAL INSULATIONS
C 570	SPEC. FOR OIL- AND RESIN-BASE CAULKING COMPOUND FOR BUILDING CONSTRUCTION
C 578	SPEC. FOR PREFORMED,BLOCK-TYPE CELLULAR POLYSTYRENE THERMAL INSULATION
C 585	REC. PRACTICE FOR INNER AND OUTER DIAMETERS OF RIGID THERMAL INSULATION FOR NOMINAL SIZE OF PIPE AND TUBING (NPS SYSTEM)
C 589	TEST FOR APPARENT IMPACT STRENGTH OF PREFORMED BLOCK-TYPE INSULATING MATERIALS
C 590	TEST FOR ACTION ON SUBSTRATES BY COATINGS, ADHESIVES AND JOINT SEALANTS USED ON OR WITH THERMAL INSULATION
C 591	SPEC. FOR RIGID PREFORMED CELLULAR URETHANE THERMAL INSULATION
C 592	SPEC. FOR MINERAL FIBER BLANKET INSULATION AND BLANKET-TYPE PIPE INSULATION (METAL MESH COVERED)(INDUSTRIAL TYPE)
C 603	TEST FOR EXTRUSION RATE AND APPLICATION LIFE OF ELASTOMERIC SEALANTS
C 610	SPEC. FOR EXPANDED PERLITE BLOCK AND PIPE THERMAL INSULATION
C 612	SPEC. FOR MINERAL FIBER BLOCK AND BOARD THERMAL INSULATION
C 634	DEF. OF TERMS RELATING TO ENVIRONMENTAL ACOUSTICS
C 635	SPEC. FOR METAL SUSPENSION SYSTEMS FOR ACOUSTICAL TILE AND LAY-IN PANEL CEILINGS
C 636	REC. PRACTICE FOR INSTALLATION OF METAL CEILING SUSPENSION SYSTEMS FOR ACOUSTICAL TILE AND LAY-IN PANELS
C 639	TEST OFR RHEOLOGICAL (FLOW) PROPERTIES OF ELASTOMERIC SEALANTS
C 640	SPEC. FOR CORKBOARD AND CORK PIPE INSULATION FOR LOW TEMP. THERMAL INSULATION
C 643	TEST FOR CHANGE IN ACOUSTICAL ABSORPTION OF CEILING MATERIALS DUE TO REPAINTING
C 647	GUIDE FOR PROPERTIES AND TESTS OF MASTICS AND COATINGS FOR THERMAL INSULATION
C 653	RECOMMENDED PRACTICE FOR DETERMINATION OF THE THERMAL RESISTANCE OF LOW-DENSITY MINERAL FIBER BLANKET-TYPE BUILDING INSULATION
C 656	SPEC. FOR STRUCTURAL INSULATING BOARD, CALCIUM SILICATE
C 661	TEST FOR INDENTATION HARDNESS OF ELASTOMERIC-TYPE SEALANTS BY MEANS OF A DUROMETER
C 665	SPEC. FOR MINERAL FIBER BLANKET THERMAL INSULATION FOR WOOD FRAME AND LIGHT CONSTRUCTION BUILDINGS
C 667	REC. PRACTICE FOR PREFABRICATED REFLECTIVE INSULATION SYSTEMS FOR EQUIPMENT AND PIPE
C 669	SPEC. FOR GLAZING COMPOUNDS FOR BACK BEDDING AND FACE GLAZING OF METAL SASH OPERATING AT TEMPERATURES ABOVE AMBIENT AIR
C 677	RECOMMENDED PRACTICE FOR USE OF A STANDARD REFERENCE SHEET FOR THE MEASUREMENT OF THE TIME-AVERAGED VAPOR PRESSURE IN A CONTROLLED HUMIDITY SPACE
C 679	TEST FOR TACK-FREE TIME OF ELASTOMERIC-TYPE JOINT SEALANTS

ASTM INSULATION STANDARDS
(SORTED BY ASTM CODE)

ASTM NO.		TITLE
C	680	REC. PRACTICE FOR DETERMINATION OF HEAT GAIN OR LOSS AND SURFACE TEMPERATURES OF INSULATED PIPE AND EQUIPMENT SYSTEMS BY THE USE OF A COMPUTER PROGRAM
C	681	TEST FOR VOLATILITY OF OIL- AND RESIN-BASED, KNIFE-GRADE, CHANNEL GLAZING COMPOUNDS
C	686	TEST FOR PARTING STRENGTH OF MINERAL BATT- AND BLANKET-TYPE INSULATION
C	687	RECOMMENDED PRACTICE FOR DETERMINATION OF THE THERMAL RESISTANCE OF LOW-DENSITY FIBROUS LOOSE FILL TYPE BUILDING INSULATION
C	691	TEST FOR STEADY-STATE THERMAL TRANSMISSION PROPERTIES OF NONHOMOGENOUS PIPE INSULATION INSTALLED
C	692	EVALUATING THE INFLUENCE OF WICKING-TYPE THERMAL INSULATIONS ON THE STRESS CORROSION CRACKING TENDENCY OF AUSTENITIC STAINLESS STEEL
C	711	TEST FOR LOW-TEMPERATURE FLEXIBILITY AND TENACITY OF ONE-PART ELASTOMERIC, SOLVENT-RELEASE TYPE SEALANTS
C	712	TEST FOR BUBBLING OF ONE-PART, ELASTOMERIC SOLVENT-RELEASE TYPE SEALANTS
C	713	TEST FOR SLUMP OF AN OIL-BASE KNIFE-GRADE CHANNEL GLAZING COMPOUND
C	716	REC. PRACTICE FOR USE OF LOCK-STRIP GASKETS
C	717	DEF. OF TERMS RELATING TO BUILDING SEALS
C	718	TEST FOR UV-COLD BOX EXPOSURE OF ONE-PART, ELASTOMERIC, SOLVENT-RELEASE TYPE SEALANTS
C	719	TEST FOR ADHESION AND COHESION OF ELASTOMERIC JOINT SEALANTS UNDER CYCLIC MOVEMENT
C	720	SPEC. FOR SPRAY-APPLIED FIBROUS THERMAL INSULATION FOR ELEVATED TEMPERATURE
C	726	SPEC. FOR MINERAL FIBER ROOF INSULATION BOARD
C	727	REC. PRACTICE FOR USE OF REFLECTIVE INSULATION IN BUILDING CONSTRUCTIONS
C	728	SPEC. FOR PERLITE THERMAL INSULATION BOARD
C	731	TEST FOR EXTRUDABILITY, AFTER PACKAGE AGING, OF LATEX SEALING COMPOUNDS
C	732	TEST FOR AGING EFFECTS OF ARTIFICIAL WEATHERING ON LATEX SEALING COMPOUNDS
C	733	TEST FOR VOLUME SHRINKAGE OF LATEX SEALING COMPOUNDS
C	734	TEST FOR LOW-TEMPERATURE FLEXIBILITY OF LATEX SEALING COMPOUNDS AFTER ARTIFICIAL WEATHERING
C	736	TEST FOR EXTENSION-RECOVERY AND ADHESION OF LATEX SEALING COMPOUNDS
C	739	SPEC. FOR CELLULOSIC FIBER (WOOD BASE) LOOSE FILL THERMAL INSULATION
C	740	REC. PRACTICE FOR EVACUATED REFLECTIVE INSULATION IN CRYOGENIC SERVICE
C	741	TEST FOR ACCELERATED AGING OF WOOD SASH FACE GLAZING COMPOUND
C	742	TEST FOR DEGREE OF SET FOR WOOD SASH GLAZING COMPOUND
C	745	TEST FOR HEAT FLUX THROUGH EVACUATED INSULATIONS USING A GUARDED FLAT PLATE BOILOFF CALORIMETER
C	755	RECOMMENDED PRACTICE FOR SELECTION OF VAPOR BARRIERS FOR THERMAL INSULATIONS
C	762	REC. PRACTICE FOR APPLICATION OF SPRAY-APPLIED FIBROUS THERMAL INSULATION
C	764	SPEC. FOR MINERAL FIBER LOOSE FILL THERMAL INSULATION
C	765	TEST FOR LOW-TEMPERATURE FLEXIBILITY OF PREFORMED SEALING TAPES
C	766	TEST FOR ADHESION AFTER IMPACT OF PREFORMED SEALING TAPES
C	771	TEST FOR WEIGHT LOSS AFTER HEAT AGING OF PREFORMED SEALING TAPES
C	772	TEST FOR OIL MIGRATION OR PLASTICIZER BLEED-OUT OF PREFORMED SEALING TAPES
C	782	TEST FOR SOFTNESS OF PREFORMED SEALING TAPES
C	790	REC. PRACTICES FOR USE OF LATEX SEALING COMPOUNDS
C	792	TEST FOR EFFECTS OF HEAT AGING ON WEIGHT LOSS, CRACKING, AND CHALKING OF ELASTOMERIC SEALANTS
C	793	TEST FOR EFFECTS OF ACCELERATED WEATHERING ON ELASTOMERIC JOINT SEALANTS
C	794	TEST FOR ADHESION-IN-PEEL OF ELASTOMERIC JOINT SEALANTS
C	795	SPEC. FOR WICKING-TYPE THERMAL INSULATION FOR USE OVER AUSTENITIC STAINLESS STEEL
C	797	REC. PRACTICES AND TERMINOLOGY FOR USE OF OIL- AND RESIN-BASED PUTTY AND GLAZING COMPOUNDS

ASTM INSULATION STANDARDS
(SORTED BY ASTM CODE)

ASTM NO.		TITLE
C	800	SPEC. FOR GLASS FIBER BLANKET INSULATION (AIRCRAFT TYPE)
C	804	REC. PRACTICES FOR USE OF SOLVENT-RELEASE TYPE SEALANTS
C	834	SPEC. FOR LATEX SEALING COMPOUNDS
C	835	TEST FOR TOTAL HEMISPHERICAL EMITTANCE OF SURFACES FROM 20 TO 1400 C (68 TO 2550 F)
C	836	SPEC. FOR HIGH SOLIDS CONTENT, COLD LIQUID-APPLIED ELASTOMERIC WATERPROOFING MEMBRANE FOR USE WITH SEPARATE WEARING COURSE
C	846	SPEC. FOR APPLICATION OF STRUCTURAL INSULATING BOARD (FIBERBOARD) SHEATHING
C	854	TEST FOR RESISTANCE TO EXTERNAL LOADS ON METAL REFLECTIVE PIPE INSULATION
C	855	THERMAL RESISTANCE FACTORS FOR PREFORMED ABOVE-DECK ROOF INSULATION
C	864	SPEC. FOR DENSE ELASTOMERIC COMPRESSION SEAL GASKETS, SETTING BLOCKS, AND SPACERS
C	870	REC. PRACTICE FOR CONDITIONING OF THERMAL INSULATING MATERIALS
C	871	CHEMICAL ANALYSIS OF THERMAL INSULATION MATERIALS FOR LEACHABLE CHLORIDE, SILICATE, AND SODIUM IONS
C	879	TESTING RELEASE PAPERS USED WITH PREFORMED SEALING TAPES
C	892	SPEC. FOR HIGH-TEMPERATURE FIBER BLANKET THERMAL INSULATION
C	898	GUIDE FOR USE OF HIGH SOLIDS CONTENT, COLD LIQUID-APPLIED ELASTOMERIC WATERPROOFING MEMBRANE WITH SEPARATE WEARING COURSE
C	907	TEST FOR TENSILE ADHESIVE STRENGTH OF PREFORMED SEALING TAPES BY DISK METHOD
C	908	TEST FOR YIELD STRENGTH OF PREFORMED SEALING TAPES
C	910	TEST FOR BOND AND COHESION OF ONE-PART ELASTOMERIC SOLVENT RELEASE-TYPE SOLVENTS
C	916	SPEC. FOR ADHESIVES FOR DUCT THERMAL INSULATION
C	919	PRACTICE FOR USE OF SEALANTS IN ACOUSTICAL APPLICATIONS
C	920	SPEC. FOR ELASTOMERIC JOINT SEALANTS
C	921	PRACTICE FOR DETERMINING THE PROPERTIES OF JACKETING MATERIALS FOR THERMAL INSULATION
C	929	PRACTICE FOR HANDLING, TRANSPORTING, SHIPPING, STORAGE, RECEIVING, AND APPLICATION OF THERMAL INSULATION MATERIALS TO BE USED OVER AUSTENITIC STAINLESS STEEL
C	930	CLASSIFICATION FOR IDENTIFYING AND CATEGORIZING POTENTIAL HEALTH AND SAFETY HAZARDS OF THERMAL INSULATION MATERIALS AND ACCESSORIES DURING INSTALLATION
C	945	TEST FOR WATER VAPOR TRANSMISSION OF MATERIALS (PENDING)
D	56	TEST FOR FLASH POINT BY TAG CLOSED TESTER
D	70	TEST FOR SPECIFIC GRAVITY OF SEMI SOLID BITIMINOUS MATERIALS
D	71	TEST FOR SPECIFIC GRAVITY OF SOLID PITCH AND ASPHALT (DISPLACEMENT METHOD)
D	92	TEST FOR FLASH AND FIRE POINTS BY CLEVELAND OPEN CUP
D	93	TEST FOR FLASH POINT BY PENSKY-MARTENS CLOSED TESTER
D	529	REC. PRACTICE FOR ACCELERATED WEATHERING TEST OF BITUMINOUS MATERIALS
D	543	TEST FOR RESISTANCE OF PLASTICS TO CHEMICAL REAGENTS
D	568	TEST FOR RATE OF BURNING AND/OR EXTENT AND TIME OF BURNING OF FLEXIBLE PLASTICS IN A VERTICAL POSITION
D	635	TEST FOR RATE OF BURNING AND/OR EXTENT AND TIME OF BURNING FOR SELF-SUPPORTING PLASTICS IN A HORIZONTAL POSITION
D	638	TEST FOR TENSILE PROPERTIES OF PLASTICS
D	658	TEST FOR ABRASION RESISTANCE OF COATINGS OF PAINT,VARNISH,LACQUER AND RELATED PRODUCTS WITH THE AIR BLAST ABRASION TESTER
D	747	TEST FOR STIFFNESS OF PLASTICS BY MEANS OF A CANTILEVER BEAM
D	777	TEST FOR FLAMMABILITY OF TREATED PAPER AND CARDBOARD
D	781	TEST FOR PUNCTURE AND STIFFNESS OF PAPERBOARD,CORRUGATED AND SOLID FIBERBOARD
D	790	TEST FOR FLEXURAL PROPERTIES OF PLASTICS AND ELECTRICAL INSULATING MATERIALS
D	828	TEST FOR TENSILE BREAKING STRENGTH OF PAPER AND PAPERBOARD
D	882	TEST FOR TENSILE PROPERTIES OF THIN PLASTIC SHEETING
D	903	TEST FOR PEEL OR STRIPPING STRENGTH OF ADHESIVE BONDS

ASTM INSULATION STANDARDS
(SORTED BY ASTM CODE)

ASTM NO.	TITLE
D 968	TEST FOR ABRASION RESISTANCE OF COATINGS OF PAINT, VARNISH, LACQUER AND RELATED PRODUCTS BY THE FALLING SAND METHOD
D 1204	TEST FOR LINEAR DIMENSIONAL CHANGES OF NONRIGID THERMOPLASTIC SHEETING OR FILM AT ELEVATED TEMPERATURES
D 1310	TEST FOR FLASH POINT OF LIQUIDS BY TAG OPEN CUP APPARATUS
D 1525	TEST FOR VICAT SOFTENING TEMPERATURE OF PLASTICS
D 1621	TEST FOR COMPRESSIVE PROPERTIES OF RIGID CELLULAR PLASTICS
D 1622	TEST FOR APPARENT DENSITY OF RIGID CELLULAR PLASTICS
D 1640	TEST FOR DRYING, CURING OR FILM FORMATION OF ORGANIC COATINGS AT ROOM TEMPERATURE
D 1654	EVALUATION OF PAINTED OR COATED SPECIMENS SUBJECTED TO CORROSIVE ENVIRONMENTS
D 1729	VISUAL EVALUATION OF COLOR DIFFERENCES OF OPAQUE MATERIALS
D 1823	TEST FOR APPARENT VISCOSITY OF PLASTISOLS AND ORGANOSOLS AT HIGH SHEAR RATES BY CASTOR-SEVERS VISCOMETER
D 1824	TEST FOR APPARENT VISCOSITY OF PLASTISOLS AND ORGANOSOLS AT LOW SHEAR RATES BY BROOKFIELD VISCOMETER
D 1929	TEST FOR IGNITION PROPERTIES OF PLASTICS
D 2015	GROSS CALORIFIC VALUE OF SOLID FUEL BY THE ADIABATIC BOMB CALORIMETER TESTER
D 2020	TEST FOR MILDEW FUNGUS RESISTANCE OF PAPER AND PAPERBOARD
D 2117	TEST FOR MELTING POINT OF SEMICRYSTALLINE POLYMERS
D 2126	TEST FOR RESPONSE OF RIGID CELLULAR PLASTICS TO THERMAL AND HUMID AGING
D 2202	TEST FOR SLUMP OF CAULKING COMPOUNDS AND SEALANTS
D 2203	TEST FOR STAINING OF CAULKING COMPOUNDS AND SEALANTS
D 2249	PREDICTING THE EFFECT OF WEATHERING OF FACE GLAZING AND BEDDING COMPOUNDS ON METAL SASH
D 2376	TEST FOR SLUMP OF FACE GLAZING AND BEDDING COMPOUNDS ON METAL SASH
D 2377	TEST FOR TACK-FREE TIME OF CAULKING COMPOUNDS AND SEALANTS
D 2450	TEST FOR BOND OF OIL- AND RESIN-BASE CAULKING COMPOUNDS
D 2451	TEST FOR DEGREE OF SET FOR GLAZING COMPOUNDS ON METAL SASH
D 2452	TEST FOR EXTRUDABILITY OF OIL- AND RESIN-BASE CAULKING COMPOUNDS
D 2453	TEST FOR SHRINKAGE AND TENACITY OF OIL- AND RESIN-BASE CAULKING COMPOUNDS
D 2842	TEST FOR WATER ABSORPTION OF RIGID CELLULAR PLASTICS
D 2843	TEST FOR DENSITY OF SMOKE FROM BURNING OR DECOMPOSITION OF PLASTICS
D 2863	TEST FOR FLAMMABILITY OF PLASTICS USING OXYGEN INDEX METHOD
D 3014	FLAME HEIGHT, TIME OF BURNING, AND WEIGHT LOSS OF RIGID CELLULAR PLASTICS IN A VERTICAL POSITION
D 3211	TEST FOR RELATIVE DENSITY OF BLACK SMOKE (RINGELMANN METHOD)
E 69	TEST FOR COMBUSTION PROPERTIES OF TREATED WOOD BY THE FIRE-TUBE APPARATUS
E 72	CONDUCTING STRENGTH TESTS OF PANELS FOR BUILDING CONSTRUCTION
E 84	TEST FOR SURFACE BURNING CHARACTERISTICS OF BUILDING MATERIALS
E 90	LABORATORY MEASUREMENT OF AIRBORNE-SOUND TRANSMISSION LOSS OF BUILDING PARTITIONS
E 96	TESTS FOR WATER VAPOR TRANSMISSION OF MATERIALS IN SHEET FORM
E 136	TEST FOR BEHAVIOR OF MATERIALS IN A VERTICAL TUBE FURNACE AT 750 C
E 154	TESTING MATERIALS FOR USE ANS VAPOR BARRIERS UNDER CONCRETE SLABS AND AS GROUND COVER IN CRAWL SPACES
E 162	TEST FOR SURFACE FLAMMABILITY OF MATERIALS USING A RADIANT HEAT ENERGY SOURCE
E 176	DEF. OF TERMS RELATING TO FIRE TESTS OF BUILDING CONSTRUCTION AND MATERIALS
E 283	TEST FOR RATE OF AIR LEAKAGE THROUGH EXTERIOR WINDOWS, CURTAIN WALLS, AND DOORS
E 286	TEST FOR SURFACE FLAMMABILITY OF BUILDING MATERIALS USING AN 8-FT(2.44 M)TUNNEL FURNACE
E 324	TEST FOR RELATIVE INITIAL AND FINAL MELTING POINTS AND THE MELTING RANGE OF ORGANIC CHEMICALS

ASTM INSULATION STANDARDS
(SORTED BY ASTM CODE)

ASTM NO.		TITLE
E	331	TEST FOR WATER PENETRATION OF EXTERIOR WINDOWS, CURTAIN WALLS, AND DOORS BY UNIFORM STATIC AIR PRESSURE DIFFERENCE
E	336	TEST FOR MEASUREMENT OF AIRBORNE SOUND INSULATION IN BUILDINGS
E	398	REC. PRACTICE FOR DYNAMIC MEASUREMENT OF WATER VAPOR TRANSFER
E	413	CLASSIFICATION FOR DETERMINATION OF SOUND TRANSMISSION CLASS
E	477	TESTING DUCT LINER MATERIALS AND PREFABRICATED SILENCERS FOR ACOUSTICAL AND AIRFLOW PERFORMANCE
E	492	LABORATORY MEASUREMENT OF IMPACT SOUND TRANSMISSION THROUGH FLOOR-CEILING ASSEMBLIES USING THE TAPPING MACHINE
E	514	TEST FOR WATER PERMEANCE OF MASONRY
E	518	TEST FOR FLEXURAL BOND STRENGTH OF MASONRY
E	519	TEST FOR DIAGONAL TENSION (SHEAR) IN MASONRY ASSEMBLAGES
E	529	FLEXURAL TESTS ON BEAMS AND GIRDERS FOR BUILDING CONSTRUCTION
E	535	REC. PRACTICE FOR PREPARATION OF FIRE TEST STANDARDS
E	546	TEST FOR FROST POINT OF SEALED INSULATING GLASS UNITS
E	547	TEST FOR WATER PENETRATION OF EXTERIOR WINDOWS, CURTAIN WALLS, AND DOORS BY CYCLIC STATIC AIR PRESSURE DIFFERENTIAL
E	564	STATIC LOAD TEST FOR SHEAR RESISTANCE OF FRAMED WALLS FOR BUILDINGS
E	575	REC. PRACTICE FOR REPORTING DATA FROM STRUCTURAL TESTS OF BUILDING CONSTRUCTIONS ELEMENTS, CONNECTIONS, AND ASSEMBLIES
E	576	TEST FOR DEW/FROST POINT OF SEALED INSULATING GLASS UNITS IN VERTICAL POSITION
E	596	LABORATORY MEASUREMENT OF THE NOISE REDUCTION OF SOUND-ISOLATING ENCLOSURES
E	605	TESTS FOR THICKNESS AND DENSITY OF SPRAYED FIRE-RESISTIVE MATERIAL APPLIED TO STRUCTURAL MEMBERS
E	662	TEST FOR SPECIFIC OPTICAL DENSITY OF SMOKE GENERATED BY SOLID MATERIALS
E	737	PRACTICE FOR INSTALLATION OF STORM WINDOWS, REPLACEMENT WINDOWS, MULTI-GLAZING STORM DOORS, AND REPLACEMENT DOORS
E	741	PRACTICE FOR MEASURING AIR LEAKAGE RATE BY THE TRACER DILUTION METHOD
G	20	TEST FOR CHEMICAL RESISTANCE OF PIPELINE COATINGS
G	21	REC. PRACTICE FOR DETERMINING RESISTANCE OF SYNTHETIC POLYMERIC MATERIALS TO FUNGI
G	22	REC. PRACTICE FOR DETERMINATION OF RESISTANCE OF PLASTICS TO BACTERIA

Appendix B

Manufacturers and Associations

MANUFACTURERS

Aerolite SPE Corporation
8025 Dixie Highway
Florence, Kentucky 41042
(606) 371-2030

Accessible Products Company
University Industrial Park
2122 West 5th Place
Tempe, Arizona 85281
(602) 967-8888

Advanced Foam Systems
2865 Thornhills S.E.
Grand Rapids, Michigan 49506
(616) 942-4780

All-Foam Division
Donray Products Company
500 SOM Center Road
Cleveland, Ohio 44143
(216) 449-6450

Alpha Associates
Two Amboy Avenue
Woodbridge, New Jersey 07095
(201) 634-5700

American Thermcell, Inc.
15546 Cleveland Street
P.O. Box 782
Elk River, Minnesota 55330
(612) 421-5600

Anderson Door Company
19106 Miles Avenue
Cleveland, Ohio 44128
(216) 475-5700

Apache Foam Products Company
2025 East Linden Avenue
Linden, New Jersey 07036
(201) 485-6723

Approved Insulation
2 Wilson Boulevard
C. Islip, New York 11722
(516) 582-4493

Arkansas Plastics, Inc.
Box 165
Sulphur Springs, Arkansas 72768
(501) 298-3224

Armm Industries, Inc.
90 N.E. 20th Street
P.O. Box 122
Lawton, Oklahoma 73502
(405) 248-7430

Armstrong World Industries, Inc.
P.O. Box 3001
Lancaster, Pennsylvania 17604
(717) 397-0611

ASG Industries, Inc.
P.O. Box 929
Kingsport, Tennessee 37662
(615) 245-0211

Automation Industries, Inc.
Flexible Tubing Division
Greenville, South Carolina 29606
(803) 288-7175

BASF Wyandotte Corporation
1609 Biddle Avenue
Wyandotte, Michigan 41814
(313) 282-3300

The polystyrene board manufacturers using BASF Wyandotte Beads are listed below and on the following six pages.

<u>Alabama</u>

Mahoney Plastics
Decatur, Alabama 35601
(205) 353-0476

<u>Alaska</u>

Western Insulfoam Coproation
Anchorage, Alaska 99504
(907) 279-9407

<u>Arizona</u>

Arizona Diversified Products
Phoenix, Arizona 85004
(602) 253-3191

<u>Arkansas</u>

Arkansas Plastics
Sulphur Springs, Arkansas 72736
(501) 298-3224

Drew International
Monticello, Arkansas 71655
(501) 367-6245

Insul Bead Corporation
Gravette, Arkansas 72736
(501) 787-5991

Stanark Plastics Company
North Little Rock, Arkansas 72114
(501) 945-1114

<u>California</u>

Falcon Manufacturing of California
Los Angeles, California 90061
(213) 329-4152

Far Western Foam Products, Inc.
Santa Fe Springs, California 90670
(213) 863-4845

W. R. Grace & Company
South Gate, California 90280
(213) 567-7764

Marko Foam Products, Inc.
Santa Ana, California 92705
(714) 835-6441

Vertex, Inc.
Vernon, California 90058
(213) 582-0751

Vertex, Inc.
Oakland, California 94604
(415) 763-2070

Western Insulfoam Corporation
Dixon, California 95620
(916) 753-4010

Western Insulfoam Corporation
Westminster, California 92683
(714) 893-6567

<u>Colorado</u>

Advanced Foam Plastics
Broomfield, Colorado 80020
(303) 466-1997

Drew Foam of Colorado
Denver, Colorado 80204
(303) 534-2342

Rocky Mountain Foam-Form, Inc.
Ft. Collins, Colorado 80529
(303) 221-5422

<u>Connecticut</u>

Foam Plastic of New England
Prospect, Connecticut 06712
(203) 758-6411

The Gilman Brothers Company
Gilman, Connecticut 06336
(203) 889-8444

Plastifoam Corporation
Rockville, Connecticut 06066
(203) 857-6274

Preferred Plastic Company
Putnam, Connecticut 06260
(203) 928-7795

Florida

Dyplast of Florida
Miami, Florida 33144
(305) 261-4637

W. R. Grace & Company
Boca Raton, Florida 33432
(305) 395-2424

Panel Foam, Inc.
Longwood, Florida 32750
(305) 339-2200

Penn-Plast, Inc.
St. Petersburg, Florida 33714
(813) 527-2163

Pioneer Plastics
Pensacola, Florida 32504
(904) 476-9572

The Plasti Kraft Corporation
Ozona, Florida 33560
(813) 784-1434

Southeastern Foam Products, Inc.
Ocala, Florida 32670
(904) 687-2852

Southern Foam Products, Inc.
Live Oak, Florida 32060
(904) 362-3286

Georgia

Foam Industries, Inc.
Conyers, Georgia 30207
(404) 922-4074

Georgia Foam
Gainsville, Georgia 30501
(404) 536-8888

W. R. Grace & Company
Atlanta, Georgia 30306
(404) 448-5880

Insulaire, Inc.
Gainesville, Georgia 30501
(404) 983-7291

Integrated Insulation Systems
Decatur, Georgia 33035
(404) 981-7160

Southeastern Foam Products, Inc.
Conyers, Georgia 30207
(404) 483-4491

Woolley & Company
Doraville, Georgia 30040
(404) 448-8473

Hawaii

Pacific Allied Products, Inc.
Kaneohe, Hawaii 96744
(808) 682-2038

Illinois

Approved Styrene Works
Chicago, Illinois 60639
(312) 523-0510

EPS Industries, Inc.
Dixon, Illinois 61021
(815) 284-6678

Litteral Life Corporation
Paris, Illinois 61944
(217) 466-0370

Indiana

EFP Corporation
Elkhart, Indiana 46514
(219) 295-4690

Southeastern Foam Products, Inc.
Bargersville, Indiana 46106
(317) 422-9271

Iowa

Holland Industries
Gilman, Iowa 50106
(515) 498-7404

Iowa Manufacturing
Indianola, Iowa 50125
(505) 961-7403

Polycell Industries, Inc.
Marion, Iowa 52302
(319) 377-9495

Kansas

Contour Packaging
Lenexa, Kansas 66051
(913) 888-4848

EPS Industries, Inc.
Wichita, Kansas 67201
(316) 942-1494

Star Foam, Inc.
Independence, Kansas 67301
(316) 331-0470

Kentucky

Day Star Corporation
Somerset, Kentucky 42501
(606) 679-4836

Drew Foam of Kentucky
Winchester, Kentucky 40391

Louisiana

Drew Foam of Louisiana
Hammond, Louisiana 70401
(504) 345-0040

Maryland

Amotex Plastics
Baltimore, Maryland 21205
(301) 485-8585

Foam Industries, Inc.
Frederick, Maryland 21701
(301) 662-3626

Polystyrene Products Company
Baltimore, Maryland 21220
(301) 335-2666

Southeastern Foam Products, Inc.
Adamstown, Maryland 21710
(301) 874-5484

Massachusetts

Dyrelite Corporation
New Bedford, Massachusetts 02744
(617) 993-9955

W. R. Grace & Company
Cambridge, Massachusetts 02140
(617) 876-1400

Insulation Technology, Inc.
Bridgewater, Massachusetts 02324

Michigan

Drew Foam of Michigan
Fenton, Michigan 48430
(313) 629-1531

Falcon Manufacturing of Mighigan, Inc.
Byron Center, Michigan 49315
(616) 878-1568

Jacobs Plastics
Adrian, Michigan 49221
(517) 263-3890

Mar-Foam Inc.
Marlette, Michigan 48453
(517) 635-6801

Marne Industries
Marne, Michigan 49435
(616) 677-3501

Michigan Foam Products
Grand Rapids, Michigan 49509
(616) 452-9611

Pacolite Plastics
Saginaw, Michigan 48604
(517) 754-3366

Robinson Industries
Coleman, Michigan 48618
(517) 465-6111

Minnesota

McArthur Company
St. Paul, Minnesota 55114
(612) 646-2773

Minnesota Diversified Products
Arden Hills, Minnesota 55112

Minnesota Diversified Products
Rockford, Minnesota 55373
(612) 477-5854

Minnesota Diversified Products
St. Paul, Minnesota 55114
(612) 645-8952

Poly Foam Incorporated
Lester Prairie, Minnesota 55354
(612) 395-2551

Mississippi

Century Insulation Manufacturing Company
Union, Mississippi 39365
(601) 774-8285

Drew Foam of Mississippi
Pearl, Mississippi 39208
(601) 939-5238

Southeastern Foam Products, Inc.
Grenada, Mississippi 38901
(601) 226-7085

Value Foam
Pearl, Mississippi 39208
(601) 939-0056

Missouri

Diversified Plastics
Nixa, Missouri 65714
(417) 725-2622

Foam Products
St. Louis, Missouri 63107
(314) 521-1711

Imperial Foam
Camdenton, Missouri 65020
(314) 873-5210

Lar-Roy Foam Company
Cape Girardeau, Missouri 63701
(314) 334-1844

N.P.S. Corporation
Perryville, Missouri 63775
(314) 547-8389

Southeastern Foam Products, Inc.
Wentzville, Missouri 63385
(314) 327-5191

Montana

Big Sky Insulation Unlimited
Belgrade, Montana 59714
(406) 388-4146

Nebraska

EPS Industries, Inc.
Omaha, Nebraska 68137
(402) 330-1700

Mid-America Industries
Mead, Nebraska 68041
(402) 624-6611

New Hampshire

Avilite Industries
Marlborough, New Hampshire 03455
(603) 876-3313

New Jersey

Poly Molding Corporation
Haskell, New Jersey 07420
(201) 835-7161

U.S. Mineral Products Company
Stanhope, New Jersey 07874
(210) 347-1200

New Mexico

Southwest Insulbead
Albuquerque, New Mexico 87102
(505) 243-0666

New York

Poly Fab Products, Inc.
Menants, New York 12204

Polystyrene Molders, Inc.
Newfield, New York 14867
(607) 564-7035

Thermal Foams, Inc.
Buffalo, New York 14207
(716) 874-6470

North Carolina

Foam Industries
Graham, North Carolina 27253
(919) 226-9873

Foam Molding, Inc.
Asheboro, North Carolina 27203
(919) 629-1495

Southeastern Foam Products, Inc.
Burlington, North Carolina 27215
(919) 227-9041

Ohio

Clark Industries
Columbus, Ohio 43201
(614) 294-3761

Foam Master, Inc.
Cincinnati, Ohio 45241
(513) 771-2266

Foam Master, Inc.
Twinsburg, Ohio 44087
(216) 425-3188

Pacemaker Plastics
Dover, Ohio 44622
(216) 364-8862

Southeastern Foam Products Company
New Middleton, Ohio 44442
(216) 542-2964

Southern Ohio Foam
Lebanon, Ohio 45036
(513) 932-7755

Stolle Corporation
Sidney, Ohio 45365
(513) 492-1111

Strata Foam Corporation
Akron, Ohio 44309
(216) 929-1811

Oklahoma

Lin Manufacturing
Clinton, Oklahoma 73601
(405) 323-3010

Sequoyah Foam Company
Sallisaw, Oklahoma 74955
(918) 775-9741

Tri State Foam Company
Tulsa, Oklahoma 74116
(918) 835-8241

Pennsylvania

EFP Corporation
Lancaster, Pennsylvania 17604
(717) 397-2165

Foam Products Corporation
York Haven, Pennsylvania 17370
(717) 266-3671

French Creek Products
Royersford, Pennsylvania 19468
(215) 948-6770

W. R. Grace & Company
New Castle, Pennsylvania 16102
(412) 654-7721

Insul-Board
Erie, Pennsylvania 16505
(814) 833-7400

Southeastern Foam Proudcts, Inc.
Fogelsville, Pennsylvania 18051
(215) 398-1177

Toyad Corporation
Latrobe, Pennsylvania 15650
(412) 537-9000

South Carolina

Dyplast of South Carolina
Starr, South Carolina 29684
(803) 296-3424

South Dakota

Webster Industries
Webster, South Dakota 57274
(605) 345-3131

Tennessee

Amotex Plastics
Nashville, Tennessee 37212
(615) 254-1381

Drew Foam of Memphis, Inc.
Memphis, Tennessee 38103
(901) 525-1569

W. F. Martin Company
Knoxville, Tennessee 37917
(615) 523-0401

Southeastern Foam Products, Inc.
Jonesboro, Tennessee 37659
(615) 753-5621

U.S. Foam Company
Memphis, Tennessee 38107
(901) 523-0357

Texas

Alamo Foam Company
San Antonio, Texas 78203
(512) 222-1286

Drew Foam of Houston
Houston, Texas 77001
(713) 224-3486

Drew Tex Foam Company
Waxahachie, Texas 75165
(214) 937-6390

Emerson Plastics
Houston, Texas 77002
(713) 225-2095

W. R. Grace & Company
Houston, Texas 77008
(713) 864-2657

Insulation Materials
Ft. Worth, Texas 76117
(817) 281-5929

Therma Foam Company
Ft. Worth, Texas 76106
(817) 429-7350

United Foam Industries
Irving, Texas 75070
(214) 255-8595

Utah

Marko Foam Products, Inc.
Salt Lake City, Utah 84104
(801) 972-1354

Virginia

General Foam Plastics Corporation
Norfolk, Virginia 23502
(703) 857-0153

Radva Plastics Corporation
Radford, Virginia 24141
(703) 639-2458

Southeastern Foam Products, Inc.
Petersburg, Virginia 23803
(804) 733-1810

Washington

W. R. Grace & Company
Auburn, Washington 93002
(206) 852-5725

Western Insulfoam Corporation
Kent, Washington 98031
(206) 242-9424

Wisconsin

W. R. Grace & Company
Milwaukee, Wisconsin 53208
(414) 344-6667

Mid West Plastics, Inc.
Pembine, Wisconsin 54156
(715) 324-5555

Plymouth Foam Products
Plymouth, Wisconsin 53073
(414) 893-0535

Sandra Corporation
North Prairie, Wisconsin 53153
(414) 392-9126

Southeastern Foam Products, Inc.
Elkhorn, Wisconsin 53121
(414) 723-2580

Spectrum Manufacturing
West Allis, Wisconsin 53214
(414) 475-1215

B.F. Goodrich General Products Division
33095 Bainbridge Road
Solon, Ohio 44139
(216) 248-4391

Benoit, Inc.
635 North Prior Avenue
St. Paul, Minneapolis 55104
(800) 328-1436

Bonded Insulation Company, Inc.
77 Pauling Street
Hagaman, New York 12086
(518) 842-1470

Brouk Company
1367 South Kingshighway
St. Louis, Missouri 63110
(314) 533-9022

Carney Insulation Corporation
4930 W. 77th Street, Suite 315
Edina, Minnesota 55435
(612) 835-3717

Casco Mineral Wool Division
(formerly Midwest Insulations Division)
L.C. Cassidy & Son, Inc.
1918 S. High School Road
Indianapolis, Indiana 46241
(317) 241-6391

Cellin Manufacturing
P.O. Box 688
Springfield, Virginia 22150
(703) 550-7277

The Celotex Corporation
Building Products Division
1500 North Dale Mabry Highway
Tampa, Florida 33607
(813) 871-4418

Certain-Teed Corporation
P.O. Box 860
Valley Forge, Pennsylvania 19482
(215) 687-5000

Coastal Foam, Inc.
129 Commerce Street
Apalachicola, Florida 32320
(804) 653-8892

Compac Corporation
Old Flanders Road
Netcong, New Jersey 07857
(201) 347-3900

Consolidated Fiber Glass Products Company, Inc.
P.O. Box 5248
Bakersfield, California 93388
(805) 323-6026

Cook Paint & Varnish Company
P.O. Box 389
Kansas City, Missouri 64141
(816) 471-4800

CPR Division
The Upjohn Company
555 Alaska Avenue
Torrance, California 90503
(213) 320-3550

CY/RO Industries
Wayne, New Jersey 07470
(201) 839-4800

Diversified Insulation, Inc.
P.O. Box 188
2705 West Highway 55
Hamel, Minnesota 55340
(612) 478-6614

Dow Chemical, U.S.A.
Granville Research Center
P.O. Box 515
Granville, Ohio 43023
(614) 587-4351

Dow Corning Corporation
Midland, Michigan 48640

Drew Foam Company, Inc.
311 Godrey
Monticello, Arkansas 71655
(501) 367-6246

Dryvit System, Inc.
420 Lincoln Avenue
Warwick, Rhode Island 02888
(401) 463-7150

Dunamis
25628 Snyder Avenue
Conifer, Colorado 80433
(303) 934-2151

E. B. Kaiser Company
2114 Chesnut Avenue
Glenview, Illinois 60025
(312) 724-4500

EFP Corporation
223 Middleton Run Road
Elkhart, Indiana 46514
(219) 295-4690

E. I. du Pont de Nemours & Company, Inc.
Wilmington, Delaware 19898
(302) 774-2629

Elwin G. Smith Division
100 Walls Street
Pittsburgh, Pennsylvania 15202
(412) 761-7474

Falcon Manufacturing of Michigan, Inc.
8240 Byron Center Road
Byron Center, Michigan 49315
(616) 878-1568

Finestone Corporation
18307 Weaver Street
Detroit, Michigan 48228
(313) 837-9672

Foam Master, Inc.
2292 E. Aurora Road
P.O. Box 306
Cleveland, Ohio 44087
(216) 425-3188

Foam Plastics of New England
New Haven Road - Route 69
Prospect, Connecticut 06712
(203) 758-6411

Foam Products, Inc.
Gay Street
York Haven, Pennsylvania 17370
(717) 266-3671

Foam Systems Corporation
1980 Atlantic Avenue
P.O. Box 5347
Riverside, California 92517
(714) 684-8333

Ford Glass Division
300 Renaissance Center, Suite 2300
Detroit, Michigan 48243

Formed Fabrics
A Division of International Paper Company
International Paper Plaza
77 W. 45th Street
New York, New York 10036
(212) 536-6361

Forty-Eight Insulations, Inc.
P.O. Box 1148
Aurora, Illinois 60507
(312) 896-4800

G.A.F. Corporation
140 W. 51st Street
New York, New York 10020
(212) 582-7600

General Aluminum Supply Corporation
P.O. Box 11430
Kansas City, Missouri 64112
(913) 722-2100

General Electric Company
Silicone Products Division
RTV Products Department
Waterford, New York 12188

General Plastics Manufacturing Company
3481 South 35th Street
Tacoma, Washington 98409
(206) 383-1631

The Gilman Brothers Company
Gilman, Connecticut 06336
(203) 889-8444

Grefco, Inc.
Building Products Division
3450 Wilshire Boulevard
Los Angeles, California 90010

Guardian Industries Corporation
43043 West Nine Mile Road
Northville, Michigan 48167

Hamfab Incorporated
Bridge and 9th Streest
Lehighton, Pennsylvania 18235
(215) 377-4120

Hamilton Manufacturing & Distributing, Inc.
118 Market Street
P.O. Box 1426
Twin Falls, Indiana 83301
(208) 733-9689

H. B. Fuller Company
Foster Products Division
P.O. Box 625
Spring House, Pennsylvania 19477

Hexcel
11711 Dublin Boulevard
Dublin, California 94566

High-R Building Systems, Inc.
225 S. Price Road
Longmont, Colorado 80501
(303) 772-3516 or 499-9108

H. L. Birum Corporation
75 S. Union Street
Lambertville, New Jersey 08530
(609) 397-1750

Homasote Company
Box 7240
West Trenton, New Jersey 08628
(609) 883-3300

Hoshall Industries, Inc.
1001 Enterprise Avenue, Suite 13
Oklahoma City, Oklahoma 73128

Hurstline Sales, Inc.
Route 7, Gilbert Lane
Concord, Tennessee 37720
(615) 966-5841

Illbruck/USA
3800 Washington Avenue North
Minneapolis, Minnesota 55412
(612) 521-3555

In-Sol, Inc.
1200 E. 4th
P.O. Box 971
Taylor, Texas 76574
(512) 352-5513

Insoport Industries, Inc.
3200 Reach Road
P.O. Box 3033
Williamsport, Pennylvania 17701
(717) 326-7325

Insta Foam Products, Inc.
2050 N. Broadway
Joliet, Illinois 60435
(815) 726-6241

Insulated Panel Systems, Inc.
Drawer B, 13410 Murphy Road
Stafford, Texas 77477
(713) 499-6541

Insulation Materials Corporation of America (IMCOA)
4325 Murray Avenue
Haltom City, Texas 76117
(817) 485-5290

International United Chemical
645 E. 60th Street
Los Angeles, California 90003

Iowa Excel Corporation
P.O. Box 353
West Des Moines, Iowa 50265
(515) 225-6878

Iowa Manufacturing Specialists, Inc.
400 E. Iowa
Indianola, Iowa 50125
(515) 961-7403

Kawneer Company
Product Information, Department C
1105 N. Front Street
Niles, Michigan 49120

Keene Corporation
Building Products Division
1603 Fulford Street
Kalamazoo, Michigan 49003
(616) 343-1226

Knauf Fiber Glass GmbH
Shelbyville, Indiana 46176
(317) 398-4434

KoolShade Corporation
722 Genevieve Street
Solano Beach, California 92075
(714) 755-5126

Koppers Company, Inc.
Organic Materials Group
Pittsburgh, Pennsylvania 15219
(412) 227-2000

Libbey-Owens-Ford Company
811 Madison Avenue
Toledo, Ohio 43695
(419) 247-3731

Lion Oil Company
Lion Oil Building
El Dorado, Arkansas 71730
(501) 863-3111

Manville Corporation
Ken-Caryl Ranch
Denver, Colorado 80217
(303) 979-1000

Marathon Roofing Products, Inc.
367 Nagel Drive
Buffalo, New York 14225
(716) 685-3340

Metal Building Interior Products Company
1176 East 38th Street
Cleveland, Ohio 44114
(216) 431-6400

Mid-America Industries, Inc.
Route 1, Box 101
Mead, Nebraska 68041
(402) 624-6611

Mizell Brothers Company
151 Regal Row
Dallas, Texas 75247
(214) 638-3491

Mono-Therm
P.O. Box 934
551 S. Yosemite Avenue
Oakdale, California 95361
(209) 847-3055

The Moore Company
Marceline, Missouri 64658
(816) 376-3583

National Cellulose Corporation
12315 Robin Boulevard
Houston, Texas 77045
(713) 433-6761

National Insulation, Inc.
1601 Garfield Avneue
Bay City, Michigan 48706
(517) 894-0647

Northeast Specialty Insulations, Inc.
One Watson Place
Saxonville, Massachusetts 01701
(617) 877-0721

Olin Corporation
120 Long Ridge Road
Stamford, Connecticut 06904
(203) 366-2262

Oren Corporation
P.O. Box 2446
Muncie, Indiana 47302
(317) 288-9988

Owens-Corning Fiberglas Corporation
Fiberglass Tower
Toledo, Ohio 43659
(419) 248-8827

Panel Foam, Inc.
811 South Wilma Street
Longwood, Florida 32750
(305) 339-2200

Panelera, Inc.
1857 South 3850 West
Salt Lake City, Utah 84104
(801) 972-3994

Patten Building Supply
435 Cleveland Avenue
Winnebago, Minnesota 56098
(507) 893-3112

Pease Rolling Shutters
2001 Troy Avenue
New Castle, Indiana 47362

Peerless Products, Inc.
2534 Madison Avenue
Kansas City, Missouri 64108
(816) 421-6690

The Perlite Institute, Inc.
45 West 45th Street
New York, New York 10036
(212) 265-2145

The members of the Perlite Institute are listed below and on the next two pages.

Alabama

Southeastern Perlite, Inc.
P.O. Box 6824
Birmingham, Alabama 35210
(205) 956-9545

California

Aztec Perlite Company
1518 Simpson Way
Escondido, California 92025
(714) 741-1733

Grefco, Inc.
Dicalite Division
3450 Wilshire Boulevard
Los Angeles, California 90010
(213) 381-5081

Redco, Inc.
11831 Vose Street
North Hollywood, California 91605
(213) 875-0440

Colorado

Grefco, Inc.
P.O. Box 308
Antonito, Colorado 81120
(303) 376-5475

Johns-Manville Perlite Corporation
Ken-Caryl Ranch
Denver, Colorado 80217
(303) 979-1000

Persolite products, Inc.
P.O. Box 105
Florence, Colorado 81226
(303) 572-3222

Florida

Airlite Processing Corporation of Florida
3505 65th Street
Vero Beach, Florida 32960
(305) 562-3518

Chemrock Corporation
P.O. Box 9317 Lake Forest Station
North Edgewood Avenue
Jacksonville, Florida 32208
(904) 355-0096

Zonolite-Construction Products Division
W. R. Grace & Company
1200 N.W. 15th Avenue
Pompano Beach, Florida 33064
(305) 974-5200

Illinois

Filter Products Corporation
124 N. Buesching Road
Lake Zurich 60047
(312) 438-2363

Grefco, Inc.
Building Products Division
2905 Butterfield Road, Suite 290
Oak Brook, Illinois 60521
(312) 654-4500

Mica Pellets, Inc.
1120 Oak Street
De Kalb, Illinois 60115
(815) 756-9525

Silbrico Corporation
6300 River Road
Hodgkins, Illinois 60525
(312) 735-3322

Indiana

Chemrock Corporation
P.O. Box 465
Highway 25 & Monon RR
Lafayette, Indiana 47902
(317) 474-8413

Grefco, Inc.
P.O. Box 48
Crawfordsville, Indiana 47933
(317) 362-6000

Kentucky

Zonolite-Construction Products Division
W.R. Grace & Company
112 North Street
Wilders, Newport 41071
(606) 291-3500

Louisiana

American Perlite Products, Inc.
P.O. Box 128
Gilliam, Louisiana 71029
(318) 296-4316 or 222-3638

Filter-Media Company of Louisiana, Inc.
P.O. Box 222
Reserve, Louisiana 70084

Maine

Chemrock Corporation
P.O. Box 177
Thomaston, Maine 04861
(207) 594-8225

Massachusetts

Whittemore Perlite Company, Inc.
Dundee Park
Andover, Massachusetts 01810
(617) 470-0317

Zonolite-Construction Products Division
W.R. Grace & Company
62 Whittemore Avenue
Cambridge, Massachusetts 02140
(617) 876-1400

Missouri

Brouk Company
1367 S. Kingshighway
St. Louis, Missouri 63110
(314) 533-9022

New Jersey

The Schundler Company
P.O. Box 249
Metuchen, New Jersey 08840
(201) 287-2244

New York

Buffalo Perlite Division Of
Pine Hill Concrete Mix Corporation
100 Sugg Road
Buffalo, New York 14225
(716) 634-5600

Scolite International Corporation
P.O. Box 1
Troy, New York 12181
(518) 272-2400

North Carolina

Carolina Perlite Company, Inc.
P.O. Box 158
Gold Hill, North Carolina 28071
(704) 279-2325

Ohio

The Cleveland Gypsum Division
The Cleveland Builders Supply Company
2145 West Third Street
Cleveland, Ohio 44113
(216) 621-4300

Oregon

Supreme Perlite Company
4600 North Suttle Road
Portland, Oregon 97217
(503) 286-4333

Pennsylvania

Pennsylvania Perlite Corporation
P.O. Box 2002
Allentown, Pennsylvania 18001
(215) 264-2891

Perlite Manufacturing Company of
Pittsburgh, Inc.
P.O. Box 478
Carnegie, Pennsylvania 15106
(412) 923-1525

Tennessee

Chemrock Corporation
P.O. Box 7151
Nashville, Tennessee 37210
(615) 254-1866

Texas

Filter-Media, Inc.
P.O. Box 19156
Houston, Texas 77024
(713) 622-1520

Perlite of Houston, Inc.
6105 Beverly Hill
Houston, Texas 77057
(713) 781-5411

Sil-Flo Incorporated
34-05 North Sylvania Avenue
Fort Worth, Texas 76111
(817) 834-1944

South Texas Perlite Company
P.O. Box 27272 Valley-Hi Station
San Antonio, Texas 78227
(512) 653-1635

Zonolite-Construction Products Division
W.R. Grace & Company
2651 Manila Road
Dallas, Texas 75212
(214) 637-0900

Wisconsin

Zonolite-Construction Products Division
W.R. Grace & Company
900 North 43rd Street
Milwaukee, Wisconsin 53208
(414) 344-6667

Pittsburgh Corning Corporation
800 Presque Isle Drive
Pittsburgh, Pennsylvania 15239
(412) 327-6100

Plaskolite, Inc.
P.O. Box 1497
Columbus, Ohio 43216
(614) 294-3281

The Plastifoam Corporation
66-68 West Street
Rockville, Connecticut 06066
(203) 875-6274

Poly Blends, Inc.
12350 Merriman Road
Livonia, Michigan 48150
(313) 427-5600

Poly-Foam, Inc.
Lester Prairie, Minnesota 55354
(612) 395-2551

Polymer Development Labs, Inc.
15731 Graham Street
Huntington Beach, California 92044
(714) 898-9586

Preferred Plastics, Inc.
Park Street
Putnam, Connecticut 06260
(203) 928-7795

Protexulate, Inc.
One World Trade Center
Suite 2173
New York, New York 10048

Rapco Foam, Inc.
122 E. 42nd Street
New York, New York 10017
(212) 986-7030

Reichhold Chemicals, Inc.
RCI Buildings
White Plains, New York 10603
(914) 682-5700

R.I. Energy Corporation
275 Harborside Boulevard
Providence, Rhode Island 02905
(401) 521-7500

RMAX, Inc.
13524 Welch Road
Dallas, Texas 75234
(214) 387-4500

Rockwool Industries, Inc.
3600 South Yosemite Street
Suite 700
Denver, Colorado 80237
(303) 773-6200

Rocky Mountain Foam-Form
3034 E. Mulberry
P.O. Box 1937
Fort Collins, Colorado 80522
(303) 221-5422

Rohm and Haas Company
Independence Mall West
Philadelphia, Pennsylvania 19105
(215) 592-3000

Rovanco Corporation
I-55 and Frontage Road
Joliet, Illinois 60436
(815) 741-6700

Ryder Industries, Inc.
P.O. Box 11196
Dallas, Texas 75223
(214) 428-1622

Sealeze Corporation
8011 White Bark Terrace
Chesterfield Industrial Air Park
Richmond, Virginia 23234

The Schundler Company
P.O. Box 249
150 Whitman Avenue
Metuchen, New Jersey 08840
(201) 287-2246

Shatterproof Glass Corporation
4815 Cabot Avenue
Detroit, Michigan 48210

Shelter Insulation, Inc.
3626 Binz-Engleman Road
San Antonio, Texas 78219
(512) 224-2741

Silbrico Corpoation
6300 River Road
Hodgkins, Illinois 60525
(312) 735-3322

Silvercote Metal Building Products
Elkhart, Indiana
(219) 293-3551

Southeastern Foam Products, Inc.
P.O. Box 406
1125 Ellington Drive
Conyers, Georgia 30207
(404) 483-4491

St. Regis
Laminated & Coated Products Division
P.O. Box 959
55 Starkey Avenue
Attleboro, Massachusetts 02703
(617) 222-3500

Stanley Hardware
Division of the The Stanley Works
New Britain, Connecticut 06050

Stauffer Chemical Company
50 Galesi Drive
Wayne, New Jersey 07470

Stepan Chemical
Edens & Winnetka
Northfield, Illinois 60093
(312) 446-7500

Texas Urethanes, Inc.
9721 Highway 290 East
Austin, Texas 78766
(512) 272-5531

Thermal Systems, Inc.
3055 W. 2100 Street
Salt Lake City, Utah 84114
(801) 972-6650

Thermoguard Insulation Company
8207 E. Trent
Spokane, Washington 99206
(206) 624-3871

Thermo Products Company
2508 New Marlin Highway
Waco, Texas 76705
(817) 756-3713

Thoro System Products
Department IO 815
7800 N.W. 38th Street
Miami, Florida 33166
(305) 592-2081

3M/Energy Control Products
220-8E, 3M Center
St. Paul, Minnesota 55144

Tomko Asphalt Products, Inc.
P.O. Box 326
Phillipsburg, Kansas 67661
(913) 543-2156

Topline Products, Inc.
P.O. Box 26746
Richmond, Virginia 23261
(800) 446-2606

Toyard Corporation
P.O. Box 30
Latrobe, Pennsylvania 15650
(412) 597-7754

Tri-State Foam Company
307 E. Brady
Tulsa, Oklahoma 74120
(918) 599-9101

United Fiber Corporation
Seneca Industrial Park
irving, New York 14081
(716) 549-5400

United Foam Corporation
2626 Vista Industria
Compton, California 90221
(213) 639-5600

United McGill Corporation
200 E. Broadway
Westerville, Ohio 43081
(614) 882-7401

U.S. Gypsum Company
101 S. Wacker Drive
Chicago, Illinois 60606
(312) 321-3759

United States Mineral Products Company
Furnace Street
Stanhope, New Jersey 07874
(201) 347-1200

Water Guidance Systems, Inc.
388 East Main Street
Branford, Connecticut 06405
(203) 481-4231

Wat Pro
2517 Highway 35
P.O. Box 336
Manasquan, New Jersey 08736
(201) 528-6727

W.R. Grace & Company
Construction Products Division
62 Whittemore Avenue
Cambridge, Massachusetts 02140

Weather Guard, Inc.
4308 Riverline Drive
St. Louis, Missouri 63045
(314) 291-7335

Western Insulfoam Corporation
15040 Golden West Circle
Westminster, California 92683
(714) 893-6567

Western Weathercheck
305 Mathew Street
Santa Clara, California 95050
(408) 244-6615

Weyerhaeuser Company
(% Jay A. Johnson - WTC 1B4)
Tacoma, Washington 98003
(206) 924-6527

Williams Products, Inc.
1750 Maplelawn Boulevard
Troy, Michigan 48084
(313) 643-6400

The Wiremold Company
West Hartford, Connecticut 06110
(203) 233-6251

Witco Chemical Corporation
900 Wilmington Road
Wilmington, Delaware 19720
(302) 328-5661

Wooley & Company
6865 Wimms Drive
Doraville, Georgia 30340
(404) 448-8473

ASSOCIATIONS

American Hardboard Association (AHA)
205 W. Touhy Avenue
Park Ridge, Illinois 60068
(312) 692-5178

American Society of Heating, Refrigerating, and Air-Conditioning Engineers, Inc. (ASHRAE)
1791 Tullie Circle NE
Atlanta, Georgia 30329

American Society for Testing and Materials (ASTM)
1916 Race Street
Philadelphia, Pennsylvania 19103
(215) 299-5585

Insulation Contractors Association of America (ICAA)
905 16th Street, N.W.
Washington, D.C. 10006
(202) 347-2791

Laminated Fiberglass Insulation Producers Association (LFIPA)
c/o Thomas Associates, Inc.
1230 Kieth Building
Cleveland, Ohio 44115
(216) 241-7333

Metal Building Manufacturer's Association (MBMA)
1230 Kieth Building
Cleveland, Ohio 44115
(216) 241-7333

Midwest Insulation Contractors Association (MICA)
Omaha, Nebraska

Mineral Insulation Manufacturer's Association (MIMA)
382 Springfield Avenue
Summit, New Jersey 07901
(201) 277-1550

National Association of Home Builders Research Foundation, Inc. (NAHB)
627 Southlawn Lane
P.O. Box 1627
Rockville, Maryland 20850
(301) 762-4200

National Association of Plumbing-Heating-Cooling Contractors (NAPHCC)
1016 20th Street, N.W.
Washington, D.C. 20036
(202) 331-7675

National Building Material Distributors Association
55 E. Monroe Street, Suite 1616
Chicago, Illinois 60603
(312) 332-7127

National Bureau of Standards (NBS)
U.S. Department of Commerce
Washington, D.C. 20234

National Concrete Masonry Association (NCMA)
P.O. Box 135
McLean, Virginia 22101

National Fenestration Council (NFC)
3310 Harrison
Topeka, Kansas 66611
(913) 266-7014

National Insulation Contractors Association (NICA)
1025 Vermont Avenue Suite 410
Washington, D.C. 20005
(202) 783-6277

The Perlite Institute, Inc.
45 West 45th Street
New York, New York 10036
(212) 265-2145

Sealed Insulating Glass Manufacturer's Association (SIGMA)
111 E. Wacker Drive
Chicago, Illinois 60601
(312) 644-6610

Sheet Metal and Air Conditioning Contractors' Association, Inc. (SMACNA)
8224 Old Courthouse Road
Tysons Corner, Vienna, Virginia 22180

The Society of Plastics Industry, Inc.
Expandable Polystyrene Materials Group
3150 Des Plaines Avenue
Des Plaines, Illinois 60018
(312) 297-6150

Thermal Insulation
Manufacturer's Association (TIMA)
7 Kirby Plaza
Mount Kisco, New York 10549
(914) 241-2284

The Vermiculite Association
52 Executive Park South
Atlanta, Georgia 30345
(404) 321-7994

Appendix C

Useful Formulas and Data Tables

Table C-1
USEFUL FORMULAS

Where:
A = Area; A_s = Surface area of solids;
V = Volume; C = Circumference;
= Angle in degrees

RECTANGLE

$A = W \times L$

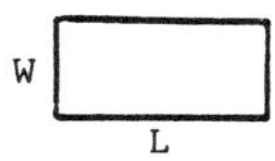

PARALLELOGRAM

$A = B \times H$

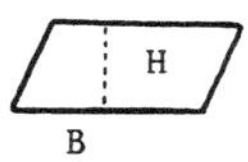

TRAPEZOID

$A = \frac{1}{2}(A + B) \times H$

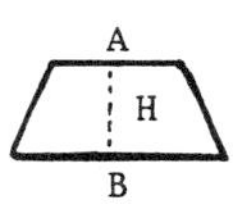

TRIANGLE

$A = \frac{B \times H}{2}$

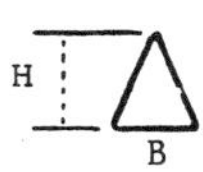

CIRCLE

$A = \pi \times R^2$

$C = \pi \times D$

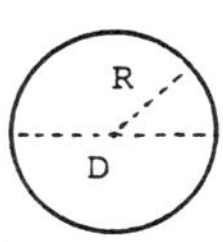

SECTOR OF CIRCLE

$A = \frac{\pi \times R^2 \times \theta}{360}$

$S = (\pi/180) \times R \times \theta$

$\theta = \frac{S}{(\pi/180) \times R}$

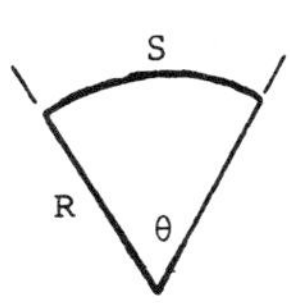

ELLIPSE

$A = \pi \times A \times B$

$C = 2\pi \times \sqrt{\frac{A^2 + B^2}{2}}$

RECTANGULAR SOLID

$A = 2(W \times L + L \times H + H \times W)$

$V = W \times L \times H$

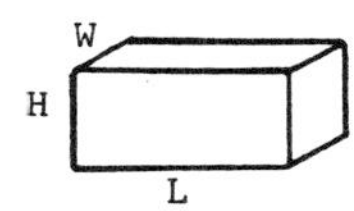

CONE

$A_1 = \pi \times R \times S + \pi \times R^2$

$V = (\pi/3) \times R^2 \times H$

CYLINDER

$A = 2\pi \times R \times L + 2\pi \times R^2$

$V = \pi \times R^2 \times L$

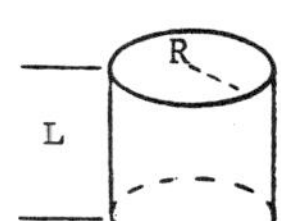

ELLIPTICAL SOLID

$V = \pi \times a \times b \times H$

$A = 2\pi \times \frac{a^2 + b^2}{2} \times H + 2\pi \times a \times b$

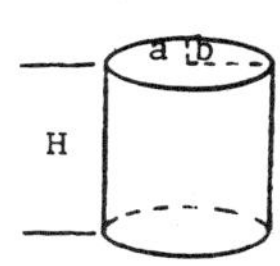

SPHERE

$A = 3\pi \times R^2$

$V = (4/3)\pi \times R^3$

TABLE C-2
HEAT LOSS PER LINEAR FOOT, BARE HORIZONTAL PIPE
(Btu/hr ft per °F temperature difference)

Nominal Pipe Size	Temperature Difference Degree Fahrenheit Between Pipe Surface And Surrounding Air (Air At 80°F)																			
In	50	100	150	200	250	300	350	400	450	500	550	600	650	700	750	800	850	900	950	1000
1/2	.466	.546	.616	.682	.752	.823	.895	.983	1.07	1.16	1.26	1.36	1.47	1.58	1.71	1.85	1.99	2.13	2.29	2.46
3/4	.572	.688	.754	.836	.921	1.01	1.10	1.21	1.32	1.43	1.55	1.68	1.82	1.97	2.12	2.29	2.46	2.64	2.84	3.05
1	.702	.819	.925	1.03	1.14	1.24	1.36	1.49	1.62	1.77	1.92	2.08	2.25	2.43	2.63	2.83	3.05	3.29	3.53	3.79
1-1/4	.870	1.02	1.15	1.27	1.41	1.54	1.69	1.86	2.03	2.21	2.40	2.60	2.81	3.05	3.89	3.55	3.82	4.12	4.43	4.76
1-1/2	.984	1.15	1.30	1.44	1.59	1.75	1.91	2.10	2.30	2.50	2.72	2.95	3.20	3.46	3.74	4.04	4.35	4.69	5.04	5.41
2	1.21	1.41	1.59	1.77	1.96	2.15	2.35	2.59	2.84	3.09	3.37	3.65	3.96	4.29	4.63	5.00	5.40	5.82	6.26	6.73
2-1/2	1.45	1.68	1.90	2.12	2.34	2.58	2.82	3.10	3.40	3.70	4.04	4.38	4.75	5.19	5.61	6.06	6.54	7.05	7.58	8.15
3	1.73	2.02	2.28	2.54	2.81	3.09	3.38	3.74	4.09	4.46	4.86	5.29	5.73	6.22	6.73	7.27	7.85	8.47	9.12	9.81
3-1/2	1.96	2.28	2.58	2.87	3.18	3.50	3.83	4.24	4.64	5.07	5.52	6.00	6.52	7.07	7.65	8.28	8.94	9.64	10.39	11.17
4	2.18	2.54	2.87	3.20	3.55	3.91	4.29	4.74	5.18	5.67	6.18	6.73	7.30	7.92	8.58	9.27	10.02	10.81	11.65	12.53
4-1/2	3.41	2.80	3.17	3.53	3.91	4.32	4.73	5.24	5.73	6.27	6.83	7.44	8.08	8.76	9.49	10.28	11.10	11.98	12.91	13.89
5	2.83	3.29	3.71	4.14	4.59	5.07	5.55	6.14	6.73	7.36	8.04	8.73	9.51	10.33	11.18	12.09	13.06	14.10	15.20	16.36
6	3.12	3.64	4.11	4.59	5.10	5.62	6.16	6.83	7.49	8.18	8.95	9.73	10.58	11.50	12.47	13.49	14.58	15.74	16.98	18.28
7	3.57	4.15	4.69	5.25	5.81	6.41	7.25	7.80	8.56	9.36	10.24	11.14	12.12	13.17	14.27	15.47	16.73	18.06	19.48	20.98
8	4.00	4.65	5.26	5.87	6.48	7.16	7.86	8.72	9.57	10.50	11.47	12.49	13.59	14.77	16.03	17.36	18.79	20.30	21.90	23.60
9	4.44	5.17	5.82	6.53	7.23	7.99	8.77	9.73	10.68	11.72	12.80	13.94	15.17	16.48	17.89	19.38	20.97	22.65	24.44	26.33
10	4.92	5.71	6.47	7.23	8.02	8.86	9.74	10.80	11.88	13.00	14.21	15.50	16.88	18.35	19.92	21.58	23.36	25.24	27.24	29.35
12	5.77	6.71	7.58	8.48	9.45	10.41	11.45	12.72	13.99	15.32	16.76	18.29	19.89	21.63	23.50	25.47	27.57	29.81	32.18	34.68
14	6.30	2.33	8.28	9.27	10.30	11.40	12.50	13.89	15.28	16.75	18.33	20.01	21.77	23.71	25.73	27.89	30.20	32.66	35.26	38.00
16	7.12	8.29	9.38	10.51	11.69	12.90	14.20	15.79	17.34	19.06	20.86	22.75	24.80	26.98	29.28	31.79	34.39	37.20	40.17	43.31
18	7.97	9.25	10.47	11.75	13.07	14.48	15.90	17.69	19.43	21.37	23.40	25.52	27.83	30.28	32.88	35.66	38.63	41.79	45.14	48.68
20	8.80	10.21	11.57	12.93	14.40	15.97	17.59	19.53	21.52	23.61	25.87	28.22	30.79	33.51	36.39	39.48	42.78	46.29	50.00	53.88
24	10.43	12.13	13.76	15.39	17.15	18.97	20.92	23.25	25.57	28.15	30.79	33.68	36.69	39.96	43.48	47.19	51.14	55.29	59.75	64.46

Table C-3
PIPE INSULATION SURFACE AREAS
PER LINEAR FOOT (ft^2/ft)

Nominal Pipe Size	Outside Pipe Diameter	Nominal Insulation Thickness, Inches											
		Bare	1	1-1/2	2	2-1/2	3	3-1/2	4	4-1/2	5	5-1/2	6
1/2	0.840	0.220	0.75	1.05	1.31	1.73	2.00						
3/4	1.050	0.275	0.75	1.05	1.31	1.73	2.00						
1	1.315	0.344	0.92	1.18	1.46	1.73	2.00						
1-1/4	1.66	0.434	0.92	1.31	1.46	1.73	2.00	2.26	2.52				
1-1/2	1.90	0.497	1.05	1.31	1.73	2.00	2.26	2.32	2.81				
2	2.37	0.622	1.18	1.46	1.73	2.00	2.26	2.52	2.81				
2-1/2	2.875	0.753	1.31	1.73	2.00	2.26	2.52	2.81	3.08				
3	3.500	0.916	1.46	1.73	2.00	2.26	2.52	2.81	3.08	3.34			
3-1/2	4.00	1.047	1.73	2.00	2.26	2.52	2.81	3.08	3.34	3.67			
4	4.50	1.178	1.73	2.00	2.26	2.52	2.81	3.08	3.34	3.67			
5	5.563	1.456	2.00	2.26	2.52	2.81	3.08	3.34	3.67	3.93	4.19		
6	6.625	1.735	2.26	2.52	2.81	3.08	3.34	3.67	3.93	4.19	4.45		
8	8.625	2.258	2.81	3.08	3.34	3.67	3.93	4.19	4.45	4.71	4.97	5.24	
10	10.75	2.814	3.34	3.67	3.93	4.19	4.45	4.71	4.97	5.24	5.50	5.76	
12	12.75	3.338	3.93	4.19	4.45	4.71	4.97	5.24	5.50	5.76	6.02	6.28	6.54
14	14.00	3.665	4.19	4.45	4.71	4.97	5.24	5.50	5.76	6.02	6.28	6.54	6.81
16	16.00	4.189	4.71	4.97	5.24	5.50	5.76	6.02	6.28	6.54	6.81	7.07	7.33
18	18.00	4.712	4.97	5.50	5.76	6.02	6.28	6.54	6.81	7.07	7.33	7.59	7.85
20	20.00	5.236	5.76	6.02	6.28	6.54	6.81	7.07	7.33	7.59	7.85	8.12	8.38
24	24.00	6.283	6.81	7.07	7.33	7.59	7.85	8.12	8.38	8.64	8.90	9.16	9.42

Ref. ANSI/ASTM C-585-76

Table C-4
ACTUAL THICKNESSES FOR PIPE INSULATION

Nominal Pipe Size, Inches	Nominal Insulation Thickness, Inches						
	1	1-1/2	2	2-1/2	3	3-1/2	4
1/2	1.01	1.57	2.07	2.88	3.38	3.88	4.38
3/4	0.90	1.46	1.96	2.78	3.28	3.78	4.28
1	1.08	1.58	2.12	2.64	3.14	3.64	4.14
1-1/4	0.91	1.66	1.94	2.47	2.97	3.47	3.97
1-1/2	1.04	1.54	2.35	2.85	3.35	3.85	4.42
2	1.04	1.58	2.10	2.60	3.10	3.60	4.17
2-1/2	1.04	1.86	2.36	2.86	3.36	3.92	4.42
3	1.02	1.54	2.04	2.54	3.04	3.61	4.11
3-1/2	1.30	1.80	2.30	2.80	3.36	3.86	4.36
4	1.04	1.54	2.04	2.54	3.11	3.61	4.11
5	0.99	1.49	1.99	2.56	3.06	3.56	4.18
6	0.96	1.46	2.02	2.52	3.02	3.65	4.15
8		1.52	2.02	2.65	3.15	3.65	4.15
10		1.58	2.08	2.58	3.08	3.58	4.08
11		1.58	2.08	2.58	3.08	3.58	4.08
12		1.58	2.08	2.58	3.08	3.58	4.08
14		1.46	1.96	2.46	2.96	3.46	3.96
16		1.46	1.96	2.46	2.96	3.46	3.96
18		1.46	1.96	2.46	2.96	3.46	3.96

Ref. ANSI/ASTM C585-76

Table C-5
INSULATION AREA OF ROUND DUCT
(FT^2/LINEAR FT)

Duct Diameter, Inches	Insulation Thickness, Inches		
	1	1-1/2	2
4	1.57	1.83	2.09
5	1.83	2.09	2.36
6	2.09	2.36	2.62
7	2.36	2.62	2.88
8	2.62	2.88	3.14
9	2.88	3.14	3.40
10	3.14	3.40	3.67
12	3.67	3.93	4.19
14	4.19	4.45	4.71
16	4.71	4.97	5.24
18	5.24	5.50	5.76
20	5.76	6.02	6.28
22	6.28	6.54	6.81
24	6.81	7.07	7.33
26	7.33	7.59	7.85
28	7.85	8.12	8.38
30	8.38	8.64	8.90
32	8.90	9.16	9.42
34	9.42	9.69	9.95
36	9.95	10.21	10.47
38	10.47	10.73	11.00
40	11.00	11.26	11.52
42	11.52	11.78	12.04
44	12.04	12.30	12.57
46	12.57	12.83	13.09
48	13.09	13.35	13.61

Appendix D

Conversion Factors

MULTIPLY	BY	TO OBTAIN
1. AREA		
cm^2	0.00108	ft^2
cm^2	0.1550	in^2
ft^2	929.03	cm^2
ft^2	0.0929	m^2
in^2	6.4516	cm^2
m^2	10000.00	cm^2
m^2	10.7650	ft^2
m^2	1550.0	in^2
2. DENSITY AND SPECIFIC VOLUME		
cm^3/g	0.01602	ft^3/lbm
cm^3/g	27.68	in^3/lbm
cm^3/g	0.001	m^3/kg
ft^3/lbm	62.43	cm^3/g
ft^3/lbm	0.06243	m^3/kg
g/cm^3	62.43	lbm/ft^3
g/cm^3	0.03613	lbm/in^3
g/cm^3	1000.	kg/m^3
kg/m^3	0.06243	lbm/ft^3
kg/m^3	3.613×10^{-5}	lbm/in^3
lbm/ft^3	0.01602	g/cm^3
lbm/ft^3	5.787×10^{-4}	lbm/in^3
lbm/in^3	27.68	g/cm^3
lbm/in^3	1728	lbm/ft^3
m^3/kg	16.02	ft^3/lbm
m^3/kg	2.768×10^4	in^3/lbm
sp-gr	1	g/cm^3
sp-gr	62.43	lbm/ft^3

MULTIPLY	BY	TO OBTAIN
3. ENERGY, WORK, AND HEAT		
Btu	1.055×10^{10}	erg (dyn cm)
Btu	1054.8	J (W sec)
Btu	0.2520	kg-cal
Btu	778.3	ft lbf
Btu	3.929×10^{-4}	hp hr
Btu	107.6	kgf m
Btu	2.930×10^{-4}	kW hr
Btu	252.0	g-cal
erg	1	dyn cm
erg	9.480×10^{-11}	Btu
ft lbf	1.285×10^{-3}	Btu
ft lbf	5.050×10^{-7}	hp hr
ft lbf	3.239×10^{-4}	kg-cal
ft lbf	0.1383	kgf m
ft lbf	3.766×10^{-7}	kW hr
g-cal	3.969×10^{-3}	Btu
g-cal	4.186	J
gf cm	9.297×10^{-8}	Btu
hp hr	2545.	Btu
hp hr	1.98×10^6	ft lbf
hp hr	641.3	kg-cal
hp hr	2.737×10^5	kgf m
hp hr	0.7457	kW hr
J	1×10^7	erg
J	2.389×10^{-4}	kg-cal
J	0.1020	kgf m
J	2.778×10^{-4}	W hr
J	0.7376	ft lbf
J	9.480×10^{-4}	Btu
J	1	N m

MULTIPLY	BY	TO OBTAIN
3. ENERGY, WORK, AND HEAT (contd)		
kg-cal	3.969	Btu
kg-cal	3087.	ft lbf
kg-cal	1.559×10^{-3}	hp hr
kg-cal	1.163×10^{-3}	W hr
kg-cal	426.9	kgf m
kgf m	7.233	ft lbf
kgf m	9.297×10^{-3}	Btu
kgf m	2.343×10^{-3}	kg-cal
kgf m	2.724×10^{-6}	kW hr
kW hr	3412.9	Btu
kW hr	2.655×10^{6}	ft lbf
kW hr	1.341	hp hr
kW hr	860.0	kg-cal
kW hr	3.671×10^{5}	kgf m
W hr	3.4129	Btu
W hr	2655.	ft lbf
W hr	1.341×10^{-3}	hp hr
W hr	0.8600	kg-cal
W hr	367.1	kgf m
W hr	1000.	kW hr

MULTIPLY	BY	TO OBTAIN
4. FORCE		
dyn	2.245×10^{-6}	lbf
dyn	$1. \times 10^{-5}$	N
kgf	9.807	N
kgf	2.205	lbf
N	1	J/m
N	1	kg m/s^2
N	0.2248	lbf
lbf	4.448×10^{5}	dyn
lbf	4.448	N

MULTIPLY	BY	TO OBTAIN
5. HEAT TRANSFER COEFFICIENT		
Btu/hr ft^2°F	4.883	kg-cal/hr m^2°C
Btu/hr ft^2°F	1.356×10^{-4}	g-cal/sec cm^2°C
Btu/hr ft^2°F	2.035×10^{-3}	W/in^2°F
Btu/hr ft^2°F	3.663×10^{-3}	W/in^2°C
Btu/hr ft^2°F	5.678×10^{-4}	W/cm^2°C
g-cal/sec cm^2°C	7375.	Btu/hr ft^2°F
g-cal/sec cm^2°C	36000.	kg-cal/hr m^2°C
g-cal/sec cm^2°C	15.00	W/in^2°F
g-cal/sec cm^2°C	4.187	W/cm^2°C
kg-cal/hr m^2°C	0.2048	Btu/hr ft^2°F
kg-cal/hr m^2°C	2.778×10^{-5}	g-cal/sec cm^2°C
kg-cal/hr m^2°C	4.168×10^{-4}	W/in^2°F
kg-cal/hr m^2°C	1.163×10^{-4}	W/cm^2°C
W/in^2°F	491.4	Btu/hr ft^2°F
W/in^2°F	2399.	kg-cal/hr m^2°C
W/in^2°F	0.06667	g-cal/sec cm^2°C
W/in^2°F	0.2790	W/cm^2°C
W/in^2°C	273.0	Btu/hr ft^2°F
W/cm^2°C	1761.	Btu/hr ft^2°F
W/cm^2°C	8598.	kg-cal/hr m^2°C
W/cm^2°C	0.2388	g-cal/sec cm^2°C
W/cm^2°C	3.584	W/in^2°F

MULTIPLY	BY	TO OBTAIN
6. LENGTH		
cm	0.03281	ft
cm	0.3937	in
ft	30.48	cm
ft	0.3048	m
in	2.540	cm
m	100.0	cm
m	3.281	ft
m	39.37	in
micron	$1. \times 10^{-6}$	m

MULTIPLY	BY	TO OBTAIN
7. MASS		
g	0.0022046	lbm
kg	1000.	g
kg	2.2046	lbm
lbm	453.59	g
lbm	0.03108	slug
slug	32.1725	lbm
8. POWER AND HEAT TRANSFER RATE		
Btu/min	0.02357	hp
Btu/min	0.01758	kW
Btu/min	17.58	W
Btu/sec	1.414	hp
Btu/sec	1.055	kW
Btu/sec	778.2	ft lbf/sec
Btu/sec	252.0	g-cal/sec
Btu/sec	3600.	Btu/hr
Btu/hr	3.927×10^{-4}	hp
ft lbf/min	1.285×10^{-3}	Btu/min
ft lbf/min	0.01667	ft lbf/sec
ft lbf/min	3.030×10^{-5}	hp
ft lbf/min	3.239×10^{-4}	kg-cal/min
ft lbf/min	2.260×10^{-5}	kW
ft lbf/sec	7.712×10^{-2}	Btu/min
ft lbf/sec	1.818×10^{-3}	hp
ft lbf/sec	1.943×10^{-2}	kg-cal/min
ft lbf/sec	1.356×10^{-3}	kW
ft lbf/sec	0.3239	g-cal/sec
ft lbf/sec	4.627	Btu/hr

MULTIPLY	BY	TO OBTAIN
8. POWER AND HEAT TRANSFER RATE (contd)		
hp	42.41	Btu/min
hp	33000.	ft lbf/min
hp	550.0	ft lbf/sec
hp	10.69	kg-cal/min
hp	0.7457	kW
hp	745.7	W
hp	2545.	Btu/hr
hp	178.2	g-cal/sec
hp(boiler)	13.14	hp
kg-cal/min	51.44	ft lbf/sec
kg-cal/min	0.09355	hp
kg-cal/min	0.06977	kW
kW	56.89	Btu/min
kW	4.426×10^{4}	ft lbf/min
kW	737.6	ft lbf/sec
kW	1.341	hp
kW	14.33	kg-cal/min
kW	1000	W
kW	238.9	g-cal/sec
kW	3.4129	Btu/hr
hp(metric)	0.9863	hp
W	0.05689	Btu/min
W	44.26	ft lbf/min
W	0.7376	ft lbf/sec
W	1.341 x 10-3.	hp
W	0.01433	kg-cal/min
W	0.0010	kW

MULTIPLY	BY	TO OBTAIN
9. PRESSURE		
atm	760.0	mm mercury
atm	29.92	in mercury
atm	33.90	ft water
atm	10332	kgf/m^2
atm	14.70	lbf/in^2
atm	101325	Pa
ft water	0.02950	atm
ft water	0.8826	in mercury
ft water	304.8	kgf/m^2
ft water	62.43	lbf/ft^2
ft water	0.4335	lbf/in^2
ft water	2989	Pa
in mercury	0.03342	atm
in mercury	1.133	ft water
in mercury	345.3	kgf/m^2
in mercury	70.73	lbf/ft^2
in mercury	0.4912	lbf/in^2
in mercury	3386.4	Pa
in water	0.002458	atm
in water	0.07355	in mercury
in water	25.40	kgf/m^2
in water	0.5781	oz/in^2
in water	5.204	lbf/ft^2
in water	0.03613	lbf/in^2
in water	1.868	mm mercury
in water	249.08	Pa
kgf/m^2	9.678×10^{-5}	atm
kgf/m^2	3.281×10^{-3}	ft water
kgf/m^2	2.896×10^{-3}	in mercury
kgf/m^2	0.2048	lbf/ft^2
kgf/m^2	1.422×10^{-3}	lbf/in^2
kgf/m^2	7.356×10^{-2}	mm mercury
kgf/m^2	9.80665	Pa
kgf/mm^2	$1. \times 10^{6}$	kgf/m^2

MULTIPLY	BY	TO OBTAIN
9. PRESSURE (contd)		
lbf/ft^2	0.01602	ft water
lbf/ft^2	6.945×10^{-3}	lbf/in^2
lbf/ft^2	47.88	Pa
lbf/in^2	0.06803	atm
lbf/in^2	2.307	ft water
lbf/in^2	2.036	in mercury
lbf/in^2	703.1	kgf/m^2
lbf/in^2	51.71	mm mercury
lbf/in^2	144.0	lbf/ft^2
lbf/in^2	6894.8	Pa
mm mercury	0.001316	atm
mm mercury	0.04462	ft water
mm mercury	13.60	kgf/m^2
mm mercury	2.785	lbf/ft^2
mm mercury	0.01934	lbf/in^2
mm mercury	133.3	Pa
Pa	0.0075	mm mercury
Pa	3.346×10^{-4}	ft water
Pa	2.953×10^{-4}	in mercury
Pa	4.015×10^{-3}	in water
Pa	0.10197	kgf/m^2
Pa	0.020885	lbf/ft^2
Pa	1.4504×10^{-4}	lbf/in^2
Pa	1	N/m^2
10. SPECIFIC HEAT		
Btu/lbm°F	4.184	J/g°C
J/g°C	0.2390	Btu/lbm°F
kg-cal/kg°C	1.000	Btu/lbm°F

11. TEMPERATURE

°K = °C + 273.16
°R = °F + 459.69
°C = (°F − 32.)/1.8
°F = 1.8°C + 32.

MULTIPLY	BY	TO OBTAIN
12. THERMAL CONDUCTIVITY		
Btu ft/hr ft²°F	12.00	Btu in/hr ft °F²
Btu ft/hr ft²°F	0.01730	W cm/cm °C²
Btu ft/hr ft²°F	1.730	W/m°C
Btu ft/hr ft °F²	0.004134	g-cal cm/sec cm² °C
Btu in/hr ft²°F	0.08333	Btu ft/hr ft²°F
Btu in/hr ft²°F	0.001442	W cm/cm²°C
Btu in/hr ft²°F	0.1442	W/m°C
Btu in/hr ft °F²	0.0003445	g-cal cm/sec cm² °C
Btu in/hr ft²°F	12.40	kg-cal cm/hr m²°C
g-cal cm/sec cm² °C	241.9	Btu ft/hr ft²°F
g-cal cm/sec cm² °C	2903.	Btu in/hr ft²°F
g-cal cm/sec cm² °C	4.186	W cm/cm²°C
W cm/cm²°C	57.80	Btu ft/hr ft²°F
W cm/cm²°C	693.6	Btu in/hr ft²°F
W cm/cm²°C	0.2389	g-cal cm/sec cm²°C
W/m°C	0.5780	Btu ft/hr ft²°C
13. THERMAL RESISTANCE		
°C/W	0.5274	°F hr/Btu
°F hr/Btu	1.896	°C/W

MULTIPLY	BY	TO OBTAIN
14. VELOCITY AND VOLUME FLOW RATE		
ft/min	0.5080	cm/sec
ft/min	0.01667	ft/sec
ft/min	0.01829	km/hr
ft/min	0.3048	m/min
ft/min	0.005080	m/sec
ft³/min	472.0	cm³/sec
ft³/min	0.1247	gal/sec
ft³/min	7.481	gal/min
ft/sec	30.48	cm/sec
ft/sec	18.29	m/min
gal/min	2.228×10^{-3}	ft³/sec
gal/min	8.021	ft³/hr
l/min	5.885×10^{-4}	ft³/sec
l/min	4.403×10^{-3}	gal/sec
m/min	1.667	cm/sec
m/min	3.281	ft/min
m/min	0.05468	ft/sec
m/sec	196.8	ft/min
m/sec	3.281	ft/sec

15. VISCOSITY (ABSOLUTE)

MULTIPLY	BY	TO OBTAIN
centipoise	0.010	poise (g/sec cm)
centipoise	2.419	lbm/hr ft
centipoise	6.720×10^{-4}	lbm/sec ft
centipoise	3.600	kg/hr m
lbm/sec ft	1	English second
lbm/sec ft	3600.	lbm/hr ft
lbm/sec ft	14.88	poise
lbm/sec ft	1488.	centipoise
lbm/hr ft	4.133×10^{-3}	poise
lbm/hr ft	0.4133	centipoise
lbf sec/ft^2	32.18	lbm/sec ft
lbf sec/ft^2	47.85×10^3	centipoise
lbf hr/ft^2	32.18	lbm/hr ft
lbf hr/ft^2	13.29	centipoise
poise	1	g/sec cm
centisokes	sp gr	centipoise

16. VISCOSITY (KINEMATIC)

MULTIPLY	BY	TO OBTAIN
centistoke	1.076×10^{-5}	ft^2/sec
centistoke	0.010	stoke
ft^2/sec	9.290×10^4	centistoke
ft^2/sec	929.0	stoke
stoke	100.0	centistoke
stoke	1.076×10^{-3}	ft^2/sec
stoke	1	cm^2/sec

17. VOLUME

MULTIPLY	BY	TO OBTAIN
cm^3	3.531×10^{-5}	ft^3
cm^3	0.06102	in^3
ft^3	2.832×10^4	cm^3
ft^3	7.481	gal
gal	0.1337	ft^3
gal	3.785×10^{-3}	m^3
gal(imp)	1.20095	gal(US)
L	1000	cm^3
L	0.03531	ft^3
L	61.02	in^3
L	0.001	m^3
L	0.2642	gal
m^3	35.31	ft^3
m^3	6.102×10^4	in^3
m^3	1000	L

18. WATER VAPOR PERMEABILITY

MULTIPLY	BY	TO OBTAIN
perm-in	1.459×10^{-12}	kg/Pa s m
perm-in	1.459	ng/Pa s m
perm-in	1.675	g cm/(mmHg)(24 hr)m^2
kg/Pa s m	6.853×10^{11}	perm-in
ng/Pa s m	0.6853	perm-in
g cm/(mmHg) (24hr)m^2	0.5968	perm-in

19. WATER VAPOR PERMEANCE

MULTIPLY	BY	TO OBTAIN
perm	5.745×10^{-11}	kg/Pa s m^2
perm	57.45	ng/Pa s m^2
perm	1.516	g/(mmHg)(24hr)m^2
kg/Pa s m^2	1.740×10^{10}	perm
ng/Pa s m^2	0.0174	perm
g/(mmHg)(24hr)m^2	0.6596	perm

Appendix E

Sea Level Psychrometric Chart

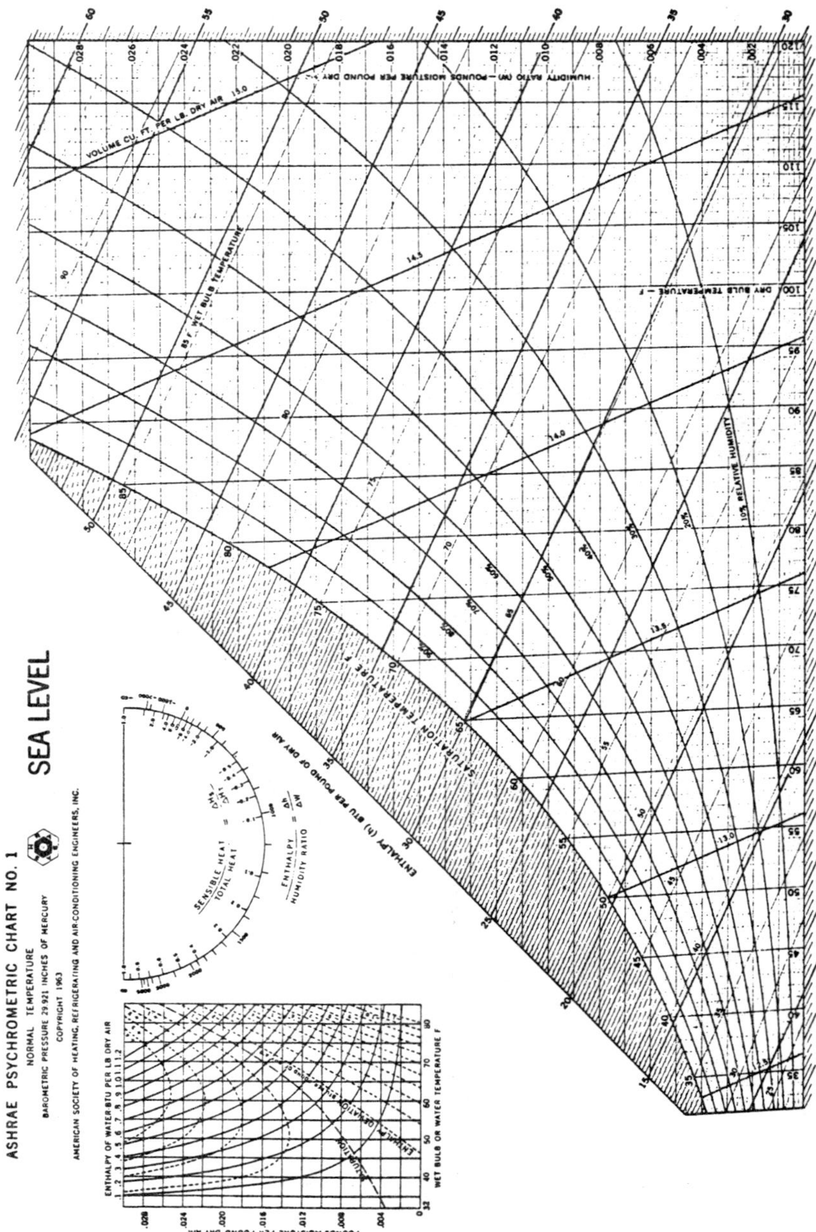

(Reprinted By Permission From ASHRAE)

Appendix F

Computer Program Listings

HTMCP - FORTRAN

```
C  THIS FORTRAN PROGRAM CAN BE USED TO DETERMINE THE OVERALL HEAT
C  TRANSFER COEFFICIENT OF A COMPOSITE SLAB OR A COMPOSITE
C  CYLINDER.  IN ADDITION THE PROGRAM WILL DETERMINE THE TEMPERATURES
C  AT THE INTERFACES BETWEEN CONSECUTIVE LAYERS IN THE COMPOSITE.
C  FOR SLAB COMPOSITES THE PROGRAM WILL ALSO COMPUTE THE AVERAGE
C  SLAB MASS AND ITS THERMAL MASS(CAPACITY).
C
C  INPUT TO THE PROGRAM CONSISTS OF THE FOLLOWING DATA:
C
C  CARD(1) ITYPE** FLAG FOR COMPOSITE TYPE (I2)
C          =1  SLAB COMPOSITE
C          =2  CYLINDER COMPOSITE
C          LYRS** NUMBER OF LAYERS IN THE COMPOSITE,INCLUDING THE
C                 INSIDE AND OUTSIDE FILM LAYERS. (I2)
C          T(1),T(K)** T(1) IS THE INSIDE TEMPERATURE AND T(K)
C                 IS THE OUTSIDE TEMPERATURE
C
C  CARD(I+1) I,XL(I),TH(I),DEN(I),SPHT(I),RES(I),DES(I,J),J=1,20
C   WHERE:  INPUT LAYERS FROM INSIDE TO OUTSIDE
C
C
C            I=THE LAYER NUMBER   (DIMENSIONLESS)
C            XL(I)=THE LAYER THICKNESS (SLAB) OR THE RADIUS OF THE
C                  LAYER TO THE OUTSIDE OF THE LAYER (CYLINDER) (INCHES)
C            TK(I)=THERMAL CONDUCTIVITY OF THE LAYER
C                        (BTU-IN/HR-FT2-DEG F)
C            DEN(I)=THE DENSITY OF THE LAYER MATERIAL
C                        (LBM/FT3)
C            SPHT(I)=THE SPECIFIC HEAT OF THE LAYER MATERIAL
C                        (BTU/LBM-DEG F)
C            RES(I)=THE THERMAL RESISTANCE OF THE LAYER
C                        (FT2-HR-DEG F/BTU)
C                   RES(I) IS ENTERED AS A POSITIVE NUMBER FOR LAYERS WHICH
C                   CONSIST OF AN AIR FILM ON AN INSIDE OR AN OUTSIDE
C                   SURFACE.  RES(I) IS ENTERED  AS ZERO WHEN THE
C                   THICKNESS AND THERMAL CONDUCTIVITY VALUES FOR THE
C                   LAYER ARE ENTERED AND AS A POSITIVE VALUE WHEN
C                   THE THICKNESS AND THERMAL CONDUCTIVITY VALUES ARE
C                   ENTERED AS ZEROES.
C            DES(I,J)=THE DESCRIPTION OF THE LAYER (DIMENSIONLESS)
C
C  LAST CARD = 0/  (WILL SIGNAL END OF ALL INPUT)
C
C
C
C
      DIMENSION XL(20),TK(20),DEN(20),SPHT(20),RES(20),DES(20)
     1,DES1(20,20)
      DIMENSION T(22),RAD(20),RR(20)
      DIMENSION DD(20),TDD(20)
      DIMENSION TITLE(30)
      INTEGER*2 TITLE,DES,DES1
      IRD=9
      IWRT=1
```

```
C  READ DATA FROM INPUT FILE
 2     READ(IRD,*) LYRS
       K=LYRS+1
       IF(LYRS.EQ.0) GO TO 90
       READ(IRD,*) ITYPE,TITLE,T(1),T(K)
       K=LYRS+1
       RO=0.0
       CO=0.0
       QO=0.0
       DO 100 I=1,LYRS
       READ(IRD,*)I,XL(I),TK(I),DEN(I),SPHT(I),RES(I),DES
       DO 101 J=1,20
       DES1(I,J)=DES(J)
  101 CONTINUE
        XL(I)=XL(I)/12.
       TK(I)=TK(I)/12.
C  IF THE COEFFICIENTS ARE BEING DETERMINED FOR A CYLINDER COMPOSITE
C  THE VALUE OF XL(I) IS THE RADIUS TO THE OUTSIDE OF LAYER I
       IF(ITYPE.EQ.2) RAD(I)=XL(I)
       IF(RES(I).NE.0.0) GO TO 20
       IF(XL(I).NE.0.0) GO TO 25
       GO TO 100
C  DETERMINE THE RESISTANCE OF LAYER I
 20    RR(I)=RES(I)
       GO TO 30
 25    RR(I)=XL(I)/TK(I)
       IF(ITYPE.EQ.2) RR(I)=ALOG(RAD(I)/RAD(I-1))/(2.0*3.14159*TK(I))
C  IF THE COMPOSITE SECTION IS A SLAB, DETERMINE THE MASS AND THE
C  THERMAL MASS OF THE COMPOSITE
 30    IF(ITYPE.EQ.1) DD(I)=XL(I)*DEN(I)
       IF(ITYPE.EQ.2) DD(I)=(RAD(I)**2-RAD(I-1)**2)*3.14159*DEN(I)
       TDD(I)=DD(I)*SPHT(I)
       DO=DO+DD(I)
       TDO=TDO+TDD(I)
       RO=RO+RR(I)
 100   CONTINUE
       CO=1.0/RO
       QO=(T(1)-T(K))/RO
       DO 200 J=2,K
       T(J)=T(J-1)-(QO*RR(J-1))
 200   CONTINUE
       WRITE(IWRT,56)
       WRITE(IWRT,55)  TITLE
       WRITE(IWRT,35)
       DO 50 I=1,LYRS
       XLMEL= XL(I) * 12.
       TKMEL=TK(I)*12.
       WRITE(IWRT,40) I,XLMEL,TKMEL,DEN(I),SPHT(I),RR(I),
      1(DES1(I,J),J=1,20)
 50    CONTINUE
        IF(ITYPE.EQ.2) GO TO 80
       WRITE(IWRT,44)
       WRITE(IWRT,45) CO,QO,DO
       WRITE(IWRT,46) TDO
```

```
      GO TO 85
80    WRITE(IWRT,47)
      WRITE(IWRT,42) CO,QO,DO
85    WRITE(IWRT,48)
      DO 86 I=1,K
      WRITE(IWRT,49) I,T(I)
86    CONTINUE
      GO TO 2
35    FORMAT(/,1X,'LAYER',1X,'THICKNESS',2X,'THERMAL',3X,'DENSITY',2X,
     1'SPECIFIC',2X,'RESISTANCE',2X,'DESCRIPTION',/,8X,'(RADIUS)',2X,
     2'COND (K)',12X,'HEAT',/,9X,'(IN)',4X,'(BTU-IN/   (LB/FT3)',2X,
     3'(BTU/   (FT2-HR-F/',/,17X,'FT2-HR-F)',13X,'LB-F)',7X,'BTU)',
     4/,1X,'_____',2X
     5,'_______  ________  ________  ________  ________  ____________')
40    FORMAT(2X,I2,3X,2(F8.4,2X),F8.4,2X,2(F8.4,2X),20A2)
44    FORMAT(/,/,/,25X,'*** FOR SLAB COMPOSITES ***')
45    FORMAT(/,20X,'COMPOSITE THERMAL AND PHYSICAL PROPERTIES',/,10X,
     1'THE OVERALL HEAT TRANSFER COEFFICIENT IS *',F8.3,' BTU/HR-FT2-F'
     2,/,10X,'THE UNIT HEAT TRANSFER THROUGH THE SLAB IS',F8.3,
     3' BTU/HR-FT2',/,10X,'THE COMPOSITE SYSTEM MASS IS ************',
     4F8.3,' LBM/FT2')
42    FORMAT(/,20X,'COMPOSITE THERMAL AND PHYSICAL PROPERTIES',/,10X,
     1'THE OVERALL HEAT TRANSFER COEFFICIENT IS *',F8.3,' BTU/HR-FT-F'
     2,/,10X,'THE UNIT HEAT TRANSFER THRU THE CYLINDER  ',F8.3,
     3' BTU/HR-FT',/,10X,'THE COMPOSITE SYSTEM MASS IS ************',
     4F8.3,' LBM/FT')
46    FORMAT(10X,'THE COMPOSITE SYSTEM THERMAL MASS IS *****',F8.3,
     1' BTU/FT2-F')
47    FORMAT(/,/,/,25X,'*** FOR CYLINDRICAL COMPOSITES ***')
48    FORMAT(/,/,20X,'TEMPERATURES AT THE LAYER INTERFACES ARE',/,/,19X
     1'T1 IS THE INSIDE FLUID OR MATERIAL TEMPERATURE',/,19X,
     2'T(LAYERS+1) IS THE OUTSIDE TEMPERATURE',/,19X,'T(I) IS THE TEMPER
     3ATURE BETWEEN LAYER I-1 AND LAYER I')
49    FORMAT(/,25X,'TEMPERATURE',I2,' **',F8.2,' F')
55    FORMAT(/,/,2X,'****',30A2,'****',/,/,/)
56    FORMAT('1',/,/,/,5X,'NAVAL CIVIL ENGINEERING LABORATORY',20X,
     1'3/23/82',/,'____________________________________________________
     2_________________',/,5X,'HTMCP-FORTRAN',30X,'EMC ENGINEERS,INC.
     3',/,/'*************************************************************
     4**************')
90    STOP
      END
```

HTMCP · BASIC

```
40  REM ******************************************************************
42  REM *  THIS BASIC PROGRAM IS DESIGNED TO CALCULATE THE OVERALL HEAT   *
43  REM * TRANSFER COEFFICIENT-OVERALL HEAT TRANSFER,COMPOSITE MASS,COMP- *
44  REM * OSITE THERMAL MASS(FOR A SLAB) AND THE TEMPERATURE GRADIENT AT  *
45  REM * LAYER INTERFACES FOR EITHER A SLAB OR A CYLINDER. IT IS LIMITED *
46  REM * TO A MAXIMUM OF 20 LAYERS,INCLUDING INSIDE AND OUTSIDE FILM CO- *
47  REM * EFFICEINTS. DATA IS INPUTED,AS DESCRIBED BELOW, AT THE END      *
48  REM * OF THE PROGRAM STARTING AT LINE 2500(LINE NUMBERS MUST ALSO BE  *
49  REM * INPUT).                                                         *
50  REM ******************************************************************
100 REM DEFINITION OF VARIABLE NAMES:
101 REM  B= TEMPERATURE(DEG F)
102 REM C1= TOTAL HEAT TRANSFER COEFFICIENT OF COMPOSITE(BTU/HR-FT2-F)
103 REM  D= DENSITY OF EACH LAYER(LBM/FT3)
104 REM E$= DESCRIPTION OF EACH LAYER(DIMENSIONLESS)
105 REM  L= THERMAL MASS OF EACH LAYER(BTU/FT2-F)
106 REM L2= THERMAL MASS OF COMPOSITE(BTU/FT2-F)
107 REM  M= MASS OF EACH LAYER(LBM/FT2)
108 REM M2= TOTAL MASS OF COMPOSITE(LBM/FT2)
109 REM  N= NUMBER OF LAYERS(MAXIMUM OF 20 LAYERS)
110 REM P1= RESISTANCE TO EACH INTERFACE-INSIDE TO OUTSIDE(HR-FT2-F/BTU)
111 REM Q1= OVERALL HEAT TRANSFER(BTU/HR-FT2)
112 REM  R= THERMAL RESISTANCE OF EACH LAYER(HR-FT2-F/BTU)
113 REM R2= TOTAL THERMAL RESISTANCE OF COMPOSITE(HR-FT2-F/BTU)
114 REM  S= SPECIFIC HEAT OF EACH LAYER(BTU/LBM-F)
115 REM T2= FLAG FOR SLAB(=1) OR CYLINDER(=2)
116 REM  X= LAYER THICKNESS(FT)
117 REM X1= ARRAY COUNTER
118 REM  Z= THERMAL RESISTANCE OF EACH LAYER FROM INPUT DATA(HR-FT2-F/BTU)
119 REM NOTE: DATA INPUT FORMAT IS IN THE FORM OF:
120 REM        DATA LINE#1: T2,N,B(INSIDE),B(OUTSIDE)
121 REM        DATA LINE#2: X(1),T(1),D(1),S(1),Z(1),E$(1)
122 REM      - WHERE (1) DENOTES THE THE FIRST LAYER OF THE COMPOSITE(WHICH
123 REM        MUST ALWAYS BE GOING IN THE DIRECTION OF THE INSIDE TO THE
124 REM        OUTSIDE, WITH THE INSIDE FILM COEFFICIENT BEING LAYER 1 AND
125 REM        THE OUTSIDE FILM COEFFICIENT BEING THE LAST LAYER).
126 REM  ******************************************************************
127 REM DATA INPUT
200 DIM X(20),T(20),D(20),S(20),Z(20),B(20),E$(20)
210 READ T2,N
220 K=N+1
230 READ B(0),B(K)
235 REM BEGINNING OF SLAB CALCULATIONS
240 FOR X1=1 TO N
250 READ X(X1),T(X1),D(X1),S(X1),Z(X1),E$(X1)
260 NEXT X1
270 IF T2=2 THEN 490
275 REM MASS AND THERMAL MASS CALCULATIONS
280 FOR X1=1 TO N
```

```
290 M(X1)=X(X1)*D(X1)/12
300 L(X1)=M(X1)*S(X1)
310 M2=M(X1)+M2
320 L2=L2+L(X1)
330 IF X(X1)=0 THEN 360
340 R2=R2+(X(X1)/T(X1))
350 GOTO 370
360 R2=R2+Z(X1)
370 NEXT X1
380 FOR X1=1 TO N
390 IF X(X1)=0 THEN  R(X1)=Z(X1)
400 IF X(X1)=0 THEN 420
410 R(X1)=X(X1)/T(X1)
420 NEXT X1
425 REM OVERALL HEAT TRANSFER COEFFICIENT,TRANSFER AND TEMPERATURE GRADIENT
430 C1=1/R2
440 Q1=C1*(B(0)-B(K))
450 FOR X1=0 TO K
460 P1=P1+R(X1)
470 B(X1)=B(0)-(Q1*P1)
480 NEXT X1
490 A$='SLAB COMPOSITE'
500 IF T2=2 THEN  A$='CYLINDER COMPOSITE'
510 IF T2=1 THEN 770
515 REM BEGINNING OF CYLINDER CALCULATIONS
517 REM RESISTANCE FOR LAYERS AND COMPOSITE
600 FOR X1=1 TO N
610 IF X1=1 THEN  R(X1)=Z(X1)
615 IF X1=N THEN  R(X1)=Z(X1)
620 IF X1=1 THEN 640
625 IF X1=N THEN 640
627 T(X1)=T(X1)/12
630 R(X1)=LOG(X(X1)/X(X1-1))/(2*3.14159*T(X1))
640 R2=R2+R(X1)
645 T(X1)=T(X1)*12
650 NEXT X1
655 REM OVERALL HEAT TRANSFER COEFFICIENT,TRANSFER AND TEMPERATURE GRADIENT
660 C1=1/R2
670 Q1=C1*(B(0)-B(K))
680 FOR X1=0 TO K
690 P1=P1+R(X1)
700 B(X1)=B(0)-(Q1*P1)
710 NEXT X1
720 REM CYLINDER MASS CALCULATION
730 FOR X1=2 TO N
735 X(X1)=X(X1)/12
736 X(X1-1)=X(X1-1)/12
740 M=((3.14159*X(X1)*X(X1)))-((3.14159*X(X1-1)*X(X1-1)))
750 M2=M2+(M*D(X1))
755 X(X1)=X(X1)*12
756 X(X1-1)=X(X1-1)*12
```

```
760  NEXT X1
770  PRINT '  NAVAL CIVIL ENGINEERING LABORATORY                        3/23/82'
775  PRINT '  ____________________________________________________________'
780  PRINT '  HTMCP-BASIC                                      EMC ENGINEERS,INC.'
785  PRINT '  ************************************************************'
790  PRINT
791  PRINT
800  PRINT TAB(30):A$
805  IF T2=1 THEN  Y$='--------------'
806  IF T2=2 THEN  Y$='------------------'
810  PRINT TAB(30):Y$
811  PRINT
812  PRINT
813  PRINT
820  PRINT 'LAYER   THICKNESS   THERMAL   DENSITY  SPECIFIC  THERMAL':
822  PRINT ' DESCRIPTION'
830  PRINT '        (OR RADIUS)  COND (K)             HEAT   RESISTANCE'
840  PRINT '         (INCHES)    (BTU-IN/  (LB/FT3)  (BTU/   (HR-FT2-F/'
850  PRINT '                     HR-FT2-F)           LB-F)      BTU)'
851  PRINT '-----  ----------  ---------  -------  -------  ---------':
852  PRINT '---------'
854  FOR X1=1 TO N
855  PRINT TAB(3):X1:TAB(11):X(X1):TAB(24):T(X1):TAB(34):D(X1):
856  PRINT TAB(43):S(X1):TAB(51):R(X1):TAB(60):E$(X1)
858  NEXT X1
870  PRINT
880  PRINT
900  PRINT TAB(23):'*** FOR A ':A$:' ***'
901  PRINT
904  IF T2=1 THEN  Q$='BTU/HR-FT2-F'
906  IF T2=2 THEN  Q$='BTU/HR-FT-F'
910  W7=10
920  PRINT TAB(W7):'OVERALL HEAT TRANSFER COEFFICIENT IS *** ':
923  PRINT USING '###.###',C1:
925  PRINT TAB(61):Q$
927  IF T2=1 THEN  G$='BTU/HR-FT2'
928  IF T2=2 THEN  G$='BTU/HR-FT'
930  PRINT TAB(W7):'OVERALL HEAT TEANSFER IS *************** ':Q1:'':G$
940  IF T2=1 THEN  B$='SYSTEM'
950  IF T2=2 THEN  B$='CYLINDER'
955  IF T2=1 THEN  C$='LBM/FT2'
956  IF T2=2 THEN  C$='LBM/FT.'
957  IF T2=1 THEN  H$='******** '
958  IF T2=2 THEN  H$='****** '
960  PRINT TAB(W7):'THE COMPOSITE ':B$:' MASS IS ':H$:M2:TAB(61):C$
970  IF T2=2 THEN 985
980  PRINT TAB(W7):'THE COMPOSITE SYSTEM THERMAL MASS IS *** ':L2:
982  PRINT TAB(61):'BTU/FT2-F'
```

```
985  PRINT
986  PRINT
990  PRINT TAB(W7):'THE TEMPERATURES AT THE ':B$:' INTERFACES ARE:'
1050 FOR X1=0 TO N
1051 PRINT
1060 PRINT TAB(W7):'TEMPERATURE ':X1+1:' *= ':B(X1):' F':
1070 IF X1=0 THEN  PRINT ' (INSIDE TEMPERATURE)'
1080 IF X1>0 THEN  IF X1<N THEN  PRINT ' (BETWEEN LAYER ':X1:' AND ':X1+1:')'
1090 IF X1=N THEN  PRINT '(OUTSIDE TEMPERATURE)'
2000 NEXT X1
2499 REM INPUT DATA SHOULD START HERE
2500 REM EXAMPLE OF INPUT DATA
2510 REM LINE #1*T2,N,B(INSIDE),B(OUTSIDE)
2520 DATA 2,4,350,80
2530 REM LINE #2: LAYER ONE DATA(INSIDE FILM COEFFICIENT)
2540 DATA .95,0,0,0,.0833, INSIDE FILM'
2550 REM THE REST OF THE DATA LINES DESCRIBE THE THREE REMAINING LAYERS
2570 DATA 2,.44,12,0,0,'CA.SILICATE'
2580 DATA 4,.26,6,0,0,'GLASS FIBER'
2590 DATA 4,0,0,0,.5,'OUTSIDE FILM'
```

Appendix G

Listing of Manufacturers' Products by Trade Name and Generic Description

This appendix presents two tables describing insulation products. Each table gives the product trade name, manufacturer's name, and a generic product description for each item. The product descriptions are summarized from manufacturer's literature. Table G-1 is sorted by trade name, and Table G-2 is sorted by product description. In both tables, the products are divided into the following major catagories:

- Frame wall insulation
- Masonry wall insulation
- Metal building wall insulation
- Ceiling insulation
- Roof insulation
- Windows and window treatments
- Doors
- Weatherstripping and caulking
- Sealants
- Pipe insulation
- Equipment insulation
- Duct insulation

A number of product trade names are commonly used by architects and engineers as examples of generic materials, or as simple one- or two-word descriptions of more complicated, prefabricated assemblies of materials. These tables provide a convenient method of converting trade names into generic descriptions. Such descriptions could be used in specifications or drawings where use of a specific trade name would be inappropriate. The tables also allow one to identify products of different manufacturers that are similar in material, form, or function.

Although these tables list and describe more than 250 individual products, they are by no means complete; compilation of a complete list would be a never-ending task. To identify additional products, the reader may wish to consult Sweet's Catalog File, the Thomas Register, as well as a local telephone directory.

In the tables presented in this appendix, product trade names marked with an asterisk (*) are not necessarily trademarked or copyrighted names, but are generic names commonly used by the particular manufacturer to identify the product.

Table G-1
INSULATION MANUFACTURERS PRODUCTS
(SORTED BY PRODUCT TRADE NAME)

FRAME WALL INSULATION

PRODUCT TRADENAME	MANUFACTURER	PRODUCT DESCRIPTION	PLATES
ADVANCED FOAM AFS "38"	ADVANCED FOAM SYSTEMS, INC	UREA-FORMALDEHYDE FOAMED-IN-PLACE INSULATION. FOR NEW OR RETROFIT SIDEWALL APPLICATIONS.	2A
FIBERGLAS SHEATHING	OWENS-CORNING FIBERGLAS CORP	GLASS FIBER RIGID BOARD, FACING IS NOT A VAPOR BARRIER BUT RESISTS RAIN PENETRATION AND AIR INFILTRATION.	2C-D, 3B-C, 4B
FLAME RESISTANT BATTS *	MANVILLE CORP	GLASS FIBER BATTS WITH FOIL-SCRIM-KRAFT (FSK) FACING. FACING EXTENDS PAST BATT EDGE FOR STAPLING FLANGE. MAY BE LEFT EXPOSED WHERE CODES PERMIT.	1B-C, 2B-D 3A-B, 4A-C
FLAME SPREAD 25	OWENS-CORNING FIBERGLAS CORP	GLASS FIBER BATT WITH SPECIAL FOIL/KRAFT LAMINATE FACING. RECOMMENDED BY MFR. FOR USE IN WALLS AND CEILINGS WHERE INSULATION FACING WILL BE EXPOSED. SUITABLE FOR LOW-ABUSE AREAS.	
FOIL-FACED BATTS *	MANVILLE CORP	GLASS FIBER BATT WITH ALUMINUM FOIL FACING. FACING EXTENDS PAST BATT EDGE FOR STAPLING FLANGE.	1B-C, 2B-D 3A-B, 4A-C
HI-R PANELS	HI-R BUILDING SYSTEMS, INC	POLYURETHANE FOAM CORE PREFABRICATED FRAMING SYSTEM. PANELS ARE ASSEMBLED INTO FINISHED WALLS AT THE JOB SITE.	
HIGH-R SHEATHING	OWENS-CORNING FIBERGLAS CORP	POLYISOCYANURATE FOAM RIGID BOARD, GLASS FIBER REINFORCED AND ALUMINUM FOIL FACED. FOR USE AS NONSTRUCTURAL FRAME SHEATHING, OR BEHIND INTERIOR WALLBOARD.	2C-D, 3B-C, 4B
INSUL-SHEATH TG	FALCON MANUFACTURING OF MICHIGAN INC.	POLYSTYRENE FOAM BOARD, HIGH-DENSITY. EDGES AND TONGUE AND GROOVE. USED AS NONSTRUCTURAL EXTERIOR SHEATHING.	2C-D, 3B-C, 4B
KRAFT-FACED BATTS *	MANVILLE CORP	GLASS FIBER BATT WITH KRAFT PAPER FACING. FACING EXTENDS PAST BATT EDGE FOR STAPLING FLANGE.	1B-C, 2B-D 3A-B, 4A-C

INSULATION MANUFACTURERS PRODUCTS
(SORTED BY PRODUCT TRADE NAME)

PRODUCT TRADENAME	MANUFACTURER	PRODUCT DESCRIPTION	PLATES

FRAME WALL INSULATION (Cont.)

PRODUCT TRADENAME	MANUFACTURER	PRODUCT DESCRIPTION	PLATES
KRAFT FACED BUILDING INSULATION *	OWENS-CORNING FIBERGLAS CORP	GLASS FIBER BATT INSULATION WITH ASPHALTED KRAFT PAPER FACING, FLANGED FOR STAPLING.	1B-C, 2B-D 3A-B, 4A-C
RETROFIL	MANVILLE CORP	GLASS FIBER LOOSE FILL DESIGNED FOR RETROFIT INSULATION OF SIDEWALLS.	2A
ROLLED BATTS *	OWENS-CORNING FIBERGLAS CORP	GLASS FIBER BATTS, UNFACED.	1D, 2B-D, 3A-B 4A-C
STYROFOAM IB	DOW CHEMICAL CO	POLYSTYRENE FOAM EXTRUDED BOARD WITH CUT-CELL SURFACE FOR PLASTERING BASE.	
STYROFOAM SM	DOW CHEMICAL CO	POLYSTYRENE FOAM EXTRUDED BOARD WITH NATURAL SKIN SURFACE. USED AS NONSTRUCTURAL EXTERIOR SHEATHING.	2C-D, 3B-C, 4B
STYROFOAM TG	DOW CHEMICAL CO	POLYSTYRENE FOAM EXTRUDED BOARD WITH HIGH-DENSITY SKIN. EDGES ARE TONGUE AND GROOVE. USED AS NONSTRUCTURAL EXTERIOR SHEATHING.	2C-D, 3B-C, 4B
SUPER "R" PLUS	FALCON MANUFACTURING OF MICHIGAN INC.	POLYSTYRENE FOAM RIGID BOARD WITH REFLECTIVE FOIL-KRAFT PAPER LAMINATED FACINGS. FACING PERFORATIONS DISALLOW VAPOR BARRIER. EDGE OF BOARD HAS FOIL OVERLAP FOR SEALING BUTT JOINTS. USED AS NONSTRUCTURAL EXTERIOR SHEATHING.	2C-D, 3B-C, 4B
TG-3000	THERMAL SYSTEMS, INC.	POLYURETHANE FOAM RIGID BOARD WITH ALUMINUM FOIL-SCRIM-FOIL LAMINATE FACERS. USED AS NONSTRUCTURAL EXTERIOR SHEATHING.	2C-D, 3B-C, 4B
THERMASOTE SIDEWALL PANELS	HOMASOTE CO	COMPOSITE OF ASBESTOS-FREE INSULATING BUILDING BOARD AND RIGID POLYURETHANE FOAM. FOAM SIDE IS FACED WITH ASPHALT SATURATED FELT OR FIBERGLASS. FOR USE AS NONSTRUCTURAL FRAME SHEATHING UNDER SIDING OR SHINGLES, OR AS A BASE FOR PAINT, STAIN OR STUCCO.	2C-D, 3B-C, 4B

INSULATION MANUFACTURERS PRODUCTS
(SORTED BY PRODUCT TRADE NAME)

FRAME WALL INSULATION (Cont.)

PRODUCT TRADENAME	MANUFACTURER	PRODUCT DESCRIPTION	PLATES
THERMATITE INSULATING SHEATHING	MANVILLE CORP	POLYISOCYANURATE FOAM RIGID BOARD,GLASS FIBER REINFORCED AND ALUMINUM FOIL FACED. A SHEATHING APPLIED OVER FRAMING MEMBERS IN NEW CONSTRUCTION OR TO EXTERIOR WALLS BEFORE INSTALLING NEW SIDING.	2C-D,3B-C,4B
THERMAX	CELOTEX CORP	POLYISOCYANURATE FOAM RIGID BOARD,GLASS FIBER REINFORCED AND ALUMINUM FOIL FACED. FOR USE AS NON-STRUCTURAL FRAME SHEATHING,OR BEHIND INTERIOR WALLBOARD.	2C-D,3B-C,4B
UNFACED BATTS *	MANVILLE CORP	GLASS FIBER BATTS,UNFACED. SEPARATE VAPOR BARRIER MAY BE USED.	1D,2B-D 3A-B,4A-C

MASONRY WALL INSULATION

PRODUCT TRADENAME	MANUFACTURER	PRODUCT DESCRIPTION	PLATES
ADVANCED FOAM AFS "38"	ADVANCED FOAM SYSTEMS,INC	UREA-FORMALDEHYDE FOAMED-IN-PLACE INSULATION. FOR NEW OR RETROFIT APPLICATIONS IN CONCRETE BLOCK AND MASONRY CAVITY WALLS.	5B-D,6A-D 7A-B,8B,9B
FLAME-RESISTANT BATTS *	MANVILLE CORP	GLASS FIBER BATTS WITH FOIL-SCRIM-KRAFT (FSK) FACING. FACING EXTENDS PAST BATT EDGE FOR STAPLING FLANGE. MAY BE LEFT EXPOSED WHERE CODES PERMIT.	6C
FLAME SPREAD 25	OWENS-CORNING FIBERGLAS CORP	GLASS FIBER BATT WITH SPECIAL FOIL/KRAFT LAMINATE FACING,RECOMMENDED BY MFR FOR USE IN WALLS AND CEILINGS WHERE INSULATION FACING WILL BE EXPOSED. SUITABLE FOR LOW-ABUSE AREAS.	6C
FOAM-FORM	ROCKY MOUNTAIN FOAM-FORM	POLYSTYRENE FOAM,EXPANDED,BLOCKS FOR POURED CONCRETE WALLS.	11D
FOIL FACED BATTS *	MANVILLE CORP	GLASS FIBER BATT WITH ALUMINUM FOIL FACING. FACING EXTENDS PAST BATT EDGE FOR STAPLING FLANGE.	6C

INSULATION MANUFACTURERS PRODUCTS
(SORTED BY PRODUCT TRADE NAME)

MASONRY WALL INSULATION (Cont.)

PRODUCT TRADENAME	MANUFACTURER	PRODUCT DESCRIPTION	PLATES
HIGH-R SHEATHING	OWENS-CORNING FIBERGLAS CORP	POLYISOCYANURATE FOAM RIGID BOARD,GLASS FIBER REINFORCED AND ALUMINUM FOIL FACED. FOR INTERIOR BASEMENT,MASONRY WALL,OR CAVITY WALL INSULATION.	6B,D,7A-B 8C,9C
INSULWAL	PANELERA	COMPOSITE OF POLYURETHANE FOAM AND GYPSUM BOARD WITH REFLECTIVE FOIL SKIN OVER FOAM. ATTACHED TO MASONRY WALL WITH SPECIAL CLIPS.	11C
KRAFT-FACED BATTS *	MANVILLE CORP	GLASS FIBER BATT WITH KRAFT PAPER FACING. FACING EXTENDS PAST BATT EDGE FOR STAPLING FLANGE.	6C
KRAFT-FACED BUILDING INSULATION *	OWENS-CORNING FIBERGLAS CORP	GLASS FIBER BATT INSULATION WITH ASPHALTED KRAFT PAPER FACING,FLANGED FOR STAPLING.	6C
MASONRY WALL BATTS *	MANVILLE CORP	GLASS FIBER BATTS,FOR INSTALLATION BETWEEN FURRING STRIPS ON MASONRY WALL INTERIORS.	6B
MASONRY WALL INSULATION *	OWENS-CORNING FIBERGLAS CORP	GLASS FIBER BATT,UNFACED,FOR INSTALLATION BETWEEN FURRING STRIPS ON MASONRY WALL INTERIORS.	6B
ROLLED BATTS *	OWENS-CORNING FIBERGLAS CORP	GLASS FIBER BATTS,UNFACED.	6C
SEMI-RIGID INSULATION BOARD *	OWENS-CORNING FIBERGLAS CORP	GLASS FIBER SEMI-RIGID BOARD,FACED OR UNFACED. FOR INTERIOR MASONRY WALLS BETWEEN FURRING STRIPS AND FOR MASONRY CAVITY WALLS.	6B,8C,9C
STYROFOAM SM	DOW CHEMICAL CO	POLYSTYRENE FOAM EXTRUDED BOARD WITH NATURAL SKIN SURFACE. USED ON INTERIOR MASONRY WALLS WITHOUT FURRING AND IN MASONRY CAVITY WALLS.	6D,7A-D,8C 9C,10B,11A
TG-3000	THERMAL SYSTEMS,INC.	POLYURETHANE FOAM RIGID BOARD WITH ALUMINUM FOIL-SCRIM-FOIL LAMINATE FACERS FOR INTERIOR BASEMENT,MASONRY WALL OR CAVITY WALL INSULATION.	6D,7A-B

INSULATION MANUFACTURERS PRODUCTS
(SORTED BY PRODUCT TRADE NAME)

MASONRY WALL INSULATION (Cont.)

PRODUCT TRADENAME	MANUFACTURER	PRODUCT DESCRIPTION	PLATES
THERMOCURVE	THORO SYSTEMS PRODUCTS	POLYSTYRENE FOAM PANELS WITH A UNIQUE CURVED DESIGN AND PROTRUDING SPACERS. PANELS FIT SNUGLY WITHIN POURED CONCRETE WALL FORMS. AVAILABLE IN THREE WIDTHS FOR ALL STANDARD POURED-IN-PLACE STRUCTURES.	
THOROWALL INSULATING THOROWALL	THORO SYSTEMS PRODUCTS	INSULATING PLASTER CONTAINING POLYSTYRENE BEADS, HYDRAULIC BINDERS, AND CHEMICAL ADDITIVES. LIGHTWEIGHT, PACKAGED AS A POWDER, ADD WATER TO APPLY. MAY BE APPLIED TO MASONRY WALL EXTERIORS WITH TROWEL OR SPRAY GUN.	
THERMASOTE SIDEWALL PANELS	HOMASOTE CO	COMPOSITE OF ASBESTOS-FREE INSULATING BUILDING BOARD AND RIGID POLYURETHANE FOAM. FOAM SIDE IS FACED WITH ASPHALT SATURATED FELT OR FIBERGLASS. FOR MASONRY WALL EXTERIOR RETROFITS.	5D, 7C-D
THERMAX INSULATION BOARD	CELOTEX CORP	POLYISOCYANURATE FOAM RIGID BOARD, GLASS FIBER REINFORCED AND ALUMINUM FOIL FACED. ONE FACING HAS A WHITE VINYL COATING, PROVIDING A WASHABLE INTERIOR FINISH.	11B
THERMAX SHEATHING	CELOTEX CORP	POLYISOCYANURATE FOAM RIGID BOARD, GLASS FIBER REINFORCED AND ALUMINUM FOIL FACED. FOR INTERIOR BASEMENT, MASONRY OR CAVITY WALL INSULATION.	6D, 7A-B 8C, 9C
T/LINER	SILVERCOTE METAL BUILDING PRODUCTS	SUPPORTS, EXTRUDED PLASTIC, FOR ATTACHING PREFINISHED INSULATION TO MASONRY WALL INTERIORS.	11B
UNFACED BATTS *	MANVILLE CORP	GLASS FIBER BATTS, UNFACED. SEPARATE VAPOR BARRIER MAY BE USED.	6C
WALL INSULATION *	KNAUF FIBER GLASS GMBH	GLASS FIBER SEMI-RIGID BOARD, UNFACED OR FSK FACED. FOR MASONRY AND CAVITY WALLS.	6B, 8C, 9C

INSULATION MANUFACTURERS PRODUCTS
(SORTED BY PRODUCT TRADE NAME)

METAL BUILDING WALL INSULATION

PRODUCT TRADENAME	MANUFACTURER	PRODUCT DESCRIPTION	PLATES
FIBERGLAS BUILDING INSULATION *	OWENS-CORNING FIBERGLAS CORP	GLASS FIBER BATTS, UNFACED. USED BETWEEN METAL STUDS.	12D
FLAME-RESISTANT BATTS *	MANVILLE CORP	GLASS FIBER BATTS WITH FOIL-SCRIM-KRAFT (FSK) FACING. FACING EXTENDS PAST BATT EDGE FOR STAPLING FLANGE. MAY BE LEFT EXPOSED WHERE CODES PERMIT.	12C
FLAME SPREAD 25	OWENS-CORNING FIBERGLAS CORP	GLASS FIBER BATT WITH SPECIAL FOIL/ KRAFT LAMINATE FACING, RECOMMENDED BY MFR FOR USE IN WALLS AND CEILINGS WHERE INSULATION FACING WILL BE EXPOSED. SUITABLE FOR LOW-ABUSE AREAS.	12C
FOIL FACED BATTS *	MANVILLE CORP	GLASS FIBER BATT WITH ALUMINUM FOIL FACING. FACING EXTENDS PAST BATT EDGE FOR STAPLING FLANGE.	12C
INSULATED "RAIN SCREEN"	H.L. BIRUM CORP	UNSPECIFIED COMPOSITION, LAMINATED THERMAL PANELS.	13D
INSULATED WALL & SOFFIT SYSTEM	FINESTONE CORP	VARIOUS RIGID INSULATIONS (GLASS FIBER, POLYSTYRENE BEAD BOARD, POLYURETHANE FOAM OR PHENOLIC FOAM) IN A PANEL COMPOSED OF SUBSTRATE INSULATION METAL BOTH MODIFIED PORTLAND CEMENT, AND ARCHITECTURAL FINISH.	13D
KRAFT-FACED BATTS *	MANVILLE CORP	GLASS FIBER BATT WITH KRAFT PAPER FACING. FACING EXTENDS PAST BATT EDGE FOR STAPLING FLANGE.	12C
KRAFT FACED BUILDING INSULATION *	OWENS-CORNING FIBERGLAS CORP	GLASS FIBER BATT INSULATION WITH ASPHALTED KRAFT PAPER FACING, FLANGED FOR STAPLING.	12C
METAL BUILDING PANEL INSULATION *	CERTAINTEED CORP	GLASS FIBER BLANKET, UNFACED.	13C
MICROLITE "L"	MANVILLE CORP	GLASS FIBER BLANKET LAMINATED WITH VARIOUS CUSTOM FACINGS.	13A-B
PAN-INSUL	MANVILLE CORP	GLASS FIBER BATT, UNFACED. FOR INTERLOCKING PRE-ENGINEERED METAL BLDG WALL CAVITIES.	13C

INSULATION MANUFACTURERS PRODUCTS
(SORTED BY PRODUCT TRADE NAME)

METAL BUILDING WALL INSULATION (Cont.)

PRODUCT TRADENAME	MANUFACTURER	PRODUCT DESCRIPTION	PLATES
PEBS BLANKET	MANVILLE CORP	GLASS FIBER BLANKET, UNFACED. FOR PRE-ENGINEERED METAL BLDG WALLS.	13C
RIGID ROLL	MANVILLE CORP	GLASS FIBER SEMI-RIGID ROLLED INSULATION WITH A TEXTURED VINYL, FSK, OR WHITE FSK FACING.	13A-B
ROLLED BATTS *	OWENS-CORNING FIBERGLAS CORP	GLASS FIBER BATTS, UNFACED.	12C
ROLL-IN	MANVILLE CORP	GLASS FIBER SEMI-RIGID ROLLED INSULATION WITH A DECORATIVE VINYL FACING.	13A-B
SNAP-IN	MANVILLE CORP	GLASS FIBER SEMI-RIGID BOARD WITH A DECORATIVE VINYL FACING.	13A-B
THERMAX INSULATION BOARD	CELOTEX CORP	POLISOCYANURATE ROAM RIGID BOARD, GLASS FIBER REINFORCED AND ALUMINUM FOIL FACED. ONE FACE HAS A WHITE VINYL COATING, PROVIDING A WASHABLE INTERIOR FINISH.	13A-B
THERMOCORE	MANVILLE CORP	PERLITE BOARD ARCHITECTURAL PANEL OF EXPANDED PERLITE PARTICLES, FIBERS AND BINDERS, LAMINATED TO MINERAL FIBER CEMENT FACINGS.	13D
TRANSIFOAM	MANVILLE CORP	POLYSTYRENE FOAM ARCHITECTURAL PANEL EXPANDED BEAD BOARD, LAMINATED TO MINERAL FIBER CEMENT FACINGS.	13D
TRANSITOP	MANVILLE CORP	WOOD FIBER AND ASPHALTIC COMPOUNDS HYDRAULICALLY PRESSED INTO RIGID ARCHITECTURAL PANELS, LAMINATED TO MINERAL FIBER CEMENT FACINGS.	13D
UNFACED BATTS *	MANVILLE CORP	GLASS FIBER BATTS, UNFACED. SEPARATE VAPOR BARRIER MAY BE USED.	12C

INSULATION MANUFACTURERS PRODUCTS
(SORTED BY PRODUCT TRADE NAME)

PRODUCT TRADENAME	MANUFACTURER	PRODUCT DESCRIPTION	PLATES

CEILING INSULATION

PRODUCT TRADENAME	MANUFACTURER	PRODUCT DESCRIPTION	PLATES
BLOWING WOOL *	MANVILLE CORP	GLASS FIBER LOOSE FILL FOR ATTICS AND OVERHEAD SPACES.	14B-D,15B
FIBERGLAS BLOWING WOOL	OWENS-CORNING FIBERGLAS CORP	GLASS FIBER LOOSE FILL FOR ATTICS AND OVERHEAD SPACES.	14B-D,15B
FIBERGLAS CUBED BLOWING WOOL	OWENS-CORNING FIBERGLAS CORP	GLASS FIBER LOOSE FILL FOR ATTICS AND OVERHEAD SPACES.	14B-D,15B
FLAME-RESISTANT BATTS *	MANVILLE CORP	GLASS FIBER BATT WITH FOIL-SCRIM-KRAFT (FSK) FACING. FACING EXTENDS PAST BATT EDGE FOR STAPLING FLANGE. MAY BE LEFT EXPOSED WHERE CODES PERMIT.	14A,C,15A-B, D,16A-B, 17A-B,18C, 19D,20B
FLAME SPREAD 25	OWENS-CORNING FIBERGLAS CORP	GLASS FIBER BATT WITH SPECIAL FOIL/KRAFT LAMINATE FACING,RECOMMENDED BY MFR FOR USE IN WALLS AND CEILINGS WHERE INSULATION FACING WILL BE EXPOSED. SUITABLE FOR LOW-ABUSE AREAS.	14A,C,15A-B, D,16A-B 17A-B,18C 19D,20B
FOIL FACED BATTS *	MANVILLE CORP	GLASS FIBER BATT WITH ALUMINUM FOIL FACING. FACING EXTENDS PAST BATT EDGE FOR STAPLING FLANGE.	14A,C,15A-B, D,16A-B,17A-B,18C 19D,20B
HIGH-R SHEATHING	OWENS-CORNING FIBERGLAS CORP	POLYISOCYANURATE FOAM RIGID BOARD,GLASS FIBER REINFORCED AND ALUMINUM FOIL FACED. FOR CATHEDRAL AND A-FRAME OVERDECK APPLICATIONS.	17B,D
K-13	NATIONAL CELLULOSE CORP	CELLULOSE FIBER,SPRAYED-ON FOR EXPOSED INTERIOR APPLICATIONS.	20C
KRAFT-FACED BATTS *	MANVILLE CORP	GLASS FIBER BATT WITH KRAFT PAPER FACING. FACING EXTENDS PAST BATT EDGE FOR STAPLING FLANGE.	14A,C,15A-B, D,16A-B 17A-B,18C, 19D,20B
KRAFT FACED BUILDING INSULATION *	OWENS-CORNING FIBERGLAS CORP	GLASS FIBER BATT INSULATION WITH ASPHALTED KRAFT PAPER FACING,FLANGED FOR STAPLING.	14A,C,15A-B, D,16A-B,17A-B,18C,19D, 20B

INSULATION MANUFACTURERS PRODUCTS
(SORTED BY PRODUCT TRADE NAME)

CEILING INSULATION (Cont.)

PRODUCT TRADENAME	MANUFACTURER	PRODUCT DESCRIPTION	PLATES
RETROFIT INSULATION	METAL BUILDING INTERIOR PRODUCTS CO	GLASS FIBER BLANKET INSULATION WITH WHITE VINYL FACING FOR INSTALLATION BETWEEN PURLINS.	
ROLLED BATTS *	OWENS-CORNING FIBERGLAS CORP	GLASS FIBER BATTS,UNFACED.	14A,C,15A-B, D,16A-B,17A-B,18C,19D, 20B
STYROFOAM SM	DOW CHEMICAL CO	POLYSTYRENE FOAM EXTRUDED BOARD WITH NATURAL SKIN SURFACE. FOR ROOF OVER-DECK APPLICATIONS (CATHEDRAL CEIL-INGS.)	17D
STYROFOAM TG	DOW CHEMICAL CO	POLYSTYRENE FOAM EXTRUDED BOARD WITH HIGH-DENSITY SKIN. EDGES ARE TONGUE AND GROOVE. FOR ROOF OVERDECK APPLIC-ATIONS (CATHEDRAL CEILINGS).	17D
SUPER BATT INSULATION	OWENS-CORNING FIBERGLAS CORP	GLASS FIBER BATTS,EXTRA THICK,FACED OR UNFACED.	15C
THERMAX INSULATION BOARD	CELOTEX CORP	POLYISOCYANURATE FOAM RIGID BOARD,GLASS FIBER REINFORCED AND ALUMINUM FOIL FACED. ONE FACER HAS A WHITE VINYL COATING.	19B-D,20A-B
THERMAX SHEATHING	CELOTEX CORP	POLYISOCYANURATE FOAM RIGID BOARD,GLASS FIBER REINFORCED AND ALUMINUM FOIL FACED. FOR CATHEDRAL AND A-FRAME OVERDECK APPLICATIONS.	17B,D
THERMOCON	THERMOCON SYSTEMS,INC	CELLULOSE FIBER,SPRAYED-ON FOR EXPOSED INTERIOR APPLICATIONS.	20C
UNFACED BATTS *	MANVILLE CORP	GLASS FIBER BATTS,UNFACED. SEPARATE VAPOR BARRIER MAY BE USED.	14A,C,15A-B, D,16A-B 17A-B,18C, 19D,20B

ROOF INSULATION

PRODUCT TRADENAME	MANUFACTURER	PRODUCT DESCRIPTION	PLATES
ALL-WEATHER CRETE	SILBRICO CORP	INSULATING CONCRETE FOR CONCRETE OR METAL DECKS,SLOPED TO PROVIDE DRAINAGE.	29A-D,30A-C

INSULATION MANUFACTURERS PRODUCTS
(SORTED BY PRODUCT TRADE NAME)

ROOF INSULATION (Cont.)

PRODUCT TRADENAME	MANUFACTURER	PRODUCT DESCRIPTION	PLATES
ELASTIZELL CONCRETE	ELASTIZELL CORP OF AMERICA	INSULATING CONCRETE CONTAINING DISCREET AIR CELLS (NO EXPANDED FILLERS). FOR USE WITH ANY DECK OR OVER EXISTING BUR.	29A-D,30A-C
EPS	ARCO POLYMERS, INC	POLYSTYRENE FOAM, EXPANDED, UNFACED. FOR NEW BURS OR SINGLE-PLY ROOFS, OR FOR RETROFIT DIRECTLY OVER OLD BUR.	22A-D,23C-D 24C-D,25A-D 26A
EXELTHERM XTRA	KOPPERS CO, INC	PHENOLIC FOAM RIGID BOARD FOR BURS OR SINGLE-PLY ROOFS. FOR ANY DECK.	22A-D,24C-D 23A,25A-D
FESCO BOARD	MANVILLE CORP	PERLITE (EXPANDED PARTICLES) BLENDED HOMOGENEOUSLY WITH SELECTED FIBERS AND BINDERS. RIGID BOARD TOP SURFACED WITH TOP-LOC COATING TO COATING TO RECEIVE BUR FOR ANY DECK.	22A-D
FESCO FOAM	MANVILLE CORP	COMPOSITE OF PERLITE BOARD (FESCO BOARD) -POLYURETHANE FOAM BOARD WITH ASPHALT ROOFING FELT FACING FOR BUR. FOR ANY DECK.	22A-B
FESCO RE-ROOF BOARD	MANVILLE CORP	PERLITE BOARD (FESCO BOARD) FOR USE IN APPLYING A NEW BUR DIRECTLY OVER ON OLD ROOF.	23C-D
FESCO TAPERED DRI-DECK SYSTEM	MANVILLE CORP	PERLITE BOARD (FESCO BOARD) FACTORY-TAPERED TO PROVIDE DRAINAGE ON A FLAT ROOF.	23A-B
FIBERGLAS ROOF INSULATION	OWENS-CORNING FIBERGLAS CORP	GLASS FIBER BOARD WITH RESINOUS BINDER, TOP-SURFACED WITH GLASS FIBER REINFORCED ASPHALT AND KRAFT FOR BUR BASE. FOR FLAT AND LOW-SLOPE NAILABLE, NONNAILABLE, AND METAL DECKS.	22A-D
FIBERGLAS/URETHANE ROOF INSULATION	OWENS-CORNING FIBERGLAS CORP	COMPOSITE OF GLASS FIBER BOARD POLYURETHANE FOAM BOARD, SURFACED WITH FIBROUS GLASS MAT FOR BUR BASE. FOR FLAT AND LOW-SLOPE NAILABLE, NON-NAILABLE, AND METAL DECKS.	22A-B
FS1000	THERMAL SYSTEMS, INC	POLYURETHANE FOAM BOARD EXTRUDED BETWEEN KRAFT, ALUMINUM FOIL, OR ASPHALT SATURATED FELT MEMBRANES. FOR ALL DECKS.	22A-D
GAFTEMP ISOTHERM	GAF CORP	POLISOCYANURATE FOAM RIGID BOARD WITH ASPHALT SATURATED ASBESTOS FACINGS.	22A-D

INSULATION MANUFACTURERS PRODUCTS
(SORTED BY PRODUCT TRADE NAME)

ROOF INSULATION (Cont.)

PRODUCT TRADENAME	MANUFACTURER	PRODUCT DESCRIPTION	PLATES
GAFTEMP PERLITE	GAF CORP	PERLITE BOARD COMPOSED OF EXPANDED PERLITE PARTICLES HOMOGENOUSLY BLENDED WITH SELECTED BINDERS AND FIBERS. TOP SURFACE SEALED WITH A SPECIAL COATING FOR BUR BASE.	22A-D
GAFTEMP URETHANE /PERLITE	GAF CORP	POLYURETHANE FOAM RIGID BOARD WITH ASPHALT-SATURATED FELT FACINGS.	22A-D
GAFTEMP URETHANE /PERLITE	GAF CORP	COMPOSITE OF PERLITE BOARD POLYURETHANE FOAM BOARD WITH ASPHALT SATURATED FELT TOP SURFACE.	22A-B
INSULATED PANEL *	INSULATED PANEL SYSTEMS, INC	POLYURETHANE FOAM CORE STANDING SEAM ROOF PANEL WITH GALVANIZED STEEL SKINS. SKINS AVAILABLE WITH STUCCO-EMBOSSED FINISHES AND VARIOUS COLORS.	31B
ITP MONEY CLIP BOARD	MANVILLE CORP	GLASS FIBER BOARD WITH GLASS-REINFORCED WHITE, METALLIZED POLYESTER FACING. FOR UNDERDECK INSTALLATION BETWEEN PURLINS IN NEW METAL BUILDINGS. FACING PROVIDES FINISHED INTERIOR SURFACE.	
ITP MONEY CLIP BATT	MANVILLE CORP	GLASS FIBER BATT FOR USE WITH ITP MONEY CLIP BOARD.	18C, 19D
MAX-I	RMAX, INC	POLYISOCYANURATE FOAM RIGID BOARD WITH ASPHALT-COATED FIBERGLASS MAT FACINGS. MAY BE USED DIRECTLY OVER STEEL ROOF DECKS AS BUR BASE.	22A-D
MICROLITE "L"	MANVILLE CORP	GLASS FIBER BLANKET, UNFACED, FOR NEW METAL BUILDINGS. INSTALLED OVER PURLINS.	20A-B
MI-T-R	MIZELL BROS. CO	GLASS FIBER BATT AND BLANKET SYSTEM FOR METAL BUILDINGS.	31A
NAILABLE BASE	RMAX, INC	COMPOSITE OF POLYISOCYANURATE FOAM RIGID BOARD WITH A BOTTOM SKIN OF ROOFING FELT AND A TOP LAYER OF NAILABLE BASE MATERIAL. FOR ROOFS WHERE A TOP NAILABLE SURFACE IS REQUIRED.	17D
NEWCOMP	SHELTER INSULATION, INC	COMPOSITE OF PERLITE BOARD-POLYURETHANE /POLYISOCYANURATE FOAM BOARD. FOAM IS FACED WITH ASPHALT-SATURATED FELT. PRODUCT IS INSTALLED FOAM-SIDE DOWN OVER NONCOMBUSTIBLE DECKS; PERLITE SIDE IS THE BUR BASE.	22A-B

INSULATION MANUFACTURERS PRODUCTS
(SORTED BY PRODUCT TRADE NAME)

ROOF INSULATION (Cont.)

PRODUCT TRADENAME	MANUFACTURER	PRODUCT DESCRIPTION	PLATES
PEBS BLANKET	MANVILLE CORP	GLASS FIBER BLANKET,UNFACED. GENERAL PURPOSE PRODUCT FOR NEW AND RETROFIT INSTALLATIONS IN METAL BUILDINGS.	12C-D,13A 18C,19D,20B 31A
PERMALITE PK	GREFCO,INC	COMPOSITE OF PERLITE BOARD (PERMALITE) POLYURETAHANE FOAM BOARD. ASPHALT SATURATED FELT FACING FOR BUR BASE.	22A-B
PERMALITE PK PLUS	GREFCO,INC	COMPOSITE OF PERLITE BOARD (PERMALITE) TOP AND BOTTOM LAYERS WITH A RIGID POLYURETHANE FOAM CORE.	22A-B
PERMALITE SEALSKIN	GREFCO,INC	PERLITE BOARD FORMED OF EXPANDED HERMETICALLY SEALED PERLITE BEADS WATERPROOFING AGENTS AND CELLULOSE BINDER. INTEGRAL SURFACE TREATMENT FOR BUR BASE.	22A-D
PERMALITE URETHANE	GREFCO,INC	POLYURETHANE FOAM BOARD WITH ASPHALT SATURATED FEET FACINGS ON BOTH SURFACES.	22A-D
PLYFOAM STYRENE	WATER GUIDANCE SYSTEMS,INC	POLYSTYRENE FOAM BOARD FOR SINGLE-PLY ROOFS.	24C-D,25A-D 26A
PLYFOAM URETHANE	WATER GUIDANCE SYSTEMS,INC	POLYURETHANE FOAM PANELS WITH ASPHALT FELT,FIBERGLASS,ALUMINUM FOIL,OR POLYETHYLENE COATED KRAFT PAPER FACINGS. FOR SINGLE-PLY ROOFS.	24C-D,25A-D 26A
PLYFOAM URETHANE /COMPOSITE	WATER GUIDANCE SYSTEMS,INC	COMPOSITE OF POLYURETHANE FOAM PANEL WITH PERLITE BOTTOM LAYER. FOR SINGLE-PLY ROOFS.	24C-D,25A-D 26A
PLY-I	RMAX,INC	POLYISOCYANURATE FOAM RIGID BOARD WITH FIBERGLASS REINFORCED ALUMINUM FOIL FACINGS. FOR SINGLE-PLY ROOFS OVER ANY DECK. MECHANICAL FASTENERS RECOMMENDED FOR ATTACHMENT.	24C-D,25A-D 26A
POLYCON-POSITE	CONSOLIDATED FIBER GLASS PRODUCTS CO	COMPOSITE OF PERLITE BOARD POLYURE-THANE FOAM BOARD.	22A-B
POLYCON-STANDARD	CONSOLIDATED FIBERGLASS PRO-DUCTS CO	POLYURETHANE FOAM RIGID BOARD WITH ASPHALT-SATURATED FELT FACINGS.	22A-D
RIGID-ROLL	MANVILLE CORP	GLASS FIBER SEMI-RIGID BLANKET WITH TEXTURED VINYL,FSK OR WHITE FSK FACING FOR NEW METAL BUILDINGS. INSTALLED OVER PURLINS.	31A

INSULATION MANUFACTURERS PRODUCTS
(SORTED BY PRODUCT TRADE NAME)

ROOF INSULATION (Cont.)

PRODUCT TRADENAME	MANUFACTURER	PRODUCT DESCRIPTION	PLATES
SILICONE ROOFING SYSTEM	GENERAL ELECTRIC CO	POLYURETHANE FOAM, SPRAYED-IN-PLACE OVER ANY NEW DECK OR EXISTING BUR, WITH TWO TOP COATS OF SPRAYED-IN-PLACE SILICONE RUBBER.	27A-D, 26A-B
SISCOMP	SHELTER INSULATION, INC	COMPOSITE OF POLYURETHANE FOAM ROOF INSULATION.	22A-B
SISDECK	SHELTER INDUSTRIES, INC	POLYURETHANE FOAM ROOF INSULATION FOR NONCOMBUSTIBLE DECKS.	22C-D
SISDECK GF(N)	SHELTER INSULATION, INC	POLYURETHANE FOAM ROOF INSULATION WITH INTEGRALLY BONDED NONASPHALTIC GLASS FACINGS. ESPECIALLY DESIGNED FOR SINGLE-PLY ROOF SYSTEMS.	24D-C, 25A-D 26A
SISTEEL	SHELTER INSULATION, INC	NONCOMPOSITE ROOF INSULATION FOR STEEL DECKS.	22A-B
STYROFOAM SM	DOW CHEMICAL CO	POLYSTYRENE FOAM EXTRUDED BOARD WITH NATURAL SKIN SURFACE. FOR ROOF OVER-DECK APPLICATIONS (CATHEDRAL CEIL-INGS.)	17D
STYROFOAM TG	DOW CHEMICAL CO	POLYSTYRENE FOAM EXTRUDED BOARD WITH HIGH-DENSITY SKIN. EDGES ARE TONGUE AND GROOVE. FOR ROOF OVERDECK APPLIC-ATIONS (CATHEDRAL CEILINGS).	17D
STYROFOAM RM	DOW CHEMICAL CO	POLYSTYRENE FOAM, EXTRUDED BOARD, UNFACED. CHANNELED TO PROVIDE DRAINAGE OF PROTECTED MEMBRANE (UPSIDE-DOWN) ROOFS.	23C-D
TAPERED EPS	ARCO POLYMERS, INC	POLYSTYRENE FOAM, EXPANDED, UNFACED TAPERED TO PROVIDE DRAINAGE OF A FLAT ROOF. FOR BURS OVER NEW OR EXISTING ROOFS.	23A-B
TAPERED FOAM	BENOIT, INC	POLYSTYRENE FOAM, MOLDED, TAPERED TO PROVIDE DRAINAGE OF FLAT ROOFS. BUR BASE FOR ANY DECK.	23A-B
TAPERED FOAMGLAS ROOF INSULATION	PITTSBURGH CORNING CORP	CELLULAR GLASS BOARD FOR BURS, TAPERED FOR DRAINAGE OF FLAT ROOFS	23A-B
TG1000	THERMAL SYSTEMS, INC	COMPOSITE OF PERLITE BOARD BASE, TOP SHEET OF POLYURETHANE BOARD, FACED WITH KRAFT, ALUMINUM FOIL, OR ASPHALT SATURATED FELT.	22A-B

INSULATION MANUFACTURERS PRODUCTS
(SORTED BY PRODUCT TRADE NAME)

ROOF INSULATION (Cont.)

PRODUCT TRADENAME	MANUFACTURER	PRODUCT DESCRIPTION	PLATES
THERMAL-ACOUSTICAL BATT	MANVILLE CORP	GLASS FIBER BATTS,WITH OR WITHOUT A VAPOR BARRIER,FOR CONTROL OF BOTH HEAT AND SOUND TRANSMISSION THROUGH SUSPENDED CEILINGS.	
THERMAROOF COMPOSITE	RMAX, INC	COMPOSITE OF PERLITE BASE LAYER POLY-ISOCYANURATE FOAM WITH FIBERGLASS FELT TOP FACING. FOR BURS OVER STEEL DECKS.	24C-D 26A
THERMAROOF PLUS	RMAX, INC	POLYISOCYANURATE FOAM RIGID BOARD WITH ALUMINUM FOIL FACINGS. FOR BALLASTED LOOSE-LAID SINGLE-PLY ROOFS.	24C,25A,C
THERMAROOF PLUS COMPOSITE	RMAX, INC	COMPOSITE OF PERLITE BASE LAYER POLY-ISOCYANURATE FOAM WITH ALUMINUM FOIL TOP SKIN. FOR SINGLE-PLY ROOFS.	24C-D,25A-D 26A
THERMAROOF STANDARD	RMAX, INC	POLYISOCYANURATE FOAM RIGID BOARD WITH FIBERGLASS FELT FACINGS. FOR BURS MAY BE APPLIED OVER A BASE LAYER OF PERLITE BOARD.	22A-D
XFS 4249	DOW CHEMICAL CO	COMPOSITE OF POLYSTYRENE FOAM,EXTRUDED BOARD (STYROFOAM RM) WITH FACTORY APPLIED 3/8INCH THICK LATEX MODIFIED PORTLAND CEMENT MORTAR FACING. FOR APPLICATIONS REQUIRING HIGH COMPRES-SIVE STRENGTH. (DEVELOPMENTAL PROD.)	
XFS 43001	DOW CHEMICAL CO	POLYSTYRENE FOAM EXTRUDED BOARD HAVING HIGH COMPRESSIVE STRENGTH. (DEVELOPMENTAL PRODUCT)	

WINDOWS AND WINDOW TREATMENTS

PRODUCT TRADENAME	MANUFACTURER	PRODUCT DESCRIPTION	PLATES
ACRYLITE SDP	CY/RO INDUSTRIES	ACRYLIC SHEET,DOUBLE SKINNED FOR THER-MAL INSULATION COMPARABLE TO INSUL-ATING GLASS. FOR SKYLIGHTS,COVERED WALKWAYS,CURTAINWALLS,GREENHOUSES.	
COOL-VIEW	SHATTERPROOF GLASS CORP	REFLECTIVE GLASS CONSISTING OF CLEAR, BRONZE,OR GRAY GLASS WITH A CHROME, BRONZE,OR GOLD COATING ON THE IN-TERIOR.	

INSULATION MANUFACTURERS PRODUCTS
(SORTED BY PRODUCT TRADE NAME)

WINDOWS AND WINDOW TREATMENTS (Cont.)

PRODUCT TRADENAME	MANUFACTURER	PRODUCT DESCRIPTION	PLATES
HEAVY-DUTY IN-SIDER	PLASKOLITE, INC.	INSIDE STORM WINDOW, ACRYLIC GLAZING. MOUNTS TO EXISTING WINDOW WITH VINYL MOULDING.	36B
INSULATING CURTAIN WALL	THERMAL TECHNOLOGY CORP	ROLLING WINDOW SHADE OF FOUR-LAYER FABRIC DESIGN, SELF-INFLATING. AUTO-MATICALLY ACTIVATED BY EXTERNAL TEMP-ERATURE SENSORS.	
KOOLSHADE SOLAR SCREENS	KOOLSHADE CORP	HORIZONTAL LOUVER SHADING SCREEN, FIXED POSITION. BLOCKS SUNLIGHT TO VARIOUS DEGREES DEPENDING ON SUN ELEVATION ANGLE.	
LEXAN	GENERAL ELECTRIC CO	ACRYLIC SHEET, CLEAR OR TINTED, VARIOUS THICKNESSES AND TYPES.	
LUCITE	E.I. DU PONT DE NEMOURS & COMPANY, INC.	ACRYLIC SHEET, PRODUCED FROM METHYL METHACRYLATE MONOMER, IN LINEAR (L) COMPOSITION, CROSS-LINKED (XL) COMP-OSITION FOR HIGHER SOLVENT RESIS-TANCE, AND (T-1000) FOR HIGH IMPACT RESISTANCE.	
MAGNETIC FLEXIBLE INS WINDOW	3M CO	INSIDE STORM WINDOW, NONINSULATING OR REFLECTIVE INSULATING FLEXIBLE FILM. MOUNTS TO EXISTING WINDOW WITH MAG-NETIC STRIPS.	36C
MAGNETIC INSULATING SHADE	3M CO	ROLLING WINDOW SHADE OF SUN CONTROL FILM. EDGES SEAL TO WINDOW FRAME WIT FLEXIBLE MAGNETIC STRIPS.	
MAGNETIC RIGID INSUL WINDOW	3M CO	INSIDE STORM WINDOW, ACRYLIC GLAZING. MOUNTS TO EXISTING WINDOW WITH MAG-NETIC STRIPS.	36A
MAGNETITE WINDOWS	VIKING ENERGY SYSTEM CO	INSIDE STORM WINDOW, ACRYLIC GLAZING. MOUNTS TO EXISTING WINDOW WITH MAG-NETIC STRIPS.	36A
PLEXIGLAS	ROHM AND HAAS CO	ACRYLIC SHEET, CLEAR OR TINTED, TRANS-PARENT TO SEMI-OPAQUE. VARIOUS GLA-ZING APPLICATIONS, TYPES INCLUDE A UV-ABSORBING TYPE.	
POLYCARBONATE SDP	CY/RO INDUSTRIES	POLYCARBONATE SHEET, DOUBLE SKINNED FOR THERMAL INSULATION COMPARABLE TO INSULATING GLASS. FOR SKYLIGHTS, COV-ERED WALKWAYS, CURTAINWALLS, GREEN-HOUSES.	

INSULATION MANUFACTURERS PRODUCTS
(SORTED BY PRODUCT TRADE NAME)

WINDOWS AND WINDOW TREATMENTS (Cont.)

PRODUCT TRADENAME	MANUFACTURER	PRODUCT DESCRIPTION	PLATES
REFLECTIVE GLASS	GUARDIAN INDUSTRIES CORP	REFLECTIVE GLASS,VACUUM DEPOSITED COATINGS OF COPPER OXIDE (C-SERIES) STAINLESS STEEL OXIDE (S-SERIES),OR TITANIUM OXIDE (T-SERIES). AVAILABLE IN INSULATING GLASS UNITS.	
REFLECTOVUE	ASG INDUSTRIES,INC	REFLECTIVE GLASS CONTAINING VACUUM DE-POSITED THIN METALLIC COATING OF PURE GOLD OR CHROME.	
ROLLING SHUTTER	PEASE ROLLING SHUTTERS	ROLLING WINDOW SHUTTER OF PVC VINYL, MOUNTED EXTERIOR TO EXISTING WINDOWS. EDGES SEALED WITH ALUMINUM SIDE RAILS.	
SCOTCHTINT SUN CONTROL FILMS	3M CO	SUN CONTROL FILM,EITHER CLEAR OR COLORED POLYESTER FILM,OR ALUMINUM VAPOR COATED POLYESTER FILM. REDUCES INCOM-ING GLARE,ULTRAVIOLET,AND HEAT GAIN, ALSO HEAT LOSS.	
SOLAR SHADE *	MOORE CO	VERTICAL WINDOW SHADE OF MULTIPLE HOLLOW ALUMINUM BLADES. MOUNTED EXTERIOR TO WINDOW OR BETWEEN INSIDE AND OUTSIDE WINDOWS.	
SUNGLAS WINDOW GLASS	FORD MOTOR CO,GLASS DIV	HEAT ABSORBING GLASS,SINGLE OR DOUBLE STRENGTH THICKNESSES.	
T-2001	DISCO ALUMINUM PRODUCTS CO	VENETIAN BLIND WINDOW SHADE MOUNTED BE-TWEEN SHEETS OF DOUBLE-PANE WINDOW.	
THERMALON	ARMSTRONG CORK CO	WINDOW INSULATION,OPAQUE,PERMANENTLY MOUNTED OVER EXISTING WINDOWS.	
THERMOPANE	LIBBEY-OWENS-FORD CO	INSULATING GLASS,DOUBLE- OR TRIPLE-PANE, SEALED,DESICCANT MATERIAL IN SEALED SPACE.	
TRU-THERM	ASG INDUSTRIES,INC	INSULATING GLASS,DOUBLE-PANE,SEALED WITH DESICCANT MATERIAL IN THE SEAL-ED SPACE. CLEAR OR TINTED.	
VARI-TRAN	LIBBEY-OWENS-FORD CO	HEAT ABSORBING GLASS,VACUUM DEPOSITED METALLIC COATINGS.	
WINDOW INSULATION *	SENTINAL FOAM PRODUCTS,INC	WINDOW INSULATION,TRANSLUCENT POLYETHYL-ENE FOAM,WITH ADHESIVE ON ONE SIDE FOR PERMANENT MOUNTING TO EXIS-TING WINDOWS.	

INSULATION MANUFACTURERS PRODUCTS
(SORTED BY PRODUCT TRADE NAME)

PRODUCT TRADENAME	MANUFACTURER	PRODUCT DESCRIPTION	PLATES

WINDOWS AND WINDOW TREATMENTS (Cont.)

PRODUCT TRADENAME	MANUFACTURER	PRODUCT DESCRIPTION	PLATES
WEATH-R-PROOF	SHATTERPROOF GLASS CORP	INSULATING GLASS,DOUBLE-PANE,SEALED, WITH DESICCANT MATERIAL IN THE SEALED SPACE. CLEAR OR TINTED.	

DOORS

PRODUCT TRADENAME	MANUFACTURER	PRODUCT DESCRIPTION	PLATES
EVER-STRAIT REPLACE DOOR	PEASE CO	ENTRANCE DOOR,RESIDENTIAL REPLACEMENT, EXPANDED POLYSTYRENE FOAM CORE, THERMAL BREAKS, MAGNETICALLY WEATHERSTRIPPED.	
INSULCLAD 260 DOORS	KAWNEER CO	ENTRANCE DOOR,COMMERCIAL,FULLY WEATHERSTRIPPED,THERMAL BREAKS.	
OUTSLIDER	PEERLESS PRODUCTS,INC	STORM DOOR,RESIDENTIAL PATIO SLIDING, MOUNTS OUTSIDE EXISTING PATIO DOOR. FULLY WEATHERSTRIPPED. OPTIONAL INSULATING OR TINTED GLASS.	
STORM DOOR *	PEERLESS PRODUCTS,INC	STORM DOOR,RESIDENTIAL ENTRANCE,FULLY WEATHERSTRIPPED.	
THERMACORE	INSOPORT INDUSTRIES,INC	OVERHEAD DOOR,COMMERCIAL INSULATED WITH A POLYURETHANE FOAM CORE BETWEEN EMBOSSED GALVANIZED STEEL SKINS, THERMAL BREAKS.	39A
THERMAL DOOR	ANDERSON DOOR CO	OVERHEAD DOOR,COMMERCIAL,INSULATED.	39A
THERMOSPAN	DALTON INTERNATIONAL,INC	OVERHEAD DOOR,COMMERCIAL,INSULATED WITH A POLYURETHANE FOAM CORE BETWEEN GALVANIZED STEEL SKINS,THERMAL BREAKS.	39A

WEATHERSTRIPPING AND CAULKING

PRODUCT TRADENAME	MANUFACTURER	PRODUCT DESCRIPTION	PLATES
EXTERIOR FOAM WEATHER STRIP *	3M CO	FOAM STRIP,ADHESIVE BACKED,FOR WINDOWS AND DOORS,IRREGULAR SURFACES. MOISTURE RESISTANT.	
GARAGE DOOR BOTTOM WEATHER STRIP *	3M CO	GARAGE DOOR WEATHERSTRIPPING,HIGH-QUALITY SYNTHETIC RUBBER. SEALS GARAGE DOOR TO FLOOR,CONFORMS TO IRREGULARITIES.	

INSULATION MANUFACTURERS PRODUCTS
(SORTED BY PRODUCT TRADE NAME)

PRODUCT TRADENAME	MANUFACTURER	PRODUCT DESCRIPTION	PLATES

WEATHERSTRIPPING AND CAULKING (Cont.)

PRODUCT TRADENAME	MANUFACTURER	PRODUCT DESCRIPTION	PLATES
INTERIOR FOAM WEATHER STRIP *	3M CO	FOAM STRIP,ADHESIVE BACKED,FOR WINDOWS AND DOORS,IRREGULAR SURFACES.	
PERIMETER WEATHER SEAL *	STANLEY HARDWARE	TUBULAR FINNED WEATHERSTRIPPING,EPDM RUBBER,RESISTS ULTRAVIOLET,WATER AB-SORPTION,FREEZING. FILLS GAPS BETWEEN WINDOWS,DOORS,AIR CONDITIONERS.	
PERMANENT CAULKING STRIPS *	3M CO	ROPE-TYPE CAULKING FOR PERMANENT IN-STALLATIONS,RESISTS WEATHERING AND MOISTURE,PAINTABLE.	
REUSABLE TUBULAR WEATHER STRIP *	3M CO	TUBULAR WEATHERSTRIPPING FOR SEASONAL USE AROUND WINDOWS AND DOORS. REUSABLE.	
THERM-L-BRUSH	SEALEZE CORP	BRUSH WEATHERSTRIPPING,FLEXIBLE NYLON, WEATHER RESISTANT,FILAMENTS MOVE EASILY IN ANY DIRECTION,CONFORMS TO GAPS AND MISALIGNMENTS. FOR ANY TYPE DOOR.	
TRANSPARENT WEATHER SEALING TAPE*	3M CO	TAPE,TRANSPARENT,SEALS CRACKS AROUND WINDOWS AND DOORS,MOISTURE RESISTANT.	
TRISEAL	SLOTTSEAL	TRIANGULAR WEATHERSTRIPPING,PVC,MAY BE PERMANENTLY FITTED INTO A MACHINED GROOVE IN WINDOWS AND DOORS.	
TUBESEAL	SLOTTSEAL	TUBULAR WEATHERSTRIPPING,PVC,EPDM RUB-BER OR SILICONE RUBBER. ATTACHED TO WINDOWS AND DOORS BY STAPLING THROUGH SIDE LEG.	
TYVEK	E.I. DU PONT DE NEMOURS & COMPANY, INC.	AIR INFILTRATION BARRIER,THIN SHEET OF HIGH-DENSITY POLYETHYLENE FIBERS,NOT A MOISTURE BARRIER. MAY BE WRAPPED AROUND SIDEWALLS IN NEW FRAME CON-STRUCTION OR PLACED OVER ATTIC INSUL-ATION TO REDUCE AIR INFILTRATION.	
UNISEAL	SLOTTSEAL	TUBULAR WEATHERSTRIPPING,PVC,EPDM RUB-BER,SILICONE RUBBER OR THERMO-PLASTIC RUBBER. FINNED OR TRAPEZOIDAL FOOT MAY BE INSERTED INTO A MACHINED GROOVE IN WINDOWS AND DOORS.	
V-SEAL WEATHER STRIP	3M CO	V-SHAPED WEATHERSTRIPPING,ADHESIVE-BACKED,POLYPROPYLENE,PRESCORED FOR	

INSULATION MANUFACTURERS PRODUCTS
(SORTED BY PRODUCT TRADE NAME)

PRODUCT TRADENAME	MANUFACTURER	PRODUCT DESCRIPTION	PLATES

WEATHERSTRIPPING AND CAULKING (Cont.)

WILL-SEAL	ILLBRUCK/USA	FOAM TAPE,ADHESIVE BACKED. SEALS OUT WEATHER AND WATER. FOR ANY JOINT,CONFORMS TO NONUNIFORMITIES.	
WINGSTRIP	SLOTTSEAL	V-SHAPED WEATHERSTRIPPING. ATTACHED BY STAPLING,ESPECIALLY USEFUL ON SLIDING WINDOWS.	

SEALANTS

790 BUILDING SEALANT	DOW CORNING CORP	SILICONE ONE-PART SEALANT,LOW MODULUS GOOD WEATHER RESISTANCE,DURABILITY. FOR JOINTS IN PRECAST CONCRETE PANELS CURTAINWALLS,EXPANSION JOINTS,SOLAR COLLECTOR PANELS,OR JOINTS WITH EXCESSIVE MOVEMENT.	
CONSTRUCTION 1200	GENERAL ELECTRIC CO	SILICONE ONE-PART SEALANT,WITH SUPERIOR ADHESION,WEATHER RESISTANCE AND ELASTICITY. FOR ALL GLAZING APPLICATIONS AND METAL CURTAINWALLS. ADHERES TO GLASS,CERAMICS,STEEL,WOOD, GRANITE,ALUMINUM AND MOST PLASTICS.	
D200	LION OIL CO	ELASTOMERIC TWO-PART SEALANT,WATER RESISTANT,MOVES AND RECOVERS FROM EXPANSION AND CONTRACTION CAUSED BY TEMPERATURE CHANGE. SEALS JOINTS BETWEEN CONCRETE,METAL,GLASS,OTHER SUBSTANCES,ABOVE AND BELOW GRADE.	
DYNASEAL W-100	WILLIAMS PRODUCTS,INC	POLYURETHANE ONE-PART SEALANT,COMBINES STRENGTH WITH FLEXIBILITY AND ABRASION RESISTANCE . FOR PRECAST CONCRETE,PORCELAIN,SHEET METAL,DOOR FRAMES,SKYLIGHTS,DAMP MASONRY.	
HEAT SEAL XL-770	HOSHALL INDUSTRIES INC	SEALANT,RUBBER-LIKE CHARACTERISTICS FOR EXCELLENT WEATHER AND AGING RESISTANCE. ADHERES TO ANY SURFACE.	
SILGLAZE	GENERAL ELECTRIC CO	SILCONE ONE-PART SEALANT,ESPECIALLY DESIGNED FOR SEALING BUTT AND LAP JOINTS IN GLAZING,CURTAINWALLS,AND MASONRY PERIMETERS. ADHERES TO GLASS, PLASTIC,METAL.	

INSULATION MANUFACTURERS PRODUCTS
(SORTED BY PRODUCT TRADE NAME)

SEALANTS (Cont.)

PRODUCT TRADENAME	MANUFACTURER	PRODUCT DESCRIPTION	PLATES
SILICONE RUBBER SEALANT *	DOW CORNING CORP	SILICONE ONE-PART SEALANT,GOOD WEATHER RESISTANCE,DURABILITY,STRENGTH. FOR SEALING MULLIONS,PLASTIC,GLASS,CURTAINWALL AND CONVENTIONAL JOINTS.	
SILICONE RUBBER SEALANT PAINTABLE *	DOW CORNING CORP	SILICONE ONE-PART SEALANT,GOOD WEATHER RESISTANCE,DURABILITY,PAINTABLE, STAINABLE. ADHERES TO WOOD,MASONRY, OTHER SUBSTRATES,FOR INTERIOR WALL JOINTS WINDOW AND DOOR FRAMING.	
SILPRUF	GENERAL ELECTRIC CO	SILICONE ONE-PART SEALANT. LOW MODULUS OF ELASTICITY ALLOWS EXCELLENT RECOVERY FROM COMPRESSION AND EXTENSION. DESIGNED FOR SEALING BUILDING JOINTS THAT HAVE A HIGH DEGREE OF MOVEMENT. WITHSTANDS WEATHER AND TEMPERATURE EXTREMES.	

PIPE INSULATION

PRODUCT TRADENAME	MANUFACTURER	PRODUCT DESCRIPTION	PLATES
850 SNAP*ON	CERTAINTEED CORP	GLASS FIBER RIGID MOLDED PIPE INSULATION. FOR TEMPERATURES UP TO 850 DEG F. AVAILABLE WITH FACTORY-APPLIED ALL-WEATHER JACKET.	40B-D
ACCOTHERM	ARMSTRONG WORLD INDUSTRIES, INC.	PHENOLIC RIGID FOAM,CONTINUOUSLY MOLDED WITH A LAMINATED ALL-SERVICE VAPOR BARRIER JACKET.	40B-D,42B 43A
AEROTUBE	MANVILLE CORP	ELASTOMERIC FLEXIBLE PREFORMED INSULATION.	40A,42A,44B
AEROTUBE II	MANVILLE CORP	ELASTOMERIC FLEXIBLE PREFORMED INSULATION. LOWER FLAMMABILITY RATINGS THAN STANDARD AEROTUBE.	40A,42A,44B
ARMAFLEX	ARMSTRONG WORLD INDUSTRIES, INC.	ELASTOMERIC FLEXIBLE PREFORMED INSULATION.	40A,42A,44B
ARMAFLEX II	ARMSTRONG WORLD INDUSTRIES, INC.	ELASTOMERIC FLEXIBLE PREFORMED INSULATION. LOWER FLAMMABLILTY RATINGS THAN STANDARD ARMAFLEX.	40A,42B,44B
ARMAFLEX INSULATION TAPE	ARMSTRONG WORLD INDUSTRIES, INC.	ELASTOMERIC FLEXIBLE TAPE,ADHESIVE BACKED. FOR WRAPPING PIPES AND FITTINGS.	44D

INSULATION MANUFACTURERS PRODUCTS
(SORTED BY PRODUCT TRADE NAME)

PIPE INSULATION (Cont.)

PRODUCT TRADENAME	MANUFACTURER	PRODUCT DESCRIPTION	PLATES
ARMALOK II	ARMSTRONG WORLD INDUSTRIES, INC.	POLYURETHANE RIGID FOAM,CONTINUOUSLY MOLDED,WITH A LAMINATED ALUMINUM FOIL AND WHITE KRAFT PAPER JACKET.	40B-D,42B 43A
CELOTEMP 1500	CELOTEX CORP	PERLITE,EXPANDED,WITH MOISTURE RESISTANT BINDER.	
COPPER CORE TEMP-TITE	MANVILLE CORP	POLYURETHANE FOAM PREINSULATED UNDER-GROUND PIPE. PVC CASING,TYPE K COPPER CORE PIPE, INTEGRAL COUPLING. FOR CHILLED AND HOT WATER FROM 35 TO 260 DEG F.	
DUAL PIPE	INSTA-FOAM PRODUCTS, INC	POLYURETHANE FOAM PREINSULATED PIPING WITH TWO PIPES INSIDE ONE OUTER JACKET.	41A
DUPLEX X-50	TPCO, INC	POLYURETHANE FOAM PREINSULATED PIPE, TWO CARRIER PIPES IN A SINGLE JACKET.	41A
FIBERGLAS 25 ASJ/SSL	OWENS-CORNING FIBERGLAS CORP	GLASS FIBER RIGID PIPE INSULATION,WITH ALL-SERVICE JACKET. FOR TEMP-ERATURES UP TO 650 DEG F.	40B-D
FRP 500 SERIES	E. B. KAISER CO	PREINSULATED PIPING, INDIVIDUALLY INSUL-ATED CARRIER PIPES IN A SINGLE OUTER CASING. INSULATION IS GLASS FIBER, CALCIUM SILICATE,OR AS SPECIFIED.	
HEAT-TITE	MANVILLE CORP	POLYURETHANE FOAM PREINSULATED UNDER-GROUND PIPE. PVC CASING, SCHED. 40 STEEL CORE PIPE. END SEALS PROTECT INSULATION. PIPES CONNECTED WITH AS-BESTOS-CEMENT COUPLINGS CONTAINING ELASTOMERIC SEALING RINGS. FOR HOT WATER UP TO 260 DEG F.	
HITEMP	INSTA-FOAM PRODUCTS, INC	CELLULAR GLASS PREINSULATED PIPING, IRON, STEEL,OR COPPER CARRIER PIPES,FRP JACKET. FOR TEMPERATURES UP TO 800 DEG F.	
IMCOAFLEX	INSULATING MATERIALS CORP OF AMERICA	POLYETHYLENE FOAM,CLOSED CELL,UNFACED, UV STABLIZED. FOR PIPING FROM -110 TO 210 DEG F.	40A,42A,44B
INSUL-8	ROVANCO CORP	POLYURETHANE FOAM PREINSULATED PIPING. SINGLE OR MULTIPLE CARRIER PIPES OF STEEL,STAINLESS STEEL,COPPER,ALUMINUM PVC,OR FIBERGLASS, IN AN OUTER JACKET OF STEEL,COATED STEEL,STAINLESS STEEL ALUMINUM,PVC,FIBERGLASS,POLYURETHANE.	41A

INSULATION MANUFACTURERS PRODUCTS
(SORTED BY PRODUCT TRADE NAME)

PIPE INSULATION (Cont.)

PRODUCT TRADENAME	MANUFACTURER	PRODUCT DESCRIPTION	PLATES
INSUL-TUBE	HALSTEAD INDUSTRIES, INC.	ELASTOMERIC FLEXIBLE PREFORMED INSULATION.	40A,42A,44B
INSULATION TAPE	HALSTEAD INDUSTRIES, INC.	ELASTOMERIC FLEXIBLE TAPE, ADHESIVE BACKED. FOR WRAPPING PIPES AND FITTINGS.	44D
ISONATE	UPJOHN CO, CPR DIVISION	POLYURETHANE FOAM, SPRAYED-IN-PLACE FOR UNDERGROUND PIPING SYSTEMS.	
KAYLO 10	OWENS-CORNING FIBERGLAS CORP	CALCIUM SILICATE RIGID INSULATION FOR PIPING AT TEMPERATURES UP TO 1200 DEG F.	40B-D
KOOL-KORE	MANVILLE CORP	POLYURETHANE FOAM PREINSULATED UNDERGROUND PIPE. PVC CASING AND CORE PIPES. FOR CHILLED WATER.	
METAL-ON	MANVILLE CORP	CALCIUM SILICATE MOLDED PIPE INSULATION WITH FACTORY-APPLIED ALUMINUM JACKET. FOR PIPE TEMPERATURES UP TO 1500 DEG F.	40C-D
MICRO-LOK 650	MANVILLE CORP	GLASS FIBER RIGID PIPE INSULATION. FOR TEMPERATURES UP TO 650 DEG F. AVAILABLE WITH FSK OR POLISHED METAL JACKET.	40B-D
NORTHSTAR PIPING SYSTEMS	TPCO, INC	PREINSULATED PIPING, SINGLE OR DOUBLE, WITHIN A CASING PIPE. INSULATION IS GLASS FIBER, CALCIUM SILICATE, OR AS SPECIFIED.	41B
PIPE INSULATING TAPE *	3M CO	FOAM FLEXIBLE TAPE, ADHESIVE BACKED. FOR WRAPPING AND FITTINGS.	44D
PIPE INSULATION	KNAUF FIBER GLASS GMBH	GLASS FIBER MOLDED PIPE INSULATION, UNFACED OR WITH ALL-SERVICE FSK JACKET. FOR TEMPERATURES UP TO 500 DEG F.	40B-D
PIPE WRAP INSULATION *	OWENS-CORNING FIBERGLAS CORP	GLASS FIBER SEMI-RIGID BOARD WITH A LAMINATED FSK JACKET. FOR WRAPPING AROUND LARGE-DIAMETER PIPES.	
PRE-INSULATED PIPING SYSTEMS	INSTA-FOAM PRODUCTS, INC	POLYURETHANE FOAM PREINSULATED PIPING. CARRIER PIPE MAY BE COPPER, STEEL, STAINLESS STEEL, ALUMINUM, FIBERGLASS, OR PVC, JACKET MAY BE STEEL, STAINLESS STEEL, ALUMINUM, PVC, OR ASBESTOS CEMENT. FOR TEMPERATURES FROM -350 TO 250 DEG F.	

INSULATION MANUFACTURERS PRODUCTS
(SORTED BY PRODUCT TRADE NAME)

PIPE INSULATION (Cont.)

PRODUCT TRADENAME	MANUFACTURER	PRODUCT DESCRIPTION	PLATES
PROTEXULATE	PROTEXULATE, INC	MINERAL POWDER LOOSE-FILL INSULATION FOR UNDERGROUND PIPING, WATER REPELLANT. FOR CRYOGENIC TEMPERATURES TO 480 DEG F.	
SEALASTIC	HALSTEAD INDUSTRIES, INC.	CORK FLEXIBLE TAPE, ADHESIVE BACKED. DEVELOPED ESPECIALLY FOR WRAPPING COLD PIPES TO PREVENT CONDENSATION.	44D
SOLAR-7	NORTHEAST SPECIALITY INSULATIONS, INC.	POLYISOCYANURATE FOAM RIGID INSULATION, FOR SINGLE OR DOUBLE PIPES, UV-RESISTANT PVC JACKET.	41A
STD 400 SERIES	E. B. KAISER CO	PREINSULATED PIPING, MULTIPLE CARRIER PIPES INSIDE A SINGLE, INTERNALLY INSULATED OUTER CASING. INSULATION IS GLASS FIBER, CALCIUM SILICATE, OR AS SPECIFIED.	
SUPER TEMP-TITE	MANVILLE CORP	COMBINATION OF CALCIUM SILICATE AND POLYURETHANE FOAM (THERMO-FOAM) PREINSULATED UNDERGROUND PIPE. CASING AND CORE PIPES ARE ASBESTOS-CEMENT. FOR HOT WATER AND STEAM UP TO 450 DEG F.	
TEMP-TITE	MANVILLE CORP	POLYURETHANE FOAM PREINSULATED UNDERGROUND PIPE. CASING AND CORE PIPES ARE ASBESTOS-CEMENT. END CAPS PROTECT INSULATION. PIPES ARE CONNECTED WITH RUBBER COUPLINGS. FOR HOT OR CHILLED WATER.	
THERMAZIP 150	ACCESSIBLE PRODUCTS CO	GLASS FIBER FLEXIBLE WOOL, LAMINATED TO A JACKET OF METALLIZED POLYESTER INNER FILM, FIBERGLASS SCRIM, PVC OUTER FILM. PATENTED LOCKING TRAC. FOR LIGHT-DUTY INDOOR USE TO 400 DEG F.	40A
THERMAZIP 175	ACCESSIBLE PRODUCTS CO	POLYURETHANE FLEXIBLE FOAM (ETHER TYPE) LAMINATED TO A JACKET OF METALLIZED POLYESTER INNER FILM, FIBERGLASS SCRIM PVC OUTER FILM. PATENTED LOCKING TRAC. FOR LIGHT-DUTY INDOOR USE FROM -60 TO 220 DEG F.	40A
THERMAZIP 250	ACCESSIBLE PRODUCTS CO	GLASS FIBER FLEXIBLE WOOL, PVC JACKET, LOCKING TRAC. FOR GENERAL INDOOR USE UP TO 850 DEG F.	40A

INSULATION MANUFACTURERS PRODUCTS
(SORTED BY PRODUCT TRADE NAME)

PIPE INSULATION (Cont.)

PRODUCT TRADENAME	MANUFACTURER	PRODUCT DESCRIPTION	PLATES
THERMAZIP 275	ACCESSIBLE PRODUCTS CO	POLYURETHANE FLEXIBLE FOAM (ETHER TYPE), PVC JACKET, LOCKING TRAC. FOR GENERAL INDOOR USE FROM -60 TO 220 DEG F.	40A
THERMAZIP 350	ACCESSIBLE PRODUCTS CO	GLASS FIBER FLEXIBLE WOOL, PVC IMPREGNATED POLYESTER FABRIC JACKET, LOCKING TRAC. FOR HEAVY-DUTY INDOOR AND OUTDOOR USE UP TO 850 DEG F.	40A
THERMAZIP 375	ACCESSIBLE PRODUCTS CO	POLYURETHANE FLEXIBLE FOAM (ETHER TYPE), PVC IMPREGNATED POLYESTER FABRIC JACKET, LOCKING TRAC. FOR HEAVY-DUTY INDOOR AND OUTDOOR USE FROM -60 TO 220 DEG F.	40A
THERMAZIP 450	ACCESSIBLE PRODUCTS CO	GLASS FIBER FLEXIBLE WOOL JACKET OF HYPALON IMPREGNATED FIBERGLASS LAMINATED TO TEDLAR OUTER FILM, LOCKING TRAC. FOR EXTENDED OUTDOOR USE UP TO 9850 DEG F. SUPERIOR CHEMICAL RESISTANCE.	40A
THERMAZIP 475	ACCESSIBLE PRODUCTS CO	POLYURETHANE FLEXIBLE FOAM (ETHER TYPE), JACKET OF HYPALON IMPREGNATED FIBERGLASS LAMINATED TO TEDLAR OUTER FILM, LOCKING TRAC. FOR EXTENDED OUTDOOR USE FROM -60 TO 220 DEG F. SUPERIOR CHEMICAL RESISTANCE.	40A
THERMAZIP 550	ACCESSIBLE PRODUCTS CO	GLASS FIBER FLEXIBLE WOOL, JACKET OF ALUMINIZED HEAVY FIBERGLASS FABRIC WITH A HIGH-TEMPERATURE COATING, LOCKING TRAC. FOR HIGH TEMPERATURE APPLICATIONS UP TO 850 DEG F. INDOORS AND OUTDOORS.	40A
THERMAZIP 552	ACCESSIBLE PRODUCTS CO	CERAMIC FIBER WOOL, JACKET OF ALUMINIZED HEAVY FIBERGLASS FABRIC WITH A HIGH-TEMPERATURE APPLICATIONS UP TO 1400 DEG F. INDOORS AND OUTDOORS.	40A
THERMAZIP 850	ACCESSIBLE PRODUCTS CO	GLASS FIBER FLEXIBLE WOOL, RIGID NONPOROUS PVC JACKET, LOCKING TRAC. FOR INDOOR USE IN FOOD PROCESSING AND SANITARY APPLICATIONS UP TO 850 DEG F	40A
THERMAZIP 875	ACCESSIBLE PRODUCTS CO	POLYURETHANE FLEXIBLE FOAM (ETHER TYPE), RIGID NONPOROUS PVC JACKET, LOCKING TRAC. FOR INDOOR USE IN FOOD PROCESSING AND SANITARY APPLICATIONS FROM -60 TO 220 DEG F.	40A

INSULATION MANUFACTURERS PRODUCTS
(SORTED BY PRODUCT TRADE NAME)

PIPE INSULATION (Cont.)

PRODUCT TRADENAME	MANUFACTURER	PRODUCT DESCRIPTION	PLATES
THERMO-12	MANVILLE CORP	CALCIUM SILICATE,MOLDED PIPE INSULATION. AVAILABLE WITH FACTORY-APPLIED ALUMINUM OR STAINLESS STEEL JACKET. AVAILABLE WITH EXTENDED LEGS TO ALLOW FOR 1/2-INCH TRACED LINE. FOR PIPING TEMPERATURES UP TO 1500 DEG F.	40B-D
TRYMER	UPJOHN CO,CPR DIVISION	POLYISOCYANURATE FOAM RIGID PREFORMED INSULATION,CHOICE OF JACKETING MATERIALS. FOR TEMPERATURES FROM -425 TO 300 DEG F.	40B-D
X-50	TPCO,INC	POLYURETHANE FOAM PREINSULATED PIPE, STEEL,STAINLESS STEEL,COPPER,PVC OR FRP CARRIER PIPE WITH PVC,POLYURETHANE,FRP,SPIRAL-WOUND METAL,OR VIRTUALLY ANY TUBULAR JACKET.	

EQUIPMENT INSULATION

PRODUCT TRADENAME	MANUFACTURER	PRODUCT DESCRIPTION	PLATES
1000 SERIES SPIN-GLAS	MANVILLE CORP	GLASS FIBER SEMI-RIGID BOARD. FOR FURNACES,BOILERS,HEATED VESSELS,DUCTS AND TANKS UP TO 850 DEG F.	45A-B
700 SERIES INDUSTRIAL INSULATION *	OWENS-CORNING FIBERGLAS CORP	GLASS FIBER SEMI-RIGID RECTANGULAR BOARD,FACED OR UNFACED. FOR EQUIPMENT,VESSELS,AND TANKS UP TO 450 DEG F.	45A-C
800 SERIES SPIN-GLAS BLANKET	MANVILLE CORP	GLASS FIBER FLEXIBLE BLANKET,FACED OR UNFACED. FOR INDUSTRIAL HEATING,AIR CONDITIONING,POWER AND PROCESS EQUIPMENT.	45C
AEROTUBE	MANVILLE CORP	ELASTOMERIC FLEXIBLE SHEET INSULATION.	
AEROTUBE II	MANVILLE CORP	ELASTOMERIC FLEXIBLE SHEET INSULATION. LOWER FLAMMABILITY RATINGS THAN STANDARD AEROTUBE SHEET.	
ARMAFLEX	ARMSTRONG WORLD INDUSTRIES, INC.	ELASTOMERIC FLEXIBLE SHEET. ADHERES WITH ADHESIVES TO LARGE FLAT OR CURVED METAL SURFACES.	

INSULATION MANUFACTURERS PRODUCTS
(SORTED BY PRODUCT TRADE NAME)

EQUIPMENT INSULATION (Cont.)

PRODUCT TRADENAME	MANUFACTURER	PRODUCT DESCRIPTION	PLATES
ARMAFLEX II	ARMSTRONG WORLD INDUSTRIES, INC.	ELASTOMERIC FLEXIBLE SHEET. ADHERES WITH ADHESIVES TO LARGE FLAT OR CURVED METAL SURFACES. LOWER FLAMMABILITY RATINGS THAN STANDARD ARMAFLEX.	
ELEVATED TEMPERATURE BOARD *	KNAUF FIBER GLASS GMBH	GLASS FIBER SEMI-RIGID BOARD BONDED WITH HIGH-TEMPERATURE THERMOSETTING RESIN UNFACED. FOR BOILER WALLS,PRECIPITATORS,TANKS,TOWERS,STACKS,AND OVENS UP TO 850 DEG F.	45A-B
GLAS-MAT 1200	MANVILLE CORP	GLASS FIBER MECHANICALLY BONDED BLANKET FOR INDUSTRIAL,MARINE,AND PROCESS APPLICATIONS UP TO 1200 DEG F.	
H. T. BANROC	MANVILLE CORP	MINERAL FIBER BLOCK,BONDED WITH A CLAY BINDER. FOR FURNACES,BOILERS,HEATED TANKS AND VESSELS UP TO 1900 DEG F.	45A-B
INSUL-SHEET	HALSTEAD INDUSTRIES,INC.	ELASTOMERIC FLEXIBLE SHEET. ADHERES WITH ADHESIVES TO LARGE FLAT OR CURVED METAL SURFACES.	
INSULATION BOARD *	KNAUF FIBER GLASS GMBH	GLASS FIBER SEMI-RIGID BOARD,UNFACED,OR FSK OR ALL-SERVICE JACKET FACED. FOR POWER AND PROCESS EQUIPMENT, BOILER AND STACK INSTALLATIONS UP TO 450 DEG F.	45A-C
INSUL-QUICK	OWENS-CORNING FIBERGLAS CORP	GLASS FIBER SEMI-RIGID BOARD WITH HIGH-TEMPERATURE BINDER,UNFACED OR FOIL FACED. FOR PROCESS BOILERS, PRECIPITATORS,AND HEATED EQUIPMENT UP TO 850 DEG F.	45A-C
ISONATE	UPJOHN CO,CPR DIVISION	POLYURETHANE FOAM,SPRAYED-IN-PLACE. FOR TANKS,VESSELS,HARD-TO-INSULATE INDUSTRIAL APPLICATIONS.	45D
KAYLO 10	OWENS-CORNING FIBERGLAS CORP	CALCIUM SILICATE BLOCK INSULATION. FOR BOILERS,TANKS AND VESSELS UP TO 1200 DEG F. AVAILABLE WITH V-GROOVE TO CONFORM TO CURVED SURFACES.	45A-B
METAL-ON	MANVILLE CORP	GLASS FIBER INSULATION WITH EMBOSSED ALUMINUM SHEET FACING. FOR HEATED TANKS UP TO 450 DEG F.	45A-C

INSULATION MANUFACTURERS PRODUCTS
(SORTED BY PRODUCT TRADE NAME)

EQUIPMENT INSULATION (Cont.)

PRODUCT TRADENAME	MANUFACTURER	PRODUCT DESCRIPTION	PLATES
PIPE AND TANK INSULATION *	MANVILLE CORP	GLASS FIBER SEMI-RIGID BOARD,BONDED TO A FLEXIBLE FSK JACKET,SEGMENTED AND SUPPLIED IN ROLLS. FOR APPLICATIONS TO PIPES,TANKS,DUCTS,VESSELS.	45A-C
TANK TOP INSUL	MANVILLE CORP	MINERAL FIBER RIGID BOARD. FOR FLAT TOP SURFACES OF HEATED TANKS AND VESSELS UP TO 250 DEG F.	45A-B
THERMAZIP 353 SHEET STOCK	ACCESSIBLE PRODUCTS CO	GLASS FIBER FLEXIBLE WOOL SHEET,PVC IMPREGNATED POLYESTER FABRIC JACKET FOR TANK AND VESSEL APPLICATIONS UP TO 850 DEG F.	45C
THERMAZIP 375 SHEET STOCK	ACCESSIBLE PRODUCTS CO	POLYURETHANE FLEXIBLE FOAM (ETHER TYPE), PVC IMPREGNATED POLYESTER FABRIC JACKET. FOR TANK AND VESSEL APPLICATIONS FROM -60 TO 220 DEG F.	45C
THERMAZIP HI-T BLANKET #53	ACCESSIBLE PRODUCTS CO	CERAMIC FIBER WOOL BLANKET,SERVICE TEMP LIMIT 1400 DEG F.,SANDWICHED BETWEEN TWO LAYERS OF SILICONE RUBBER COATED FIBERGLASS FABRIC (UP TO 500 DEG F.),OR SILICA CLOTH. FOR INSULATING HEAT EXCHANGERS,VALVES,FLANGES,EXPANSION JOINTS. REMOVABLE.	45C
THERMAZIP HI-T BLANKET #54	ACCESSIBLE PRODUCTS CO	CERAMIC FIBER (ALUMINA AND SILICA WOOL BLANKET,SANDWICHED BETWEEN TWO LAYER OF SILICONE COATED FIBERGLASS FABRIC (UP TO 500 DEG F.) OR SILICA CLOTH (UP TO 1800 DEG F.). FOR INSULATING HEAT EXCHANGERS,VALVES,FLANGES,EXPANSION JOINTS. REMOVABLE.	45C
THERMO-12	MANVILLE CORP	CALCIUM SILICATE,MOLDED FLAT OR RADIUS BLOCK. FOR EQUIPMENT TEMPERATURES UP TO 1500 DEG F.	45A-B
TIW,TYPE I	OWENS-CORNING FIBERGLAS CORP	GLASS FIBER FLEXIBLE WOOL WRAP. FOR INDUSTRIAL OVENS AND IRREGULAR SURFACES UP TO 1000 DEG F. LOW COMPRESSIVE STRENGTH.	
TIW,TYPE II	OWENS-CORNING FIBERGLAS CORP	GLASS FIBER WOOL BATTS FOR METAL MESH BLANKETS AND FOR BOILERS,VESSELS,AND EQUIPMENT UP TO 1000 DEG F.	
TRYMER	UPJOHN CO,CPR DIVISION	POLYISOCYANURATE FOAM BOARDSTOCK FOR INDUSTRIAL APPLICATIONS FROM -425 TO 300 DEG F.	

INSULATION MANUFACTURERS PRODUCTS
(SORTED BY PRODUCT TRADE NAME)

DUCT INSULATION

PRODUCT TRADENAME	MANUFACTURER	PRODUCT DESCRIPTION	PLATES
700 SERIES INDUSTRIAL INSULATION	OWENS-CORNING FIBERGLAS CORP	GLASS FIBER SEMI-RIGID RECTANGULAR BOARD,FACED OR UNFACED. FOR DUCT-WORK UP TO 450 DEG F.	46A
800 SERIES SPIN-GLAS	MANVILLE CORP	GLASS FIBER FLEXIBLE BLANKET,FACED OR UNFACED. FOR EXTERIOR INSULATION OF ROUND AND RECTANGULAR SHEET METAL DUCTS.	46B,47B
ACOUSTI-K27	UNITED MCGILL CORP	GLASS FIBER INSULATED RIGID ROUND DUCT,SPIRAL-WOUND METAL INNER AND OUTER SHELLS.	47D
AEROFLEX DUCT LINER	OWENS-CORNING FIBERGLAS CORP	GLASS FIBER FLEXIBLE MAT WITH FLAME-RESISTANT COATING. APPLIED TO THE INTERIOR OF DUCTS,FOR TEMPERATURES UP TO 250 DEG F.	46C
AEROFLEX DUCT LINER BOARD	OWENS-CORNING FIBERGLAS CORP	GLASS FIBER SEMI-RIGID BOARD WITH FLAME-RESISTANT COATING. APPLIED TO THE INTERIOR OF DUCTS,FOR TEMPERATURES UP TO 250 DEG F.	46C
CERTAFLEX-25&7	CERTAINTEED CORP	GLASS FIBER INSULATED FLEXIBLE ROUND DUCT,A STEEL WIRE HELIX ENCLOSED IN A DOUBLE-LAYER POLYESTER AIR BARRIER, REINFORCED METALLIZED MYLAR OUTER JACKET. FOR DUCT SYSTEMS UP TO 200 DEG F.	47C
CERTAFLEX G25	CERTAINTEED CORP	GLASS FIBER INSULATED FLEXIBLE ROUND DUCT,IMPERVIOUS WIRE REINFORCED INNER CORE WITH A POLYTHYLENE JACKET. FOR LOW-VELOCITY DUCT SYSTEMS UP TO 200 DEG F.	47C
DUCT WRAP	KNAUF FIBER GLASS GMBH	GLASS FIBER FLEXIBLE BLANKET,UNFACED OR WITH FSK OR VINYL VAPOR BARRIER. FOR HEATING AND AIR CONDITIONING DUCTS FROM 40 TO 250 DEG F.	46B,47B
ELEVATED TEMPERATURE BOARD *	KNAUF FIBER GLASS GMBH	GLASS FIBER SEMI-RIGID BOARD BONDED WITH HIGH-TEMPERATURE THERMOSETTING RESIN,UNFACED. FOR HOT DUCTWORK UP TO 850 DEG F.	46A
FLEXIBLE DUCT TYPE WG	WIREMOLD CO	GLASS FIBER INSULATED FLEXIBLE ROUND DUCT,GALVANIZED STEEL WIRE HELIX CORE WITH AIRTIGHT POLYESTER FILM,POLYOLE-FIN JACKET.	47C

INSULATION MANUFACTURERS PRODUCTS
(SORTED BY PRODUCT TRADE NAME)

DUCT INSULATION (Cont.)

PRODUCT TRADENAME	MANUFACTURER	PRODUCT DESCRIPTION	PLATES
FLEXIBLE DUCT TYPE WK	WIREMOLD CO	GLASS FIBER INSULATED FLEXIBLE ROUND DUCT,GALVANIZED STEEL WIRE HELIX CORE WITH AIRTIGHT POLYESTER FILM,REIN-FORCED ALUMINIZED JACKET.	47C
FLEXIBLE DUCT TYPE 57K	WIREMOLD CO	GLASS FIBER INSULATED FLEXIBLE ROUND DUCT,CORE OF HIGH-TEMPERATURE VINYL ORGANOSOL-COATED GLASS FABRIC MECH-ANICALLY LOCKED INTO A FLAT STEEL SPIRAL,POLYOLEFIN JACKET.	47D
FIBERGLAS DUCT BOARD	OWENS-CORNING FIBERGLAS CORP	GLASS FIBER RIGID BOARD WITH ALUMINUM FOIL VAPOR BARRIER. FOR FABRICATING RECTANGULAR DUCTWORK AND FITTINGS. FOR TEMPERATURES UP TO 250 DEG F.	46D
FIBERGLAS DUCT WRAP	OWENS-CORNING FIBERGLAS CORP	GLASS FIBER BLANKET,UNFACED OR FACED WITH A REINFORCED FOIL KRAFT FACING. FOR TEMPERATURES FROM 40 TO 250 DEG F.	46B,47B
FIBERGLAS VALUFLEX	OWENS-CORNING FIBERGLAS CORP	GLASS FIBER INSULATED FLEXIBLE ROUND DUCT,A RESILIENT INNER AIR BARRIER, WITH A POLYETHYLENE JACKET. FOR HEATING AND AIR CONDITIONING DUCTS UP TO 250 DEG F.	47C
FLEX-MET	MANVILLE CORP	GLASS FIBER INSULATED FLEXIBLE ROUND DUCT,FLEXIBLE ALUMINUM CORE,INSUL-ATION COVERED ON BOTH SIDES WITH VINYL OR ALUMINIZED MYLAR VAPOR BARRIER. FOR MANY HEATING AND AIR CONDITIONING DUCT APPLICATIONS FROM 10 TO 250 DEG F.	47D
INL-25 FLEXIBLE DUCT	OWENS-CORNING FIBERGLAS CORP	GLASS FIBER INSULATED FLEXIBLE ROUND DUCT WITH RESILIENT INNER AIR BARRIER AND REINFORCED JACKET. FOR TEMPERA-TURES UP TO 250 DEG F.	47C
INSULATION BOARD *	KNAUF FIBER GLASS GMBH	GLASS FIBER SEMI-RIGID BOARD,UNFACED,OR FSK OR ALL-SERVICE JACKET FACED. FOR HEATING AND AIR CONDITIONING DUCTS FROM -20 TO 450 DEG F.	46A
INSUL-QUICK	OWENS-CORNING FIBERGLAS CORP	GLASS FIBER SEMI-RIGID BOARD WITH HIGH-TEMPERATURE BINDER,UNFACED OR FOIL FACED. FOR DUCTWORK AND CHIMNEY LINERS UP TO 850 DEG F.	46A

INSULATION MANUFACTURERS PRODUCTS
(SORTED BY PRODUCT TRADE NAME)

DUCT INSULATION (Cont.)

PRODUCT TRADENAME	MANUFACTURER	PRODUCT DESCRIPTION	PLATES
LINACOUSTIC	MANVILLE CORP	GLASS FIBER FLEXIBLE DUCT LINER. APPLIED TO THE INTERIOR OF DUCTS FOR TEMPERATURES UP TO 250 DEG F.	46C
LINACOUSTIC R	MANVILLE CORP	GLASS FIBER PLENUM LINER BOARD WITH A BLACK MAT COATING. APPLIED TO THE INTERIOR OF PLENUMS UP TO 250 DEG F.	46C
MICROLITE	MANVILLE CORP	GLASS FIBER FLEXIBLE BLANKET,FACED OR UNFACED. FOR EXTERIOR INSULATION OF ROUND AND RECTANGULAR SHEET METAL DUCTS.	46B,47B
MICRO-AIRE DUCT BOARD	MANVILLE CORP	GLASS FIBER RIGID BOARD WITH FSK OR HEAVY-DUTY FOIL-KRAFT-SCRIM-KRAFT (HDF) FACING. PREMOLDED SLIP JOINT EDGES. FOR FABRICATING RECTANGULAR HEATING AND COOLING DUCTWORK.	46D
MICRO-AIRE HV-3	MANVILLE CORP	GLASS FIBER RIGID ROUND DUCT,SCRIM-REINFORCED FOIL JACKET,SLIP-JOINT ENDS. FOR HIGH-PRESSURE,HIGH-VELOCITY HEATING AND AIR CONDITIONING DUCTS UP TO 250 DEG F.	47A
MICRO-AIRE J/FLX SL	MANVILLE CORP	GLASS FIBER INSULATED FLEXIBLE ROUND DUCT,VINYL-COATED STEEL HELIX CORE BONDED TO POLYETHYLENE,WITH AN OUTER VAPOR BARRIER JACKET OF FIBERGLASS REINFORCED METALLIZED MYLAR/NEOPRENE LAMINATE. FOR RESIDENTIAL AND LOW-PRESSURE COMMERCIAL APPLICATIONS UP TO 250 DEG F.	47C
RIGID ROUND DUCT	MANVILLE CORP	GLASS FIBER RIGID ROUND DUCT,SCRIM-REINFORCED FOIL JACKET,SLIP-JOINT ENDS. FOR HEATING AND AIR CONDITIONING DUCTS UP TO 250 DEG F.	47A
STANDARD DUCT INSULATION *	CERTAINTEED CORP	GLASS FIBER FLEXIBLE BLANKET,FSK OR VINYL FACING. FOR WRAPPING HEATING AND COOLING DUCTWORK FROM 35 TO 250 DEG F.	46B,47B
THERMAFLEX G-KM DUCT	AUTOMATION INDUSTRIES,INC	GLASS FIBER INSULATED FLEXIBLE ROUND DUCT,STEEL WIRE HELIX CORE BONDED TO A POLYMERIC LINER,WITH FIBERGLASS REINFORCED POLYOLEFIN JACKET. FOR LOW AND MEDIUM PRESSURE SYSTEMS TO 200 DEG F.	47C

INSULATION MANUFACTURERS PRODUCTS
(SORTED BY PRODUCT TRADE NAME)

PRODUCT TRADENAME	MANUFACTURER	PRODUCT DESCRIPTION	PLATES

DUCT INSULATION (Cont.)

PRODUCT TRADENAME	MANUFACTURER	PRODUCT DESCRIPTION	PLATES
THERMAFLEX M-KE DUCT	AUTOMATION INDUSTRIES, INC	GLASS FIBER INSULATED FLEXIBLE ROUND DUCT, STEEL WIRE HELIX CORE BONDED TO A POLYMERIC LINER WITH FIBERGLASS REINFORCED METALLIZED FILM JACKET. FOR LOW AND MEDIUM PRESSURE SYSTEMS UP TO 200 DEG F.	47C

Table G-2

INSULATION MANUFACTURERS PRODUCTS
(SORTED BY PRODUCT DESCRIPTION)

PRODUCT TRADENAME	MANUFACTURER	PRODUCT DESCRIPTION	PLATES
FRAME WALL INSULATION			
THERMASOTE SIDEWALL PANELS	HOMASOTE CO	COMPOSITE OF ASBESTOS-FREE INSULATING BUILDING BOARD AND RIGID POLYURETHANE FOAM. FOAM SIDE IS FACED WITH ASPHALT SATURATED FELT OR FIBERGLASS. FOR USE AS NONSTRUCTURAL FRAME SHEATHING UNDER SIDING OR SHINGLES,OR AS A BASE FOR PAINT,STAIN OR STUCCO.	2C-D,3B-C,4B
UNFACED BATTS *	MANVILLE CORP	GLASS FIBER BATTS,UNFACED. SEPARATE VAPOR BARRIER MAY BE USED.	1D,2B-D 3A-B,4A-C
ROLLED BATTS *	OWENS-CORNING FIBERGLAS CORP	GLASS FIBER BATTS,UNFACED.	1D,2B-D 3A-B,4A-C
KRAFT-FACED BATTS *	MANVILLE CORP	GLASS FIBER BATT WITH KRAFT PAPER FACING. FACING EXTENDS PAST BATT EDGE FOR STAPLING FLANGE.	1B-C,2B-D 3A-B,4A-C
KRAFT FACED BUILDING INSULATION *	OWENS-CORNING FIBERGLAS CORP	GLASS FIBER BATT INSULATION WITH AS-PHALTED KRAFT PAPER FACING,FLANGED FOR STAPLING.	1B-C,2B-D 3A-B,4A-C
FOIL-FACED BATTS *	MANVILLE CORP	GLASS FIBER BATT WITH ALUMINUM FOIL FACING. FACING EXTENDS PAST BATT EDGE FOR STAPLING FLANGE.	1B-C,2B-D 3A-B,4A-C
FLAME-RESISTANT BATTS *	MANVILLE CORP	GLASS FIBER BATTS WITH FOIL-SCRIM-KRAFT (FSK) FACING. FACING EXTENDS PAST BATT EDGE FOR STAPLING FLANGE. MAY BE LEFT EXPOSED WHERE CODES PERMIT.	1B-C,2B-D 3A-C,4A-C
FLAME SPREAD 25	OWENS-CORNING FIBERGLAS CORP	GLASS FIBER BATT WITH SPECIAL FOIL/KRAFT LAMINATE FACING. RECOMMENDED BY MFR. FOR USE IN WALLS AND CEILINGS WHERE INSULATION FACING WILL BE EXPOSED. SUITABLE FOR LOW-ABUSE AREAS.	
RETROFIL	MANVILLE CORP	GLASS FIBER LOOSE FILL DESIGNED FOR RETROFIT INSULATION OF SIDEWALLS.	2A

INSULATION MANUFACTURERS PRODUCTS
(SORTED BY PRODUCT DESCRIPTION)

FRAME WALL INSULATION (Cont.)

PRODUCT TRADENAME	MANUFACTURER	PRODUCT DESCRIPTION	PLATES
FIBERGLAS SHEATHING	OWENS-CORNING FIBERGLAS CORP	GLASS FIBER RIGID BOARD,FACING IS NOT A VAPOR BARRIER BUT RESISTS RAIN PENETRATION AND AIR INFILTRATION.	2C-D,3B-C,4B
THERMATITE INSULATING SHEATHING	MANVILLE CORP	POLYISOCYANURATE FOAM RIGID BOARD,GLASS FIBER REINFORCED AND ALUMINUM FOIL FACED. A SHEATHING APPLIED OVER FRAMING MEMBERS IN NEW CONSTRUCTION OR TO EXTERIOR WALLS BEFORE INSTALLING NEW SIDING.	2C-D,3B-C,4B
HIGH-R SHEATHING	OWENS-CORNING FIBERGLAS CORP	POLYISOCYANURATE FOAM RIGID BOARD,GLASS FIBER REINFORCED AND ALUMINUM FOIL FACED. FOR USE AS NONSTRUCTURAL FRAME SHEATHING,OR BEHIND INTERIOR WALLBOARD.	2C-D,3B-C,4B
THERMAX	CELOTEX CORP	POLYISOCYANURATE FOAM RIGID BOARD,GLASS FIBER REINFORCED AND ALUMINUM FOIL FACED. FOR USE AS NON-STRUCTURAL FRAME SHEATHING,OR BEHIND INTERIOR WALLBOARD.	2C-D,3B-C,4B
STYROFOAM IB	DOW CHEMICAL CO	POLYSTYRENE FOAM EXTRUDED BOARD WITH CUT-CELL SURFACE FOR PLASTERING BASE.	
STYROFOAM SM	DOW CHEMICAL CO	POLYSTYRENE FOAM EXTRUDED BOARD WITH NATURAL SKIN SURFACE. USED AS NONSTRUCTURAL EXTERIOR SHEATHING.	2C-D,3B-C,4B
STYROFOAM TG	DOW CHEMICAL CO	POLYSTYRENE FOAM EXTRUDED BOARD WITH HIGH-DENSITY SKIN. EDGES ARE TONGUE AND GROOVE. USED AS NONSTRUCTURAL EXTERIOR SHEATHING.	2C-D,3B-C,4B
INSUL-SHEATH TG	FALCON MANUFACTURING OF MICHIGAN INC.	POLYSTYRENE FOAM BOARD,HIGH-DENSITY. EDGES AND TONGUE AND GROOVE. USED AS NONSTRUCTURAL EXTERIOR SHEATHING.	2C-D,3B-C,4B
SUPER "R" PLUS	FALCON MANUFACTURING OF MICHIGAN INC.	POLYSTYRENE FOAM RIGID BOARD WITH REFLECTIVE FOIL-KRAFT PAPER LAMINATED FACINGS. FACING PERFORATIONS DISALLOW VAPOR BARRIER. EDGE OF BOARD HAS FOIL OVERLAP FOR SEALING BUTT JOINTS. USED AS NONSTRUCTURAL EXTERIOR SHEATHING.	2C-D,3B-C,4B

INSULATION MANUFACTURERS PRODUCTS
(SORTED BY PRODUCT DESCRIPTION)

FRAME WALL INSULATION (Cont.)

PRODUCT TRADENAME	MANUFACTURER	PRODUCT DESCRIPTION	PLATES
TG-3000	THERMAL SYSTEMS, INC.	POLYURETHANE FOAM RIGID BOARD WITH ALUMINUM FOIL-SCRIM-FOIL LAMINATE FACERS. USED AS NONSTRUCTURAL EXTERIOR SHEATHING.	2C-D,3B-C,4B
HI-R PANELS	HI-R BUILDING SYSTEMS, INC	POLYURETHANE FOAM CORE PREFABRICATED FRAMING SYSTEM. PANELS ARE ASSEMBLED INTO FINISHED WALLS AT THE JOB SITE.	
ADVANCED FOAM AFS "38"	ADVANCED FOAM SYSTEMS, INC	UREA-FORMALDEHYDE FOAMED-IN-PLACE INSULATION. FOR NEW OR RETROFIT SIDEWALL APPLICATIONS.	2A

MASONRY WALL INSULATION

PRODUCT TRADENAME	MANUFACTURER	PRODUCT DESCRIPTION	PLATES
THERMASOTE SIDEWALL PANELS	HOMASOTE CO	COMPOSITE OF ASBESTOS-FREE INSULATING BUILDING BOARD AND RIGID POLYURETHANE FOAM. FOAM SIDE IS FACED WITH ASPHALT SATURATED FELT OR FIBERGLASS. FOR MASONRY WALL EXTERIOR RETROFITS.	5D,7C-D
INSULWAL	PANELERA	COMPOSITE OF POLYURETHANE FOAM AND GYPSUM BOARD WITH REFLECTIVE FOIL SKIN OVER FOAM. ATTACHED TO MASONRY WALL WITH SPECIAL CLIPS.	11C
MASONRY WALL BATTS *	MANVILLE CORP	GLASS FIBER BATTS,FOR INSTALLATION BETWEEN FURRING STRIPS ON MASONRY WALL INTERIORS.	6B
MASONRY WALL INSULATION *	OWENS-CORNING FIBERGLAS CORP	GLASS FIBER BATT,UNFACED,FOR INSTALLATION BETWEEN FURRING STRIPS ON MASONRY WALL INTERIORS.	6B
UNFACED BATTS *	MANVILLE CORP	GLASS FIBER BATTS,UNFACED. SEPARATE VAPOR BARRIER MAY BE USED.	6C
ROLLED BATTS *	OWENS-CORNING FIBERGLAS CORP	GLASS FIBER BATTS,UNFACED.	6C
KRAFT-FACED BATTS *	MANVILLE CORP	GLASS FIBER BATT WITH KRAFT PAPER FACING. FACING EXTENDS PAST BATT EDGE FOR STAPLING FLANGE.	6C

INSULATION MANUFACTURERS PRODUCTS
(SORTED BY PRODUCT DESCRIPTION)

MASONRY WALL INSULATION (Cont.)

PRODUCT TRADENAME	MANUFACTURER	PRODUCT DESCRIPTION	PLATES
KRAFT-FACED BUILDING INSULATION *	OWENS-CORNING FIBERGLAS CORP	GLASS FIBER BATT INSULATION WITH ASPHALTED KRAFT PAPER FACING,FLANGED FOR STAPLING.	6C
FOIL-FACED BATTS *	MANVILLE CORP	GLASS FIBER BATT WITH ALUMINUM FOIL FACING. FACING EXTENDS PAST BATT EDGE FOR STAPLING FLANGE.	6C
FLAME-RESISTANT BATTS *	MANVILLE CORP	GLASS FIBER BATTS WITH FOIL-SCRIM-KRAFT (FSK) FACING. FACING EXTENDS PAST BATT EDGE FOR STAPLING FLANGE. MAY BE LEFT EXPOSED WHERE CODES PERMIT.	6C
FLAME SPREAD 25	OWENS-CORNING FIBERGLAS CORP	GLASS FIBER BATT WITH SPECIAL FOIL/ KRAFT LAMINATE FACING,RECOMMENDED BY MFR FOR USE IN WALLS AND CEILINGS WHERE INSULATION FACING WILL BE EXPOSED. SUITABLE FOR LOW-ABUSE AREAS.	6C
SEMI-RIGID INSULATION BOARD *	OWENS-CORNING FIBERGLAS CORP	GLASS FIBER SEMI-RIGID BOARD,FACED OR UNFACED. FOR INTERIOR MASONRY WALLS BETWEEN FURRING STRIPS AND FOR MASON-RY CAVITY WALLS.	6B,8C,9C
WALL INSULATION *	KNAUF FIBER GLASS GMBH	GLASS FIBER SEMI-RIGID BOARD,UNFACED OR FSK FACED. FOR MASONRY AND CAV-ITY WALLS.	
THOROWALL INSULATING PLASTER	THORO SYSTEMS PRODUCTS	INSULATING PLASTER CONTAINING POLY-STYRENE BEADS,HYDRAULIC BINDERS,AND CHEMICAL ADDITIVES. LIGHTWEIGHT, PACKAGED AS A POWDER,ADD WATER TO APPLY. MAY BE APPLIED TO MASONRY WALL EXTERIORS WITH TROWEL OR SPRAY GUN.	
HIGH-R SHEATHING	OWENS-CORNING FIBERGLAS CORP	POLYISOCYANURATE FOAM RIGID BOARD,GLASS FIBER REINFORCED AND ALUMINUM FOIL FACED. FOR INTERIOR BASEMENT,MASONRY WALL,OR CAVITY WALL INSULATION.	6B,D,7A-B 8C,9C

INSULATION MANUFACTURERS PRODUCTS
(SORTED BY PRODUCT DESCRIPTION)

MASONRY WALL INSULATION (Cont.)

PRODUCT TRADENAME	MANUFACTURER	PRODUCT DESCRIPTION	PLATES
THERMAX SHEATHING	CELOTEX CORP	POLYISOCYANURATE FOAM RIGID BOARD,GLASS FIBER REINFORCED AND ALUMINUM FOIL FACED. FOR INTERIOR BASEMENT,MASONRY OR CAVITY WALL INSULATION.	6B,D,7A-B 8C,9C
THERMAX INSULATION BOARD	CELOTEX CORP	POLYISOCYANURATE FOAM RIGID BOARD,GLASS FIBER REINFORCED AND ALUMINUM FOIL FACED. ONE FACING HAS A WHITE VINYL COATING,PROVIDING A WASHABLE INTERIOR FINISH.	11B
STYROFOAM SM	DOW CHEMICAL CO	POLYSTYRENE FOAM EXTRUDED BOARD WITH NATURAL SKIN SURFACE. USED ON INTERIOR MASONRY WALLS WITHOUT FURRING AND IN MASONRY CAVITY WALLS.	6D,7A-D,8C 8C,10B,11A 10A
FOAM-FORM	ROCKY MOUNTAIN FOAM-FORM	POLYSTYRENE FOAM,EXPANDED,BLOCKS FOR POURED CONCRETE WALLS.	11D
THERMOCURVE	THORO SYSTEMS PRODUCTS	POLYSTYRENE FOAM PANELS WITH A UNIQUE CURVED DESIGN AND PROTRUDING SPACERS. PANELS FIT SNUGLY WITHIN POURED CONCRETE WALL FORMS. AVAILABLE IN THREE WIDTHS FOR ALL STANDARD POURED-IN-PLACE STRUCTURES.	
TG-3000	THERMAL SYSTEMS,INC.	POLYURETHANE FOAM RIGID BOARD WITH ALUMINUM FOIL-SCRIM-FOIL LAMINATE FACERS FOR INTERIOR BASEMENT,MASONRY WALL OR CAVITY WALL INSULATION.	6D,7A-B
T/LINER	SILVERCOTE METAL BUILDING PRODUCTS	SUPPORTS,EXTRUDED PLASTIC,FOR ATTACHING PREFINISHED INSULATION TO MASONRY WALL INTERIORS.	11B
ADVANCED FOAM AFS "38"	ADVANCED FOAM SYSTEMS,INC	UREA-FORMALDEHYDE FOAMED-IN-PLACE INSULATION. FOR NEW OR RETROFIT APPLICATIONS IN CONCRETE BLOCK AND MASONRY CAVITY WALLS.	5B-D,6A-D 7A-B,8B,9B

INSULATION MANUFACTURERS PRODUCTS
(SORTED BY PRODUCT DESCRIPTION)

METAL BUILDING WALL INSULATION

PRODUCT TRADENAME	MANUFACTURER	PRODUCT DESCRIPTION	PLATES
THERMOCORE	MANVILLE CORP	ARCHITECTURAL PANEL OF EXPANDED PERLITE PARTICLES, FIBERS AND BINDERS, LAMINATED TO MINERAL FIBER CEMENT FACINGS.	13D
TRANSIFOAM	MANVILLE CORP	ARCHITECTURAL PANEL OF EXPANDED POLYSTYRENE BEAD BOARD, LAMINATED TO MINERAL FIBER CEMENT FACINGS.	13D
INSULATED WALL & SOFFIT SYSTEM	FINESTONE CORP	ARCHITECTURAL PANEL COMPOSED OF SUBSTRATE RIGID INSULATION, (GLASS FIBER, POLYSTYRENE BEAD BOARD, POLYURETHANE FOAM OR PHENOLIC FOAM), METAL LATH, MODIFIED PORTLAND CEMENT, AND ARCHITECTURAL FINISH.	13D
TRANSITOP	MANVILLE CORP	ARCHITECTURAL PANEL, HYDRAULICALLY PRESSED, WOOD FIBER AND ASPHALTIC COMPOUNDS, LAMINATED TO MINERAL FIBER CEMENT FACINGS.	13D
INSULATED "RAIN SCREEN"	H.L. BIRUM CORP	ARCHITECTURAL PANEL, LAMINATED. UNSPECIFIED COMPOSITION.	13D
UNFACED BATTS *	MANVILLE CORP	GLASS FIBER BATTS, UNFACED. SEPARATE VAPOR BARRIER MAY BE USED.	12C
ROLLED BATTS *	OWENS-CORNING FIBERGLAS CORP	GLASS FIBER BATTS, UNFACED.	12C
FIBERGLAS BUILDING INSULATION *	OWENS-CORNING FIBERGLAS CORP	GLASS FIBER BATTS, UNFACED. USED BETWEEN METAL STUDS.	12D
PEBS BLANKET	MANVILLE CORP	GLASS FIBER BLANKET, UNFACED. FOR PRE-ENGINEERED METAL BLDG WALLS.	13C
PAN-INSUL	MANVILLE CORP	GLASS FIBER BATT, UNFACED. FOR INTERLOCKING PRE-ENGINEERED METAL BLDG WALL CAVITIES.	13C
METAL BUILDING PANEL INSULATION *	CERTAINTEED CORP	GLASS FIBER BLANKET, UNFACED.	13C
KRAFT-FACED BATTS *	MANVILLE CORP	GLASS FIBER BATT WITH KRAFT PAPER FACING. FACING EXTENDS PAST BATT EDGE FOR STAPLING FLANGE.	12C

INSULATION MANUFACTURERS PRODUCTS
(SORTED BY PRODUCT DESCRIPTION)

METAL BUILDING WALL INSULATION (Cont.)

PRODUCT TRADENAME	MANUFACTURER	PRODUCT DESCRIPTION	PLATES
KRAFT FACED BUILDING INSULATION *	OWENS-CORNING FIBERGLAS CORP	GLASS FIBER BATT INSULATION WITH ASPHALTED KRAFT PAPER FACING, FLANGED FOR STAPLING.	12C
FOIL-FACED BATTS *	MANVILLE CORP	GLASS FIBER BATT WITH ALUMINUM FOIL FACING. FACING EXTENDS PAST BATT EDGE FOR STAPLING FLANGE.	12C
FLAME-RESISTANT BATTS *	MANVILLE CORP	GLASS FIBER BATTS WITH FOIL-SCRIM-KRAFT (FSK) FACING. FACING EXTENDS PAST BATT EDGE FOR STAPLING FLANGE. MAY BE LEFT EXPOSED WHERE CODES PERMIT.	12C
FLAME SPREAD 25	OWENS-CORNING FIBERGLAS CORP	GLASS FIBER BATT WITH SPECIAL FOIL/KRAFT LAMINATE FACING, RECOMMENDED BY MFR FOR USE IN WALLS AND CEILINGS WHERE INSULATION FACING WILL BE EXPOSED. SUITABLE FOR LOW-ABUSE AREAS.	12C
RIGID ROLL	MANVILLE CORP	GLASS FIBER SEMI-RIGID ROLLED INSULATION WITH A TEXTURED VINYL, FSK, OR WHITE FSK FACING.	13A-B
MICROLITE "L"	MANVILLE CORP	GLASS FIBER BLANKET LAMINATED WITH VARIOUS CUSTOM FACINGS.	13A-B
ROLL-IN	MANVILLE CORP	GLASS FIBER SEMI-RIGID ROLLED INSULATION WITH A DECORATIVE VINYL FACING.	13A-B
SNAP-IN	MANVILLE CORP	GLASS FIBER SEMI-RIGID BOARD WITH A DECORATIVE VINYL FACING.	13A-B
THERMAX INSULATION BOARD	CELOTEX CORP	POLISOCYANURATE FOAM RIGID BOARD, GLASS FIBER REINFORCED AND ALUMINUM FOIL FACED. ONE FACE HAS A WHITE VINYL COATING, PROVIDING A WASHABLE INTERIOR FINISH.	13A-B

INSULATION MANUFACTURERS PRODUCTS
(SORTED BY PRODUCT DESCRIPTION)

CEILING INSULATION

PRODUCT TRADENAME	MANUFACTURER	PRODUCT DESCRIPTION	PLATES
K-13	NATIONAL CELLULOSE CORP	CELLULOSE FIBER, SPRAYED-ON FOR EXPOSED INTERIOR APPLICATIONS.	20C
THERMOCON	THERMOCON SYSTEMS, INC	CELLULOSE FIBER, SPRAYED-ON FOR EXPOSED INTERIOR APPLICATIONS.	20C
UNFACED BATTS *	MANVILLE CORP	GLASS FIBER BATTS, UNFACED. SEPARATE VAPOR BARRIER MAY BE USED.	14A,C,15A-B, D,16A-B 17A-B,18C, 19D,20B
ROLLED BATTS *	OWENS-CORNING FIBERGLAS CORP	GLASS FIBER BATTS, UNFACED.	14A,C,15A-B, D,16A-B,17A-B,18C,19D, 20B
SUPER BATT INSULATION	OWENS-CORNING FIBERGLAS CORP	GLASS FIBER BATTS, EXTRA THICK, FACED OR UNFACED.	15C
KRAFT-FACED BATTS *	MANVILLE CORP	GLASS FIBER BATT WITH KRAFT PAPER FACING. FACING EXTENDS PAST BATT EDGE FOR STAPLING FLANGE.	14A,C,15A-B, D,16A-B 17A-B,18C, 19D,20B
KRAFT FACED BUILDING INSULATION *	OWENS-CORNING FIBERGLAS CORP	GLASS FIBER BATT INSULATION WITH ASPHALTED KRAFT PAPER FACING, FLANGED FOR STAPLING.	14A,C,15A-B, D,16A-B,17A-B,18C,19D, 20B
FLAME-RESISTANT BATTS *	MANVILLE CORP	GLASS FIBER BATT WITH FOIL-SCRIM-KRAFT (FSK) FACING. FACING EXTENDS PAST BATT EDGE FOR STAPLING FLANGE. MAY BE LEFT EXPOSED WHERE CODES PERMIT.	14A,C,15A-B, D,16A-B, 17A-B,18C, 19D,20B
FLAME SPREAD 25	OWENS-CORNING FIBERGLAS CORP	GLASS FIBER BATT WITH SPECIAL FOIL/KRAFT LAMINATE FACING, RECOMMENDED BY MFR FOR USE IN WALLS AND CEILINGS WHERE INSULATION FACING WILL BE EXPOSED. SUITABLE FOR LOW-ABUSE AREAS.	14A,C,15A-B, D,16A-B 17A-B,18C 19D,20B
RETROFIT INSULATION	METAL BUILDING INTERIOR PRODUCTS CO	GLASS FIBER BLANKET INSULATION WITH WHITE VINYL FACING FOR INSTALLATION BETWEEN PURLINS.	

INSULATION MANUFACTURERS PRODUCTS
(SORTED BY PRODUCT DESCRIPTION)

CEILING INSULATION (Cont.)

PRODUCT TRADENAME	MANUFACTURER	PRODUCT DESCRIPTION	PLATES
FOIL-FACED BATTS *	MANVILLE CORP	GLASS FIBER BATT WITH ALUMINUM FOIL FACING. FACING EXTENDS PAST BATT EDGE FOR STAPLING FLANGE.	14A,C,15A-B, D,16A-B,17A-B, 19D,20B
BLOWING WOOL *	MANVILLE CORP	GLASS FIBER LOOSE FILL FOR ATTICS AND OVERHEAD SPACES.	14B,D,15B,16C
FIBERGLAS BLOWING WOOL	OWENS-CORNING FIBERGLAS CORP	GLASS FIBER LOOSE FILL FOR ATTICS AND OVERHEAD SPACES.	14B,D,15B,16C
FIBERGLAS CUBED BLOWING WOOL	OWENS-CORNING FIBERGLAS CORP	GLASS FIBER LOOSE FILL FOR ATTICS AND OVERHEAD SPACES.	14B,D,15B,16C
THERMAX SHEATHING	CELOTEX CORP	POLYISOCYANURATE FOAM RIGID BOARD,GLASS FIBER REINFORCED AND ALUMINUM FOIL FACED. FOR CATHEDRAL AND A-FRAME OVERDECK APPLICATIONS.	17B,D
THERMAX INSULATION BOARD	CELOTEX CORP	POLYISOCYANURATE FOAM RIGID BOARD,GLASS FIBER REINFORCED AND ALUMINUM FOIL FACED. ONE FACER HAS A WHITE VINYL COATING.	19B-D,20A-B
HIGH-R SHEATHING	OWENS-CORNING FIBERGLAS CORP	POLYISOCYANURATE FOAM RIGID BOARD,GLASS FIBER REINFORCED AND ALUMINUM FOIL FACED. FOR CATHEDRAL AND A-FRAME OVERDECK APPLICATIONS.	17B,D
STYROFOAM SM	DOW CHEMICAL CO	POLYSTYRENE FOAM EXTRUDED BOARD WITH NATURAL SKIN SURFACE. FOR ROOF OVER-DECK APPLICATIONS (CATHEDRAL CEIL-INGS.)	17D
STYROFOAM TG	DOW CHEMICAL CO	POLYSTYRENE FOAM EXTRUDED BOARD WITH HIGH-DENSITY SKIN. EDGES ARE TONGUE AND GROOVE. FOR ROOF OVERDECK APPLIC-ATIONS (CATHEDRAL CEILINGS).	17D

ROOF INSULATION

PRODUCT TRADENAME	MANUFACTURER	PRODUCT DESCRIPTION	PLATES
TAPERED FOAMGLAS ROOF INSULATION	PITTSBURGH CORNING CORP	CELLULAR GLASS BOARD FOR BURS,TAPERED FOR DRAINAGE OF FLAT ROOFS.	23A-B

INSULATION MANUFACTURERS PRODUCTS
(SORTED BY PRODUCT DESCRIPTION)

ROOF INSULATION (Cont.)

PRODUCT TRADENAME	MANUFACTURER	PRODUCT DESCRIPTION	PLATES
FIBERGLAS/URETHANE ROOF INSULATION	OWENS-CORNING FIBERGLAS CORP	COMPOSITE OF GLASS FIBER BOARD POLYURETHANE FOAM BOARD, SURFACED WITH FIBROUS GLASS MAT FOR BUR BASE. FOR FLAT AND LOW-SLOPE NAILABLE, NON-NAILABLE, AND METAL DECKS.	22A-B
NAILABLE BASE	RMAX, INC	COMPOSITE OF POLYISOCYANURATE FOAM RIGID BOARD WITH A BOTTOM SKIN OF ROOFING FELT AND A TOP LAYER OF NAILABLE BASE MATERIAL. FOR ROOFS WHERE A TOP NAILABLE SURFACE IS REQUIRED.	17D
THERMAROOF COMPOSITE	RMAX, INC	COMPOSITE OF PERLITE BASE LAYER POLYISOCYANURATE FOAM WITH FIBERGLASS FELT TOP FACING. FOR BURS OVER STEEL DECKS.	24C-D 26A
THERMAROOF PLUS COMPOSITE	RMAX, INC	COMPOSITE OF PERLITE BASE LAYER POLYISOCYANURATE FOAM WITH ALUMINUM FOIL TOP SKIN. FOR SINGLE-PLY ROOFS.	24C-D, 25A-D 26A
FESCO FOAM	MANVILLE CORP	COMPOSITE OF PERLITE BOARD (FESCO BOARD) -POLYURETHANE FOAM BOARD WITH ASPHALT ROOFING FELT FACING FOR BUR. FOR ANY DECK.	22A-B
PERMALITE PK	GREFCO, INC	COMPOSITE OF PERLITE BOARD (PERMALITE) POLYURETAHANE FOAM BOARD. ASPHALT SATURATED FELT FACING FOR BUR BASE.	22A-B
PERMALITE PK PLUS	GREFCO, INC	COMPOSITE OF PERLITE BOARD (PERMALITE) TOP AND BOTTOM LAYERS WITH A RIGID POLYURETHANE FOAM CORE.	22A-B
TG1000	THERMAL SYSTEMS, INC	COMPOSITE OF PERLITE BOARD BASE, TOP SHEET OF POLYURETHANE BOARD, FACED WITH KRAFT, ALUMINUM FOIL, OR ASPHALT SATURATED FELT.	22A-B
POLYCON-POSITE	CONSOLIDATED FIBER GLASS PRODUCTS CO	COMPOSITE OF PERLITE BOARD POLYURETHANE FOAM BOARD.	22A-B
GAFTEMP URETHANE /PERLITE	GAF CORP	COMPOSITE OF PERLITE BOARD POLYURETHANE FOAM BOARD WITH ASPHALT SATURATED FELT TOP SURFACE.	22A-B

INSULATION MANUFACTURERS PRODUCTS
(SORTED BY PRODUCT DESCRIPTION)

ROOF INSULATION (Cont.)

PRODUCT TRADENAME	MANUFACTURER	PRODUCT DESCRIPTION	PLATES
NEWCOMP	SHELTER INSULATION, INC	COMPOSITE OF PERLITE BOARD-POLYURETHANE /POLYISOCYANURATE FOAM BOARD. FOAM IS FACED WITH ASPHALT-SATURATED FELT. PRODUCT IS INSTALLED FOAM-SIDE DOWN OVER NONCOMBUSTIBLE DECKS; PERLITE SIDE IS THE BUR BASE.	22A-B
SISCOMP	SHELTER INSULATION, INC	COMPOSITE OF POLYURETHANE FOAM ROOF INSULATION.	22A-B
PLYFOAM URETHANE /COMPOSITE	WATER GUIDANCE SYSTEMS, INC	COMPOSITE OF POLYURETHANE FOAM PANEL WITH PERLITE BOTTOM LAYER. FOR SINGLE-PLY ROOFS.	24C-D, 25A-D 26A
XFS 4249	DOW CHEMICAL CO	COMPOSITE OF POLYSTYRENE FOAM, EXTRUDED BOARD (STYROFOAM RM) WITH FACTORY APPLIED 3/8INCH THICK LATEX MODIFIED PORTLAND CEMENT MORTAR FACING. FOR APPLICATIONS REQUIRING HIGH COMPRES-SIVE STRENGTH. (DEVELOPMENTAL PROD.)	
ITP MONEY CLIP BATT	MANVILLE CORP	GLASS FIBER BATT FOR USE WITH ITP MONEY CLIP BOARD.	18C, 19D
THERMAL-ACOUSTICAL BATTS *	MANVILLE CORP	GLASS FIBER BATTS, WITH OR WITHOUT A VAPOR BARRIER, FOR CONTROL OF BOTH HEAT AND SOUND TRANSMISSION THROUGH SUSPENDED CEILINGS.	
MI-T-R	MIZELL BROS. CO	GLASS FIBER BATT AND BLANKET SYSTEM FOR METAL BUILDINGS.	31A
MICROLITE "L"	MANVILLE CORP	GLASS FIBER BLANKET, UNFACED, FOR NEW METAL BUILDINGS. INSTALLED OVER PUR-LINS.	20A-B
RIGID-ROLL	MANVILLE CORP	GLASS FIBER SEMI-RIGID BLANKET WITH TEXTURED VINYL, FSK OR WHITE FSK FACING FOR NEW METAL BUILDINGS. INSTALLED OVER PURLINS.	31A
PEBS BLANKET	MANVILLE CORP	GLASS FIBER BLANKET, UNFACED. GENERAL PURPOSE PRODUCT FOR NEW AND RETROFIT INSTALLATIONS IN METAL BUILDINGS.	12C-D, 13A 16C, 19D, 20B 31A

INSULATION MANUFACTURERS PRODUCTS
(SORTED BY PRODUCT DESCRIPTION)

ROOF INSULATION (Cont.)

PRODUCT TRADENAME	MANUFACTURER	PRODUCT DESCRIPTION	PLATES
FIBERGLAS ROOF INSULATION	OWENS-CORNING FIBERGLAS CORP	GLASS FIBER BOARD WITH RESINOUS BINDER, TOP-SURFACED WITH GLASS FIBER REIN-FORCED ASPHALT AND KRAFT FOR BUR BASE. FOR FLAT AND LOW-SLOPE NAILABLE,NONNAILABLE,AND METAL DECKS.	22A-D
ITP MONEY CLIP BOARD	MANVILLE CORP	GLASS FIBER BOARD WITH GLASS-REINFORCED WHITE,METALLIZED POLYESTER FACING. FOR UNDERDECK INSTALLATION BETWEEN PURLINS IN NEW METAL BUILDINGS. FACING PROVIDES FINISHED INTERIOR SURFACE.	
ALL-WEATHER CRETE	SILBRICO CORP	INSULATING CONCRETE FOR CONCRETE OR METAL DECKS,SLOPED TO PROVIDE DRAINAGE.	29A-D,30A-C
ELASTIZELL CONCRETE	ELASTIZELL CORP OF AMERICA	INSULATING CONCRETE CONTAINING DISCREET AIR CELLS (NO EXPANDED FILLERS). FOR USE WITH ANY DECK OR OVER EXISTING BUR.	29A-D,30A-C
SISTEEL	SHELTER INSULATION,INC	NONCOMPOSITE ROOF INSULATION FOR STEEL DECKS.	22A-B
FESCO BOARD	MANVILLE CORP	PERLITE (EXPANDED PARTICLES) BLENDED HOMOGENEOUSLY WITH SELECTED FIBERS AND BINDERS. RIGID BOARD TOP SUR-FACED WITH TOP-LOC COATING TO COAT-ING TO RECEIVE BUR FOR ANY DECK.	22A-D
FESCO RE-ROOF BOARD	MANVILLE CORP	PERLITE BOARD (FESCO BOARD) FOR USE IN APPLYING A NEW BUR DIRECTLY OVER ON OLD ROOF.	23C-D
FESCO TAPERED DRI-DECK SYSTEM	MANVILLE CORP	PERLITE BOARD (FESCO BOARD) FACTORY-TAPERED TO PROVIDE DRAINAGE ON A FLAT ROOF.	23A-B
PERMALITE SEALSKIN	GREFCO,INC	PERLITE BOARD FORMED OF EXPANDED HERMETICALLY SEALED PERLITE BEADS WATERPROOFING AGENTS AND CELLULOSE BINDER. INTEGRAL SURFACE TREATMENT FOR BUR BASE.	22A-D

INSULATION MANUFACTURERS PRODUCTS
(SORTED BY PRODUCT DESCRIPTION)

ROOF INSULATION (Cont.)

PRODUCT TRADENAME	MANUFACTURER	PRODUCT DESCRIPTION	PLATES
GAFTEMP PERLITE	GAF CORP	PERLITE BOARD COMPOSED OF EXPANDED PERLITE PARTICLES HOMOGENOUSLY BLENDED WITH SELECTED BINDERS AND FIBERS. TOP SURFACE SEALED WITH A SPECIAL COATING FOR BUR BASE.	22A-D
EXELTHERM XTRA	KOPPERS CO, INC	PHENOLIC FOAM RIGID BOARD FOR BURS OR SINGLE-PLY ROOFS. FOR ANY DECK.	22A-D,24C-D 25A-D,26A
GAFTEMP ISOTHERM	GAF CORP	POLISOCYANURATE FOAM RIGID BOARD WITH ASPHALT SATURATED ASBESTOS FACINGS.	22A-D
MAX-I	RMAX, INC	POLYISOCYANURATE FOAM RIGID BOARD WITH ASPHALT-COATED FIBERGLASS MAT FACINGS. MAY BE USED DIRECTLY OVER STEEL ROOF DECKS AS BUR BASE.	22A-D
THERMAROOF STANDARD	RMAX, INC	POLYISOCYANURATE FOAM RIGID BOARD WITH FIBERGLASS FELT FACINGS. FOR BURS MAY BE APPLIED OVER A BASE LAYER OF PERLITE BOARD.	22A-D
THERMAROOF PLUS	RMAX, INC	POLYISOCYANURATE FOAM RIGID BOARD WITH ALUMINUM FOIL FACINGS. FOR BALLASTED LOOSE-LAID SINGLE-PLY ROOFS.	24C,25A,C
PLY-I	RMAX, INC	POLYISOCYANURATE FOAM RIGID BOARD WITH FIBERGLASS REINFORCED ALUMINUM FOIL FACINGS. FOR SINGLE-PLY ROOFS OVER ANY DECK. MECHANICAL FASTENERS RECOMMENDED FOR ATTACHMENT.	24C-D,25A-D 26A
EPS	ARCO POLYMERS, INC	POLYSTYRENE FOAM,EXPANDED,UNFACED. FOR NEW BURS OR SINGLE-PLY ROOFS,OR FOR RETROFIT DIRECTLY OVER OLD BUR.	22A-D,23C-D 24C-D,25A-D 26A-D
TAPERED EPS	ARCO POLYMERS, INC	POLYSTYRENE FOAM,EXPANDED,UNFACED TAPERED TO PROVIDE DRAINAGE OF A FLAT ROOF. FOR BURS OVER NEW OR EXISTING ROOFS.	23A-B
STYROFOAM RM	DOW CHEMICAL CO	POLYSTYRENE FOAM,EXTRUDED BOARD,UNFACED. CHANNELED TO PROVIDE DRAINAGE OF PROTECTED MEMBRANE (UPSIDE-DOWN) ROOFS.	23C-D

INSULATION MANUFACTURERS PRODUCTS
(SORTED BY PRODUCT DESCRIPTION)

ROOF INSULATION (Cont.)

PRODUCT TRADENAME	MANUFACTURER	PRODUCT DESCRIPTION	PLATES
XFS 43001	DOW CHEMICAL CO	POLYSTYRENE FOAM EXTRUDED BOARD HAVING HIGH COMPRESSIVE STRENGTH. (DEVELOPMENTAL PRODUCT)	
PLYFOAM STYRENE	WATER GUIDANCE SYSTEMS, INC	POLYSTYRENE FOAM BOARD FOR SINGLE-PLY ROOFS.	24C-D, 25A-D 26A
TAPERED FOAM	BENOIT, INC	POLYSTYRENE FOAM, MOLDED, TAPERED TO PROVIDE DRAINAGE OF FLAT ROOFS. BUR BASE FOR ANY DECK.	23A-B
PERMALITE URETHANE	GREFCO, INC	POLYURETHANE FOAM BOARD WITH ASPHALT SATURATED FEET FACINGS ON BOTH SURFACES.	22A-D
FS1000	THERMAL SYSTEMS, INC	POLYURETHANE FOAM BOARD EXTRUDED BETWEEN KRAFT, ALUMINUM FOIL, OR ASPHALT SATURATED FELT MEMBRANES. FOR ALL DECKS.	22A-D
POLYCON-STANDARD	CONSOLIDATED FIBERGLASS PRODUCTS CO	POLYURETHANE FOAM RIGID BOARD WITH ASPHALT-SATURATED FELT FACINGS.	22A-D
GAFTEMP URETHANE	GAF CORP	POLYURETHANE FOAM RIGID BOARD WITH ASPHALT-SATURATED FELT FACINGS.	22A-D
SISDECK	SHELTER INDUSTRIES, INC	POLYURETHANE FOAM ROOF INSULATION FOR NONCOMBUSTIBLE DECKS.	22C-D
SISDECK GF(N)	SHELTER INSULATION, INC	POLYURETHANE FOAM ROOF INSULATION WITH INTEGRALLY BONDED NONASPHALTIC GLASS FACINGS. ESPECIALLY DESIGNED FOR SINGLE-PLY ROOF SYSTEMS.	24D-C, 25A-D 26A
PLYFOAM URETHANE	WATER GUIDANCE SYSTEMS, INC DUCTS CO	POLYURETHANE FOAM PANELS WITH ASPHALT FELT, FIBERGLASS, ALUMINUM FOIL, OR POLYETHYLENE COATED KRAFT PAPER FACINGS. FOR SINGLE-PLY ROOFS.	24C-D, 25A-D 26A
INSULATED PANEL *	INSULATED PANEL SYSTEMS, INC	POLYURETHANE FOAM CORE STANDING SEAM ROOF PANEL WITH GALVANIZED STEEL SKINS. SKINS AVAILABLE WITH STUCCO-EMBOSSED FINISHES AND VARIOUS COLORS.	31B

INSULATION MANUFACTURERS PRODUCTS
(SORTED BY PRODUCT DESCRIPTION)

PRODUCT TRADENAME	MANUFACTURER	PRODUCT DESCRIPTION	PLATES

ROOF INSULATION (Cont.)

PRODUCT TRADENAME	MANUFACTURER	PRODUCT DESCRIPTION	PLATES
SILICONE ROOFING SYSTEM	GENERAL ELECTRIC CO	POLYURETHANE FOAM,SPRAYED-IN-PLACE OVER ANY NEW DECK OR EXISTING BUR,WITH TWO TOP COATS OF SPRAYED-IN-PLACE SILICONE RUBBER.	27A-D,28A-B

WINDOWS AND WINDOW TREATMENTS

PRODUCT TRADENAME	MANUFACTURER	PRODUCT DESCRIPTION	PLATES
SCOTCHTINT SUN CONTROL FILMS	3M CO	SUN CONTROL FILM,EITHER CLEAR OR COLORED POLYESTER FILM,OR ALUMINUM VAPOR COATED POLYESTER FILM. REDUCES INCOMING GLARE,ULTRAVIOLET,AND HEAT GAIN, ALSO HEAT LOSS.	
VARI-TRAN	LIBBEY-OWENS-FORD CO	HEAT ABSORBING GLASS,VACUUM DEPOSITED METALLIC COATINGS.	
SUNGLAS WINDOW GLASS	FORD MOTOR CO,GLASS DIV	HEAT ABSORBING GLASS,SINGLE OR DOUBLE STRENGTH THICKNESSES.	
TRU-THERM	ASG INDUSTRIES,INC	INSULATING GLASS,DOUBLE-PANE,SEALED WITH DESICCANT MATERIAL IN THE SEALED SPACE. CLEAR OR TINTED.	
WEATH-R-PROOF	SHATTERPROOF GLASS CORP	INSULATING GLASS,DOUBLE-PANE,SEALED, WITH DESICCANT MATERIAL IN THE SEALED SPACE. CLEAR OR TINTED.	
THERMOPANE	LIBBEY-OWENS-FORD CO	INSULATING GLASS,DOUBLE- OR TRIPLE-PANE, SEALED,DESICCANT MATERIAL IN SEALED SPACE.	
REFLECTOVUE	ASG INDUSTRIES,INC	REFLECTIVE GLASS CONTAINING VACUUM DEPOSITED THIN METALLIC COATING OF PURE GOLD OR CHROME.	
REFLECTIVE GLASS	GUARDIAN INDUSTRIES CORP	REFLECTIVE GLASS,VACUUM DEPOSITED COATINGS OF COPPER OXIDE (C-SERIES) STAINLESS STEEL OXIDE (S-SERIES),OR TITANIUM OXIDE (T-SERIES). AVAILABLE IN INSULATING GLASS UNITS.	

INSULATION MANUFACTURERS PRODUCTS
(SORTED BY PRODUCT DESCRIPTION)

WINDOWS AND WINDOW TREATMENTS (Cont.)

PRODUCT TRADENAME	MANUFACTURER	PRODUCT DESCRIPTION	PLATES
COOL-VIEW	SHATTERPROOF GLASS CORP	REFLECTIVE GLASS CONSISTING OF CLEAR, BRONZE,OR GRAY GLASS WITH A CHROME, BRONZE,OF GOLD COATING ON THE INTERIOR.	
LUCITE	E.I. DU PONT DE NEMOURS & COMPANY, INC.	ACRYLIC SHEET,PRODUCED FROM METHYL METHACRYLATE MONOMER,IN LINEAR (L) COMPOSITION,CROSS-LINKED (XL) COMPOSITION FOR HIGHER SOLVENT RESISTANCE,AND (T-1000) FOR HIGH IMPACT RESISTANCE.	
ACRYLITE SDP	CY/RO INDUSTRIES	ACRYLIC SHEET,DOUBLE SKINNED FOR THERMAL INSULATION COMPARABLE TO INSULATING GLASS. FOR SKYLIGHTS,COVERED WALKWAYS,CURTAINWALLS,GREENHOUSES.	
LEXAN	GENERAL ELECTRIC CO	ACRYLIC SHEET, CLEAR OR TINTED,VARIOUS THICKNESSES AND TYPES.	
PLEXIGLAS	ROHM AND HAAS CO	ACRYLIC SHEET,CLEAR OR TINTED,TRANSPARENT TO SEMI-OPAQUE. VARIOUS GLAZING APPLICATIONS,TYPES INCLUDE A UV-ABSORBING TYPE.	
POLYCARBONATE SDP	CY/RO INDUSTRIES	POLYCARBONATE SHEET,DOUBLE SKINNED FOR THERMAL INSULATION COMPARABLE TO INSULATING GLASS. FOR SKYLIGHTS,COVERED WALKWAYS,CURTAINWALLS,GREENHOUSES.	
INSULATING CURTAIN WALL	THERMAL TECHNOLOGY CORP	ROLLING WINDOW SHADE OF FOUR-LAYER FABRIC DESIGN,SELF-INFLATING. AUTOMATICALLY ACTIVATED BY EXTERNAL TEMPERATURE SENSORS.	
MAGNETIC INSULATING SHADE	3M CO	ROLLING WINDOW SHADE OF SUN CONTROL FILM. EDGES SEAL TO WINDOW FRAME WITH FLEXIBLE MAGNETIC STRIPS.	
T-2001	DISCO ALUMINUM PRODUCTS CO	VENETIAN BLIND WINDOW SHADE MOUNTED BETWEEN SHEETS OF DOUBLE-PANE WINDOW.	

INSULATION MANUFACTURERS PRODUCTS
(SORTED BY PRODUCT DESCRIPTION)

WINDOWS AND WINDOW TREATMENTS (Cont.)

PRODUCT TRADENAME	MANUFACTURER	PRODUCT DESCRIPTION	PLATES
SOLAR SHADE *	MOORE CO	VERTICAL WINDOW SHADE OF MULTIPLE HOLLOW ALUMINUM BLADES. MOUNTED EXTERIOR TO WINDOW OR BETWEEN INSIDE AND OUTSIDE WINDOWS.	
KOOLSHADE SOLAR SCREENS	KOOLSHADE CORP	HORIZONTAL LOUVER SHADING SCREEN, FIXED POSITION. BLOCKS SUNLIGHT TO VARIOUS DEGREES DEPENDING ON SUN ELEVATION ANGLE.	
ROLLING SHUTTER	PEASE ROLLING SHUTTERS	ROLLING WINDOW SHUTTER OF PVC VINYL, MOUNTED EXTERIOR TO EXISTING WINDOWS. EDGES SEALED WITH ALUMINUM SIDE RAILS.	
THERMALON	ARMSTRONG CORK CO	WINDOW INSULATION, OPAQUE, PERMANENTLY MOUNTED OVER EXISTING WINDOWS.	
WINDOW INSULATION *	SENTINAL FOAM PRODUCTS, INC	WINDOW INSULATION, TRANSLUCENT POLYETHYLENE FOAM, WITH ADHESIVE ON ONE SIDE FOR PERMANENT MOUNTING TO EXISTING WINDOWS.	
MAGNETIC RIGID INSUL WINDOW	3M CO	INSIDE STORM WINDOW, ACRYLIC GLAZING. MOUNTS TO EXISTING WINDOW WITH MAGNETIC STRIPS.	36A
MAGNETITE WINDOWS	VIKING ENERGY SYSTEM CO	INSIDE STORM WINDOW, ACRYLIC GLAZING. MOUNTS TO EXISTING WINDOW WITH MAGNETIC STRIPS.	36A
HEAVY-DUTY IN-SIDER	PLASKOLITE, INC.	INSIDE STORM WINDOW, ACRYLIC GLAZING. MOUNTS TO EXISTING WINDOW WITH VINYL MOULDING.	36B
MAGNETIC FLEXIBLE WINDOW	3M CO	INSIDE STORM WINDOW, NONINSULATING OR REFLECTIVE INSULATING FLEXIBLE FILM. MOUNTS TO EXISTING WINDOW WITH MAGNETIC STRIPS.	36C

INSULATION MANUFACTURERS PRODUCTS
(SORTED BY PRODUCT DESCRIPTION)

PRODUCT TRADENAME	MANUFACTURER	PRODUCT DESCRIPTION	PLATES

DOORS

PRODUCT TRADENAME	MANUFACTURER	PRODUCT DESCRIPTION	PLATES
INSULCLAD 260 DOORS	KAWNEER CO	ENTRANCE DOOR, COMMERCIAL, FULLY WEATHER-STRIPPED, THERMAL BREAKS.	
EVER-STRAIT REPLACE DOOR	PEASE CO	ENTRANCE DOOR, RESIDENTIAL REPLACEMENT, EXPANDED POLYSTYRENE FOAM CORE, THERMAL BREAKS, MAGNETICALLY WEATHER-STRIPPED.	
STORM DOOR *	PEERLESS PRODUCTS, INC	STORM DOOR, RESIDENTIAL ENTRANCE, FULLY WEATHERSTRIPPED.	
OUTSLIDER	PEERLESS PRODUCTS, INC	STORM DOOR, RESIDENTIAL PATIO SLIDING, MOUNTS OUTSIDE EXISTING PATIO DOOR. FULLY WEATHERSTRIPPED. OPTIONAL INSULATING OR TINTED GLASS.	
THERMAL DOOR	ANDERSON DOOR CO	OVERHEAD DOOR, COMMERCIAL, INSULATED.	39A
THERMOSPAN	DALTON INTERNATIONAL, INC	OVERHEAD DOOR, COMMERCIAL, INSULATED WITH A POLYURETHANE FOAM CORE BETWEEN GALVANIZED STEEL SKINS, THERMAL BREAKS.	39A
THERMACORE	INSOPORT INDUSTRIES, INC	OVERHEAD DOOR, COMMERCIAL INSULATED WITH A POLYURETHANE FOAM CORE BETWEEN EMBOSSED GALVANIZED STEEL SKINS, THERMAL BREAKS.	39A

WEATHERSTRIPPING AND CAULKING

PRODUCT TRADENAME	MANUFACTURER	PRODUCT DESCRIPTION	PLATES
TYVEK	E.I. DU PONT DE NEMOURS & COMPANY, INC.	AIR INFILTRATION BARRIER, THIN SHEET OF HIGH-DENSITY POLYETHYLENE FIBERS, NOT A MOISTURE BARRIER. MAY BE WRAPPED AROUND SIDEWALLS IN NEW FRAME CONSTRUCTION OR PLACED OVER ATTIC INSULATION TO REDUCE AIR INFILTRATION.	
THERM-L-BRUSH	SEALEZE CORP	BRUSH WEATHERSTRIPPING, FLEXIBLE NYLON, WEATHER RESISTANT, FILAMENTS MOVE EASILY IN ANY DIRECTION, CONFORMS TO GAPS AND MISALIGNMENTS. FOR ANY TYPE DOOR.	

INSULATION MANUFACTURERS PRODUCTS
(SORTED BY PRODUCT DESCRIPTION)

WEATHERSTRIPPING AND CAULKING (Cont.)

PRODUCT TRADENAME	MANUFACTURER	PRODUCT DESCRIPTION	PLATES
INTERIOR FOAM WEATHER STRIP *	3M CO	FOAM STRIP,ADHESIVE BACKED,FOR WINDOWS AND DOORS,IRREGULAR SURFACES.	
EXTERIOR FOAM WEATHER STRIP *	3M CO	FOAM STRIP,ADHESIVE BACKED,FOR WINDOWS AND DOORS,IRREGULAR SURFACES. MOISTURE RESISTANT.	
WILL-SEAL	ILLBRUCK/USA	FOAM TAPE,ADHESIVE BACKED. SEALS OUT WEATHER AND WATER. FOR ANY JOINT,CONFORMS TO NONUNIFORMITIES.	
GARAGE DOOR BOTTOM WEATHER STRIP *	3M CO	GARAGE DOOR WEATHERSTRIPPING,HIGH-QUALITY SYNTHETIC RUBBER. SEALS GARAGE DOOR TO FLOOR,CONFORMS TO IRREGULARITIES.	
PERMANENT CAULKING STRIP	3M CO	ROPE-TYPE CAULKING FOR PERMANENT INSTALLATIONS,RESISTS WEATHERING AND MOISTURE,PAINTABLE.	
TRANSPARENT WEATHER SEALING TAPE *	3M CO	TAPE,TRANSPARENT,SEALS CRACKS AROUND WINDOWS AND DOORS,MOISTURE RESISTANT.	
TRISEAL	SLOTTSEAL	TRIANGULAR WEATHERSTRIPPING,PVC,MAY BE PERMANENTLY FITTED INTO A MACHINED GROOVE IN WINDOWS AND DOORS.	
REUSABLE TUBULAR WEATHER STRIP *	3M CO	TUBULAR WEATHERSTRIPPING FOR SEASONAL USE AROUND WINDOWS AND DOORS. REUSABLE.	
TUBESEAL	SLOTTSEAL	TUBULAR WEATHERSTRIPPING,PVC,EPDM RUBBER OR SILICONE RUBBER. ATTACHED TO WINDOWS AND DOORS BY STAPLING THROUGH SIDE LEG.	
UNISEAL	SLOTTSEAL	TUBULAR WEATHERSTRIPPING,PVC,EPDM RUBBER,SILICONE RUBBER OR THERMOPLASTIC RUBBER. FINNED OR TRAPEZOIDAL FOOT MAY BE INSERTED INTO A MACHINED GROOVE IN WINDOWS AND DOORS.	
PERIMETER WEATHER SEAL *	STANLEY HARDWARE	TUBULAR FINNED WEATHERSTRIPPING,EPDM RUBBER,RESISTS ULTRAVIOLET,WATER ABSORPTION,FREEZING. FILLS GAPS BETWEEN WINDOWS,DOORS,AIR CONDITIONERS.	

INSULATION MANUFACTURERS PRODUCTS
(SORTED BY PRODUCT DESCRIPTION)

WEATHERSTRIPPING AND CAULKING (Cont.)

PRODUCT TRADENAME	MANUFACTURER	PRODUCT DESCRIPTION	PLATES
WINGSTRIP	SLOTTSEAL	V-SHAPED WEATHERSTRIPPING. ATTACHED BY STAPLING, ESPECIALLY USEFUL ON SLIDING WINDOWS.	
V-SEAL WEATHER STRIP *	3M CO	V-SHAPED WEATHERSTRIPPING, ADHESIVE-BACKED, POLYPROPYLENE, PRESCORED FOR FOLDING INTO V. FOR WINDOW AND DOORS.	

SEALANTS

PRODUCT TRADENAME	MANUFACTURER	PRODUCT DESCRIPTION	PLATES
D200	LION OIL CO	ELASTOMERIC TWO-PART SEALANT, WATER RESISTANT, MOVES AND RECOVERS FROM EXPANSION AND CONTRACTION CAUSED BY TEMPERATURE CHANGE. SEALS JOINTS BETWEEN CONCRETE, METAL, GLASS, OTHER SUBSTANCES, ABOVE AND BELOW GRADE.	
DYNASEAL W-100	WILLIAMS PRODUCTS, INC	POLYURETHANE ONE-PART SEALANT, COMBINES STRENGTH WITH FLEXIBILITY AND ABRASION RESISTANCE . FOR PRECAST CONCRETE, PORCELAIN, SHEET METAL, DOOR FRAMES, SKYLIGHTS, DAMP MASONRY.	
HEAT SEAL XL-770	HOSHALL INDUSTRIES INC	SEALANT, RUBBER-LIKE CHARACTERISTICS FOR EXCELLENT WEATHER AND AGING RESISTANCE. ADHERES TO ANY SURFACE.	
SILGLAZE	GENERAL ELECTRIC CO	SILCONE ONE-PART SEALANT, ESPECIALLY DESIGNED FOR SEALING BUTT AND LAP JOINTS IN GLAZING, CURTAINWALLS, AND MASONRY PERIMETERS. ADHERES TO GLASS, PLASTIC, METAL.	
CONSTRUCTION 1200	GENERAL ELECTRIC CO	SILICONE ONE-PART SEALANT, WITH SUPERIOR ADHESION, WEATHER RESISTANCE AND ELASTICITY. FOR ALL GLAZING APPLICATIONS AND METAL CURTAINWALLS. ADHERES TO GLASS, CERAMICS, STEEL, WOOD, GRANITE, ALUMINUM AND MOST PLASTICS.	

INSULATION MANUFACTURERS PRODUCTS
(SORTED BY PRODUCT DESCRIPTION)

SEALANTS (Cont.)

PRODUCT TRADENAME	MANUFACTURER	PRODUCT DESCRIPTION	PLATES
SILPRUF	GENERAL ELECTRIC CO	SILICONE ONE-PART SEALANT. LOW MODULUS OF ELASTICITY ALLOWS EXCELLENT RECOVERY FROM COMPRESSION AND EXTENSION. DESIGNED FOR SEALING BUILDING JOINTS THAT HAVE A HIGH DEGREE OF MOVEMENT. WITHSTANDS WEATHER AND TEMPERATURE EXTREMES.	
SILICONE RUBBER SEALANT	DOW CORNING CORP	SILICONE ONE-PART SEALANT, GOOD WEATHER RESISTANCE,DURABILITY,STRENGTH. FOR SEALING MULLIONS,PLASTIC,GLASS,CURTAINWALL AND CONVENTIONAL JOINTS.	
SILICONE RUBBER SEALANT PAINTABLE	DOW CORNING CORP	SILICONE ONE-PART SEALANT,GOOD WEATHER RESISTANCE,DURABILITY,PAINTABLE, STAINABLE. ADHERES TO WOOD,MASONRY, OTHER SUBSTRATES,FOR INTERIOR WALL JOINTS WINDOW AND DOOR FRAMING.	
790 BUILDING SEALANT	DOW CORNING CORP	SILICONE ONE-PART SEALANT,LOW MODULUS GOOD WEATHER RESISTANCE,DURABILITY. FOR JOINTS IN PRECAST CONCRETE PANELS CURTAINWALLS,EXPANSION JOINTS,SOLAR COLLECTOR PANELS,OR JOINTS WITH EXCESSIVE MOVEMENT.	

PIPE INSULATION

PRODUCT TRADENAME	MANUFACTURER	PRODUCT DESCRIPTION	PLATES
KAYLO 10	OWENS-CORNING FIBERGLAS CORP	CALCIUM SILICATE RIGID INSULATION FOR PIPING AT TEMPERATURES UP TO 1200 DEG F.	40B-D
METAL-ON	MANVILLE CORP	CALCIUM SILICATE MOLDED PIPE INSULATION WITH FACTORY-APPLIED ALUMINUM JACKET. FOR PIPE TEMPERATURES UP TO 1500 DEG F.	40C-D
THERMO-12	MANVILLE CORP	CALCIUM SILICATE,MOLDED PIPE INSULATION. AVAILABLE WITH FACTORY-APPLIED ALUMINUM OR STAINLESS STEEL JACKET. AVAILABLE WITH EXTENDED LEGS TO ALLOW FOR 1/2-INCH TRACED LINE. FOR PIPING TEMPERATURES UP TO 1500 DEG F.	40B-D

INSULATION MANUFACTURERS PRODUCTS
(SORTED BY PRODUCT DESCRIPTION)

PIPE INSULATION (Cont.)

PRODUCT TRADENAME	MANUFACTURER	PRODUCT DESCRIPTION	PLATES
THERMAZIP 552	ACCESSIBLE PRODUCTS CO	CERAMIC FIBER WOOL,JACKET OF ALUMINIZED HEAVY FIBERGLASS FABRIC WITH A HIGH-TEMPERATURE APPLICATIONS UP TO 1400 DEG F. INDOORS AND OUTDOORS.	40A
SEALASTIC	HALSTEAD INDUSTRIES,INC.	CORK FLEXIBLE TAPE,ADHESIVE BACKED. DEVELOPED ESPECIALLY FOR WRAPPING COLD PIPES TO PREVENT CONDENSATION.	44D
ARMAFLEX	ARMSTRONG WORLD INDUSTRIES, INC.	ELASTOMERIC FLEXIBLE PREFORMED INSUL-ATION.	40A,42A,44B
INSUL-TUBE	HALSTEAD INDUSTRIES,INC.	ELASTOMERIC FLEXIBLE PREFORMED INSUL-ATION.	40A,42A,44B
AEROTUBE	MANVILLE CORP	ELASTOMERIC FLEXIBLE PREFORMED INSUL-ATION.	40A,42A,44B
ARMAFLEX II	ARMSTRONG WORLD INDUSTRIES, INC.	ELASTOMERIC FLEXIBLE PREFORMED INSUL-ATION. LOWER FLAMMABLILTY RATINGS THAN STANDARD ARMAFLEX.	40A,42B,44B
AEROTUBE II	MANVILLE CORP	ELASTOMERIC FLEXIBLE PREFORMED INSUL-ATION. LOWER FLAMMABILITY RATINGS THAN STANDARD AEROTUBE.	40A,42A,44B
ARMAFLEX INSULATION TAPE	ARMSTRONG WORLD INDUSTRIES, INC.	ELASTOMERIC FLEXIBLE TAPE,ADHESIVE BACKED. FOR WRAPPING PIPES AND FITTINGS.	44D
INSULATION TAPE	HALSTEAD INDUSTRIES,INC.	ELASTOMERIC FLEXIBLE TAPE,ADHESIVE BACKED. FOR WRAPPING PIPES AND FITTINGS.	44D
PIPE INSULATING TAPE *	3M CO	FOAM FLEXIBLE TAPE,ADHESIVE BACKED. FOR WRAPPING PIPES AND FITTINGS.	44D
THERMAZIP 150	ACCESSIBLE PRODUCTS CO	GLASS FIBER FLEXIBLE WOOL,LAMINATED TO A JACKET OF METALLIZED POLYESTER INNER FILM,FIBERGLASS SCRIM,PVC OUTER FILM. PATENTED LOCKING TRAC. FOR LIGHT-DUTY INDOOR USE TO 400 DEG F.	40A

INSULATION MANUFACTURERS PRODUCTS
(SORTED BY PRODUCT DESCRIPTION)

PIPE INSULATION (Cont.)

PRODUCT TRADENAME	MANUFACTURER	PRODUCT DESCRIPTION	PLATES
THERMAZIP 250	ACCESSIBLE PRODUCTS CO	GLASS FIBER FLEXIBLE WOOL,PVC JACKET, LOCKING TRAC. FOR GENERAL INDOOR USE UP TO 850 DEG F.	40A
THERMAZIP 350	ACCESSIBLE PRODUCTS CO	GLASS FIBER FLEXIBLE WOOL,PVC IMPREGNATED POLYESTER FABRIC JACKET,LOCKING TRAC. FOR HEAVY-DUTY INDOOR AND OUTDOOR USE UP TO 850 DEG F.	40A
THERMAZIP 450	ACCESSIBLE PRODUCTS CO	GLASS FIBER FLEXIBLE WOOL JACKET OF HYPALON IMPREGNATED FIBERGLASS LAMINATED TO TEDLAR OUTER FILM,LOCKING TRAC. FOR EXTENDED OUTDOOR USE UP TO 9850 DEG F. SUPERIOR CHEMICAL RESISTANCE.	40A
THERMAZIP 550	ACCESSIBLE PRODUCTS CO	GLASS FIBER FLEXIBLE WOOL,JACKET OF ALUMINIZED HEAVY FIBERGLASS FABRIC WITH A HIGH-TEMPERATURE COATING,LOCKING TRAC. FOR HIGH TEMPERATURE APPLICATIONS UP TO 850 DEG F. INDOORS AND OUTDOORS.	40A
THERMAZIP 850	ACCESSIBLE PRODUCTS CO	GLASS FIBER FLEXIBLE WOOL,RIGID NONPOROUS PVC JACKET,LOCKING TRAC. FOR INDOOR USE IN FOOD PROCESSING AND SANITARY APPLICATIONS UP TO 850 DEG F	40A
PIPE INSULATION	KNAUF FIBER GLASS GMBH	GLASS FIBER MOLDED PIPE INSULATION,UNFACED OR WITH ALL-SERVICE FSK JACKET. FOR TEMPERATURES UP TO 500 DEG F.	40B-D
MICRO-LOK 650	MANVILLE CORP	GLASS FIBER RIGID PIPE INSULATION. FOR TEMPÉRATURES UP TO 650 DEG F. AVAILABLE WITH FSK OR POLISHED METAL JACKET.	40B-D
FIBERGLAS 25 ASJ/SSL	OWENS-CORNING FIBERGLAS CORP	GLASS FIBER RIGID PIPE INSULATION,WITH ALL-SERVICE JACKET. FOR TEMPERATURES UP TO 650 DEG F.	40B-D
850 SNAP*ON	CERTAINTEED CORP	GLASS FIBER RIGID MOLDED PIPE INSULATION. FOR TEMPERATURES UP TO 850 DEG F. AVAILABLE WITH FACTORY-APPLIED ALL-WEATHER JACKET.	40B-D

INSULATION MANUFACTURERS PRODUCTS
(SORTED BY PRODUCT DESCRIPTION)

PIPE INSULATION (Cont.)

PRODUCT TRADENAME	MANUFACTURER	PRODUCT DESCRIPTION	PLATES
PIPE WRAP INSULATION *	OWENS-CORNING FIBERGLAS CORP	GLASS FIBER SEMI-RIGID BOARD WITH A LAMINATED FSK JACKET. FOR WRAPPING AROUND LARGE-DIAMETER PIPES.	
PROTEXULATE	PROTEXULATE, INC	MINERAL POWDER LOOSE-FILL INSULATION FOR UNDERGROUND PIPING, WATER REPELLANT. FOR CRYOGENIC TEMPERATURES TO 480 DEG F.	
CELOTEMP 1500	CELOTEX CORP	PERLITE, EXPANDED, WITH MOISTURE RESISTANT BINDER.	
ACCOTHERM	ARMSTRONG WORLD INDUSTRIES, INC.	PHENOLIC RIGID FOAM, CONTINUOUSLY MOLDED WITH A LAMINATED ALL-SERVICE VAPOR BARRIER JACKET.	40B-D, 42B 43A
IMCOAFLEX	INSULATING MATERIALS CORP OF AMERICA	POLYETHYLENE FOAM, CLOSED CELL, UNFACED, UV STABLIZED. FOR PIPING FROM -110 TO 210 DEG F.	40A, 42A, 44B
TRYMER	UPJOHN CO, CPR DIVISION	POLYISOCYANURATE FOAM RIGID PREFORMED INSULATION, CHOICE OF JACKETING MATERIALS. FOR TEMPERATURES FROM -425 TO 300 DEG F.	40B-D
SOLAR-7	NORTHEAST SPECIALITY INSULATION	POLYISOCYANURATE FOAM RIGID INSULATION, FOR SINGLE OR DOUBLE PIPES, UV-RESISTANT PVC JACKET.	41A
ARMALOK II	ARMSTRONG WORLD INDUSTRIES, INC.	POLYURETHANE RIGID FOAM, CONTINUOUSLY MOLDED, WITH A LAMINATED ALUMINUM FOIL AND WHITE KRAFT PAPER JACKET.	40B-D, 42B 43A
THERMAZIP 175	ACCESSIBLE PRODUCTS CO	POLYURETHANE FLEXIBLE FOAM (ETHER TYPE) LAMINATED TO A JACKET OF METALLIZED POLYESTER INNER FILM, FIBERGLASS SCRIM PVC OUTER FILM. PATENTED LOCKING TRAC. FOR LIGHT-DUTY INDOOR USE FROM -60 TO 220 DEG F.	40A
THERMAZIP 275	ACCESSIBLE PRODUCTS CO	POLYURETHANE FLEXIBLE FOAM (ETHER TYPE), PVC JACKET, LOCKING TRAC. FOR GENERAL	40A

INSULATION MANUFACTURERS PRODUCTS
(SORTED BY PRODUCT DESCRIPTION)

PIPE INSULATION (Cont.)

PRODUCT TRADENAME	MANUFACTURER	PRODUCT DESCRIPTION	PLATES
THERMAZIP 375	ACCESSIBLE PRODUCTS CO	POLYURETHANE FLEXIBLE FOAM (ETHER TYPE), PVC IMPREGNATED POLYESTER FABRIC JACKET,LOCKING TRAC. FOR HEAVY-DUTY INDOOR AND OUTDOOR USE FROM -60 TO 220 DEG F.	40A
THERMAZIP 475	ACCESSIBLE PRODUCTS CO	POLYURETHANE FLEXIBLE FOAM (ETHER TYPE), JACKET OF HYPALON IMPREGNATED FIBER-GLASS LAMINATED TO TEDLAR OUTER FILM, LOCKING TRAC. FOR EXTENDED OUTDOOR USE FROM -60 TO 220 DEG F. SUPERIOR CHEMICAL RESISTANCE.	40A
THERMAZIP 875	ACCESSIBLE PRODUCTS CO	POLYURETHANE FLEXIBLE FOAM (ETHER TYPE), RIGID NONPOROUS PVC JACKET, LOCKING TRAC. FOR INDOOR USE IN FOOD PROCESS-ING AND SANITARY APPLICATIONS FROM -60 TO 220 DEG F.	40A
ISONATE	UPJOHN CO,CPR DIVISION	POLYURETHANE FOAM,SPRAYED-IN-PLACE FOR UNDERGROUND PIPING SYSTEMS.	
SUPER TEMP-TITE	MANVILLE CORP	COMBINATION OF CALCIUM SILICATE AND POLYURETHANE FOAM (THERMO-FOAM) PREINSULATED UNDERGROUND PIPE. CASING AND CORE PIPES ARE ASBESTOS-CEMENT. FOR HOT WATER AND STEAM UP TO 450 DEG F.	
HITEMP	INSTA-FOAM PRODUCTS,INC	CELLULAR GLASS PREINSULATED PIPING,IRON, STEEL,OR COPPER CARRIER PIPES,FRP JACKET. FOR TEMPERATURES UP TO 800 DEG F.	
TEMP-TITE	MANVILLE CORP	POLYURETHANE FOAM PREINSULATED UNDER-GROUND PIPE. CASING AND CORE PIPES ARE ASBESTOS-CEMENT. END CAPS PRO-TECT INSULATION. PIPES ARE CONNECTED WITH RUBBER COUPLINGS. FOR HOT OR CHILLED WATER.	

INSULATION MANUFACTURERS PRODUCTS
(SORTED BY PRODUCT DESCRIPTION)

PIPE INSULATION (Cont.)

PRODUCT TRADENAME	MANUFACTURER	PRODUCT DESCRIPTION	PLATES
HEAT-TITE	MANVILLE CORP	POLYURETHANE FOAM PREINSULATED UNDERGROUND PIPE. PVC CASING, SCHED. 40 STEEL CORE PIPE. END SEALS PROTECT INSULATION. PIPES CONNECTED WITH ASBESTOS-CEMENT COUPLINGS CONTAINING ELASTOMERIC SEALING RINGS. FOR HOT WATER UP TO 260 DEG F.	
COPPER CORE TEMP-TITE	MANVILLE CORP	POLYURETHANE FOAM PREINSULATED UNDERGROUND PIPE. PVC CASING,TYPE K COPPER CORE PIPE,INTEGRAL COUPLING. FOR CHILLED AND HOT WATER FROM 35 TO 260 DEG F.	
KOOL-KORE	MANVILLE CORP	POLYURETHANE FOAM PREINSULATED UNDERGROUND PIPE. PVC CASING AND CORE PIPES. FOR CHILLED WATER.	
PRE-INSULATED PIPING SYSTEMS	INSTA-FOAM PRODUCTS,INC	POLYURETHANE FOAM PREINSULATED PIPING. CARRIER PIPE MAY BE COPPER,STEEL, STAINLESS STEEL,ALUMINUM,FIBERGLASS, OR PVC,JACKET MAY BE STEEL,STAINLESS STEEL,ALUMINUM,PVC,OR ASBESTOS CEMENT. FOR TEMPERATURES FROM -350 TO 250 DEG F.	
X-50	TPCO,INC	POLYURETHANE FOAM PREINSULATED PIPE, STEEL,STAINLESS STEEL,COPPER,PVC OR FRP CARRIER PIPE WITH PVC,POLYURETHANE,FRP,SPIRAL-WOUND METAL,OR VIRTUALLY ANY TUBULAR JACKET.	
INSUL-8	ROVANCO CORP	POLYURETHANE FOAM PREINSULATED PIPING. SINGLE OR MULTIPLE CARRIER PIPES OF STEEL,STAINLESS STEEL,COPPER,ALUMINUM PVC,OR FIBERGLASS,IN AN OUTER JACKET OF STEEL,COATED STEEL,STAINLESS STEEL ALUMINUM,PVC,FIBERGLASS,POLYURETHANE.	41A
DUAL PIPE	INSTA-FOAM PRODUCTS,INC	POLYURETHANE FOAM PREINSULATED PIPING WITH TWO PIPES INSIDE ONE OUTER JACKET.	41A
DUPLEX X-50	TPCO,INC	POLYURETHANE FOAM PREINSULATED PIPE, TWO CARRIER PIPES IN A SINGLE JACKET.	41A

INSULATION MANUFACTURERS PRODUCTS
(SORTED BY PRODUCT DESCRIPTION)

PRODUCT TRADENAME	MANUFACTURER	PRODUCT DESCRIPTION	PLATES

PIPE INSULATION (Cont.)

PRODUCT TRADENAME	MANUFACTURER	PRODUCT DESCRIPTION	PLATES
NORTHSTAR PIPING SYSTEMS	TPCO, INC	PREINSULATED PIPING,SINGLE OR DOUBLE, WITHIN A CASING PIPE. INSULATION IS GLASS FIBER,CALCIUM SILICATE,OR AS SPECIFIED.	41B
FRP 500 SERIES	E. B. KAISER CO	PREINSULATED PIPING,INDIVIDUALLY INSUL-ATED CARRIER PIPES IN A SINGLE OUTER CASING. INSULATION IS GLASS FIBER, CALCIUM SILICATE,OR AS SPECIFIED.	
STD 400 SERIES	E. B. KAISER CO	PREINSULATED PIPING,MULTIPLE CARRIER PIPES INSIDE A SINGLE,INTERNALLY INSULATED OUTER CASING. INSULATION IS GLASS FIBER,CALCIUM SILICATE,OR AS SPECIFIED.	

EQUIPMENT INSULATION

PRODUCT TRADENAME	MANUFACTURER	PRODUCT DESCRIPTION	PLATES
KAYLO 10	OWENS-CORNING FIBERGLAS CORP	CALCIUM SILICATE BLOCK INSULATION. FOR BOILERS,TANKS AND VESSELS UP TO 1200 DEG F. AVAILABLE WITH V-GROOVE TO CONFORM TO CURVED SURFACES.	
THERMO-12	MANVILLE CORP	CALCIUM SILICATE,MOLDED FLAT OR RADIUS BLOCK. FOR EQUIPMENT TEMPERATURES UP TO 1500 DEG F.	
THERMAZIP HI-T BLANKET #53	ACCESSIBLE PRODUCTS CO	CERAMIC FIBER WOOL BLANKET,SERVICE TEMP LIMIT 1400 DEG F.,SANDWICHED BETWEEN TWO LAYERS OF SILICONE RUBBER COAT-ED FIBERGLASS FABRIC (UP TO 500 DEG F.),OR SILICA CLOTH. FOR INSULATING HEAT EXCHANGERS,VALVES,FLANGES,EX-PANSION JOINTS. REMOVABLE.	
THERMAZIP HI-T BLANKET #54	ACCESSIBLE PRODUCTS CO	CERAMIC FIBER (ALUMINA AND SILICA) WOOL BLANKET,SANDWICHED BETWEEN TWO LAYERS OF SILICONE COATED FIBERGLASS FABRIC (UP TO 500 DEG F.) OR SILICA CLOTH (UP TO 1800 DEG F.). FOR INSULATING HEAT EXCHANGERS,VALVES,FLANGES,EX-PANSION JOINTS. REMOVABLE.	

INSULATION MANUFACTURERS PRODUCTS
(SORTED BY PRODUCT DESCRIPTION)

EQUIPMENT INSULATION (Cont.)

PRODUCT TRADENAME	MANUFACTURER	PRODUCT DESCRIPTION	PLATES
ARMAFLEX	ARMSTRONG WORLD INDUSTRIES, INC.	ELASTOMERIC FLEXIBLE SHEET. ADHERES WITH ADHESIVES TO LARGE FLAT OR CURVED METAL SURFACES.	
INSUL-SHEET	HALSTEAD INDUSTRIES, INC.	ELASTOMERIC FLEXIBLE SHEET. ADHERES WITH ADHESIVES TO LARGE FLAT OR CURVED METAL SURFACES.	
AEROTUBE	MANVILLE CORP	ELASTOMERIC FLEXIBLE SHEET INSULATION.	
ARMAFLEX II	ARMSTRONG WORLD INDUSTRIES, INC.	ELASTOMERIC FLEXIBLE SHEET. ADHERES WITH ADHESIVES TO LARGE FLAT OR CURVED METAL SURFACES. LOWER FLAMMABILITY RATINGS THAN STANDARD ARMAFLEX.	
AEROTUBE II	MANVILLE CORP	ELASTOMERIC FLEXIBLE SHEET INSULATION. LOWER FLAMMABILITY RATINGS THAN STANDARD AEROTUBE SHEET.	
800 SERIES SPIN-GLAS BLANKET	MANVILLE CORP	GLASS FIBER FLEXIBLE BLANKET, FACED OR UNFACED. FOR INDUSTRIAL HEATING, AIR CONDITIONING, POWER AND PROCESS EQUIPMENT.	
THERMAZIP 353 SHEET STOCK	ACCESSIBLE PRODUCTS CO	GLASS FIBER FLEXIBLE WOOL SHEET, PVC IMPREGNATED POLYESTER FABRIC JACKET FOR TANK AND VESSEL APPLICATIONS UP TO 850 DEG F.	
TIW, TYPE II	OWENS-CORNING FIBERGLAS CORP	GLASS FIBER WOOL BATTS FOR METAL MESH BLANKETS AND FOR BOILERS, VESSELS, AND EQUIPMENT UP TO 1000 DEG F.	
TIW, TYPE I	OWENS-CORNING FIBERGLAS CORP	GLASS FIBER FLEXIBLE WOOL WRAP. FOR INDUSTRIAL OVENS AND IRREGULAR SURFACES UP TO 1000 DEG F. LOW COMPRESSIVE STRENGTH.	
GLAS-MAT 1200	MANVILLE CORP	GLASS FIBER MECHANICALLY BONDED BLANKET FOR INDUSTRIAL, MARINE, AND PROCESS APPLICATIONS UP TO 1200 DEG F.	

INSULATION MANUFACTURERS PRODUCTS
(SORTED BY PRODUCT DESCRIPTION)

EQUIPMENT INSULATION (Cont.)

PRODUCT TRADENAME	MANUFACTURER	PRODUCT DESCRIPTION	PLATES
700 SERIES INDUSTRIAL INSULATION *	OWENS-CORNING FIBERGLAS CORP	GLASS FIBER SEMI-RIGID RECTANGULAR BOARD,FACED OR UNFACED. FOR EQUIPMENT,VESSELS,AND TANKS UP TO 450 DEG F.	45A-C
INSULATION BOARD *	KNAUF FIBER GLASS GMBH	GLASS FIBER SEMI-RIGID BOARD,UNFACED,OR FSK OR ALL-SERVICE JACKET FACED. FOR POWER AND PROCESS EQUIPMENT, BOILER AND STACK INSTALLATIONS UP TO 450 DEG F.	
METAL-ON	MANVILLE CORP	GLASS FIBER INSULATION WITH EMBOSSED ALUMINUM SHEET FACING. FOR HEATED TANKS UP TO 450 DEG F.	45A-C
PIPE AND TANK INSULATION *	MANVILLE CORP	GLASS FIBER SEMI-RIGID BOARD,BONDED TO A FLEXIBLE FSK JACKET,SEGMENTED AND SUPPLIED IN ROLLS. FOR APPLICATIONS TO PIPES,TANKS,DUCTS,VESSELS.	45A-C
ELEVATED TEMPERATURE BOARD *	KNAUF FIBER GLASS GMBH	GLASS FIBER SEMI-RIGID BOARD BONDED WITH HIGH-TEMPERATURE THERMOSETTING RESIN, UNFACED. FOR BOILER WALLS,PRECIPITATORS,TANKS,TOWERS,STACKS,AND OVENS UP TO 850 DEG F.	
1000 SERIES SPIN-GLAS	MANVILLE CORP	GLASS FIBER SEMI-RIGID BOARD. FOR FURNACES,BOILERS,HEATED VESSELS,DUCTS AND TANKS UP TO 850 DEG F.	45A-B
INSUL-QUICK	OWENS-CORNING FIBERGLAS CORP	GLASS FIBER SEMI-RIGID BOARD WITH HIGH-TEMPERATURE BINDER,UNFACED OR FOIL FACED. FOR PROCESS BOILERS, PRECIPITATORS,AND HEATED EQUIPMENT UP TO 850 DEG F.	45A-C
TANK TOP INSUL	MANVILLE CORP	MINERAL FIBER RIGID BOARD. FOR FLAT TOP SURFACES OF HEATED TANKS AND VESSELS UP TO 250 DEG F.	45A-B
H. T. BANROC	MANVILLE CORP	MINERAL FIBER BLOCK,BONDED WITH A CLAY BINDER. FOR FURNACES,BOILERS,HEATED TANKS AND VESSELS UP TO 1900 DEG F.	45A-B
TRYMER	UPJOHN CO,CPR DIVISION	POLYISOCYANURATE FOAM BOARDSTOCK FOR INDUSTRIAL APPLICATIONS FROM -425 TO 300 DEG F.	

INSULATION MANUFACTURERS PRODUCTS
(SORTED BY PRODUCT DESCRIPTION)

EQUIPMENT INSULATION (Cont.)

PRODUCT TRADENAME	MANUFACTURER	PRODUCT DESCRIPTION	PLATES
THERMAZIP 375 SHEET STOCK	ACCESSIBLE PRODUCTS CO	POLYURETHANE FLEXIBLE FOAM (ETHER TYPE), PVC IMPREGNATED POLYESTER FABRIC JACKET. FOR TANK AND VESSEL APPLICATIONS FROM -60 TO 220 DEG F.	
ISONATE	UPJOHN CO,CPR DIVISION	POLYURETHANE FOAM,SPRAYED-IN-PLACE. FOR TANKS,VESSELS,HARD-TO-INSULATE INDUSTRIAL APPLICATIONS.	

DUCT INSULATION

PRODUCT TRADENAME	MANUFACTURER	PRODUCT DESCRIPTION	PLATES
MICROLITE	MANVILLE CORP	GLASS FIBER FLEXIBLE BLANKET,FACED OR UNFACED. FOR EXTERIOR INSULAT ON OF ROUND AND RECTANGULAR SHEET METAL DUCTS.	46B,47B
800 SERIES SPIN-GLAS	MANVILLE CORP	GLASS FI.ER FLEXIBLE BLANKET,FACED OR UNFACED. FOR EXTERIOR INSULATION OF ROUND AND RECTANGULAR SHEET METAL DUCTS.	46B,47B
FIBERGLAS DUCT WRAP	OWENS-CORNING FIBERGLAS CORP	GLASS FIBER BLANKET,UNFACED OR FACED WITH A REINFORCED FOIL KRAFT FAC NG. FOR TEMPERATURES FROM 40 TO 250 DEG F.	
DUCT WRAP	KNAUF FIBER GLASS GMBH	GLASS FIBER FLEXIBLE BLANKET,UNFACED OR WITH FSK OR VINYL VAPOR BARRIER. FOR HEATING AND AIR CONDITIONING DUCTS FROM 40 TO 250 DEG F.	46B,47B
STANDARD DUCT INSULATION *	CERTAINTEED CORP	GLASS FIBER FLEXIBLE BLANKET,FSK OR VINYL FACING. FOR WRAPPING HEATING AND COOLING DUCTWORK FROM 35 TO 250 DEG F.	
700 SERIES INDUSTRIAL INSULATION	OWENS-CORNING FIBERGLAS CORP	GLASS FIBER SEMI-RIGID RECTANGULAR BOARD,FACED OR UNFACED. FOR DUCTWORK UP TO 450 DEG F.	46A

INSULATION MANUFACTURERS PRODUCTS
(SORTED BY PRODUCT DESCRIPTION)

DUCT INSULATION (Cont.)

PRODUCT TRADENAME	MANUFACTURER	PRODUCT DESCRIPTION	PLATES
INSULATION BOARD *	KNAUF FIBER GLASS GMBH	GLASS FIBER SEMI-RIGID BOARD,UNFACED,OR FSK OR ALL-SERVICE JACKET FACED. FOR HEATING AND AIR CONDITIONING DUCTS FROM -20 TO 450 DEG F.	46A
ELEVATED TEMPERATURE BOARD *	KNAUF FIBER GLASS GMBH	GLASS FIBER SEMI-RIGID BOARD BONDED WITH HIGH-TEMPERATURE THERMOSETTING RESIN,UNFACED. FOR HOT DUCTWORK UP TO 850 DEG F.	46A
INSUL-QUICK	OWENS-CORNING FIBERGLAS CORP	GLASS FIBER SEMI-RIGID BOARD WITH HIGH-TEMPERATURE BINDER,UNFACED OR FOIL FACED. FOR DUCTWORK AND CHIMNEY LINERS UP TO 850 DEG F.	46A
MICRO-AIRE DUCT BOARD	MANVILLE CORP	GLASS FIBER RIGID BOARD WITH FSK OR HEAVY-DUTY FOIL-KRAFT-SCRIM-KRAFT (HDF) FACING. PREMOLDED SLIP JOINT EDGES. FOR FABRICATING RECTANGULAR HEATING AND COOLING DUCTWORK.	46D
FIBERGLAS DUCT BOARD	OWENS-CORNING FIBERGLAS CORP	GLASS FIBER RIGID BOARD WITH ALUMINUM FOIL VAPOR BARRIER. FOR FABRICATING RECTANGULAR DUCTWORK AND FITTINGS. FOR TEMPERATURES UP TO 250 DEG F.	46D
LINACOUSTIC	MANVILLE CORP	GLASS FIBER FLEXIBLE DUCT LINER. APPLIED TO THE INTERIOR OF DUCTS FOR TEMP-ERATURES UP TO 250 DEG F.	46C
AEROFLEX DUCT LINER	OWENS-CORNING FIBERGLAS CORP	GLASS FIBER FLEXIBLE MAT WITH FLAME-RESISTANT COATING. APPLIED TO THE INTERIOR OF DUCTS,FOR TEMPERATURES UP TO 250 DEG F.	46C
LINACOUSTIC R	MANVILLE CORP	GLASS FIBER PLENUM LINER BOARD WITH A BLACK MAT COATING. APPLIED TO THE INTERIOR OF PLENUMS UP TO 250 DEG F.	46C
AEROFLEX DUCT LINER BOARD	OWENS-CORNING FIBERGLAS CORP	GLASS FIBER SEMI-RIGID BOARD WITH FLAME-RESISTANT COATING. APPLIED TO THE INTERIOR OF DUCTS,FOR TEMPERATURES UP TO 250 DEG F.	46C
INL-25 FLEXIBLE DUCT	OWENS-CORNING FIBERGLAS CORP	GLASS FIBER INSULATED FLEXIBLE ROUND DUCT WITH RESILIENT INNER AIR BARRIER	47C

INSULATION MANUFACTURERS PRODUCTS
(SORTED BY PRODUCT DESCRIPTION)

DUCT INSULATION (Cont.)

PRODUCT TRADENAME	MANUFACTURER	PRODUCT DESCRIPTION	PLATES
FIBERGLAS VALUFLEX	OWENS-CORNING FIBERGLAS CORP	GLASS FIBER INSULATED FLEXIBLE ROUND DUCT,A RESILIENT INNER AIR BARRIER, WITH A POLYETHYLENE JACKET. FOR HEATING AND AIR CONDITIONING DUCTS UP TO 250 DEG F.	47C
CERTAFLEX G25	CERTAINTEED CORP	GLASS FIBER INSULATED FLEXIBLE ROUND DUCT,INPERVIOUS WIRE REINFORCED INNER CORE WITH A POLYTHYLENE JACKET. FOR LOW-VELOCITY DUCT SYSTEMS UP TO 200 DEG F.	47C
FLEXIBLE DUCT TYPE WG	WIREMOLD CO	GLASS FIBER INSULATED FLEXIBLE ROUND DUCT,GALVANIZED STEEL WIRE HELIX CORE WITH AIRTIGHT POLYESTER FILM,POLYOLE-FIN JACKET.	47C
THERMAFLEX G-KM DUCT	AUTOMATION INDUSTRIES,INC	GLASS FIBER INSULATED FLEXIBLE ROUND DUCT,STEEL WIRE HELIX CORE BONDED TO A POLYMERIC LINER,WITH FIBER-GLASS REINFORCED POLYOLEFIN JACKET. FOR LOW AND MEDIUM PRESSURE SYSTEMS TO 200 DEG F.	47C
FLEXIBLE DUCT TYPE WK	WIREMOLD CO	GLASS FIBER INSULATED FLEXIBLE FOUND DUCT,GALVANIZED STEEL WIRE HELIX CORE WITH AIRTIGHT POLYESTER FILM,REIN-FORCED ALUMINIZED JACKET.	47C
THERMAFLEX M-KE DUCT	AUTOMATION INDUSTRIES,INC	GLASS FIBER INSULATED FLEXIBLE ROUND DUCT,STEEL WIRE HELIX CORE BONDED TO A POLYMERIC LINER WITH FIBERGLASS REINFORCED METALLIZED FILM JACKET. FOR LOW AND MEDIUM PRESSURE SYSTEMS UP TO 200 DEG F.	47C
CERTAFLEX-25&7	CERTAINTEED CORP	GLASS FIBER INSULATED FLEXIBLE ROUND DUCT,A STEEL WIRE HELIX ENCLOSED IN A DOUBLE-LAYER POLYESTER AIR BARRIER, REINFORCED METALLIZED MYLAR OUTER JACKET. FOR DUCT SYSTEMS UP TO 200 DEG F.	47C

INSULATION MANUFACTURERS PRODUCTS
(SORTED BY PRODUCT DESCRIPTION)

DUCT INSULATION (Cont.)

PRODUCT TRADENAME	MANUFACTURER	PRODUCT DESCRIPTION	PLATES
MICRO-AIRE J/FLX SL	MANVILLE CORP	GLASS FIBER INSULATED FLEXIBLE ROUND DUCT,VINYL-COATED STEEL HELIX CORE BONDED TO POLYETHYLENE,WITH AN OUTER VAPOR BARRIER JACKET OF FIBERGLASS REINFORCED METALLIZED MYLAR/NEOPRENE LAMINATE. FOR RESIDENTIAL AND LOW-PRESSURE COMMERCIAL APPLICATIONS UP TO 250 DEG F.	47C
ACOUSTI-K27	UNITED MCGILL CORP	GLASS FIBER INSULATED RIGID ROUND DUCT,SPIRAL-WOUND METAL INNER AND OUTER SHELLS.	47D
FLEXIBLE DUCT TYPE 57K	WIREMOLD CO	GLASS FIBER INSULATED FLEXIBLE ROUND DUCT,CORE OF HIGH-TEMPERATURE VINYL ORGANOSOL-COATED GLASS FABRIC MECHANICALLY LOCKED INTO A FLAT STEEL SPIRAL,POLYOLEFIN JACKET.	47D
FLEX-MET	MANVILLE CORP	GLASS FIBER INSULATED FLEXIBLE ROUND DUCT,FLEXIBLE ALUMINUM CORE,INSULATION COVERED ON BOTH SIDES WITH VINYL OR ALUMINIZED MYLAR VAPOR BARRIER. FOR MANY HEATING AND AIR CONDITIONING DUCT APPLICATIONS FROM 10 TO 250 DEG F.	47D
RIGID ROUND DUCT	MANVILLE CORP	GLASS FIBER RIGID ROUND DUCT,SCRIM-REINFORCED FOIL JACKET,SLIP-JOINT ENDS. FOR HEATING AND AIR CONDITIONING DUCTS UP TO 250 DEG F.	47A
MICRO-AIRE HV-3	MANVILLE CORP	GLASS FIBER RIGID ROUND DUCT,SCRIM-REINFORCED FOIL JACKET,SLIP-JOINT ENDS. FOR HIGH-PRESSURE,HIGH-VELOCITY HEATING AND AIR CONDITIONING DUCTS UP TO 250 DEG F.	47A

Other Noyes Publications

HANDBOOK OF COAL-BASED ELECTRIC POWER GENERATION

The Technology, Utilization, Application, and Economics of Coal for Generating Electric Power

by Robert H. Shannon, P.E.

Consulting Engineer
Rockville, Maryland

The information in this book is directed toward providing a single, basic reference work which will be of assistance in future studies and evaluations by the utility industry and the nation, as they move in the direction of increased use of coal for power generation. In view of constant changes in regulatory policy, economic and energy goals, this book should be a handy and rapidly accessed source of useful and practical data on coal-based power.

The various chapters in the book detail the resource, its exploration and development; coal mining, cleaning, and transportation; utilization of coal as a fuel; environmental aspects of coal-fired power; and wastes disposal. Fundamentals included in all of the material have been keyed to provide in-depth understanding sufficient to serve as the basis for use in coal supply/utilization and electric power generation evaluations, as well as overall studies and planning leading to management decisions and action.

A condensed table of contents listing **chapter titles and selected subtitles** is given below. The book is copiously annotated with tables and figures.

ISBN 0-8155-0907-3 (1982) **372 pages**

Other Noyes Publications

COGENERATION TECHNOLOGY AND ECONOMICS FOR THE PROCESS INDUSTRIES

Edited by D.J. De Renzo

Energy Technology Review No. 81

This book assesses the potential for industrial cogeneration development in the immediate future, for the process industries. Available technology and economics are discussed, and the details of case studies in five major energy-intensive industries—textiles, chemicals, pulp and paper, petroleum refining, and food—are presented. The case studies involve plants actually using cogeneration, as well as projected sites for this energy-saving technology. The book is based on research by *Resource Planning Associates, Inc.; Jack Faucett Associates; TRW, Inc.; Arthur D. Little Inc.; Westinghouse Electric Corp.;* and *Gibbs & Hill Inc.*

Cogeneration is the sequential production of electric or mechanical power and useful thermal energy. Much has been written and said about the energy-saving potential of co-locating facilities which generate electric power with facilities which consume large quantities of low-grade heat. Co-location of such facilities allows the by-product heat from electricity generation, which has traditionally been "thrown away," to be used productively. Properly designed and operated cogeneration facilities could, thus, save large amounts of fuel and capital.

Listed below is a condensed table of contents including **part and chapter titles.**

ISBN 0-8155-0932-4 (1983)

395 pages

Other Noyes Publications

INDUSTRIAL ENERGY CONSERVATION TECHNOLOGIES AND RESEARCH OPPORTUNITIES

Edited by D.J. De Renzo

Energy Technology Review No. 82

This study on industrial energy conservation technologies and research opportunities is based on research by *Battelle Pacific Northwest Laboratory* and *Fuel & Energy Consultants, Inc.*

Energy policy is an issue of great national concern. Planned and considered conservation techniques can minimize the costs and side effects of whatever energy source is consumed by limiting the amounts which must be purchased or supplied, and informed use of technology can often result in significant improvements in energy efficiency.

This three part book assesses and evaluates energy technologies currently available for application to industrial energy conservation. Also considered are research needs for developing more advanced technologies for efficient energy use.

Part I provides concise summaries of specific technologies, noting potential energy savings, economics, operational considerations, and applications. Part II describes a wide range of attractive technical areas for which specific research and development programs could be implemented. Industry, transportation and buildings are the broad end-use areas covered. Part III contains the background studies used to assess the research needs and opportunities for advanced energy conservation systems proposed in Part II.

A condensed table of contents listing **parts, chapters, and selected subtitles** is given below.

ISBN 0-8155-0941-3(1983) **704 pages**

Other Noyes Publications

SPILL PREVENTION AND FAIL-SAFE ENGINEERING FOR PETROLEUM AND RELATED PRODUCTS

by

J.L. Goodier **R.J. Siclari** **P.A. Garrity**

Pacific Northwest Laboratory
Battelle Memorial Institute

Pollution Technology Review No. 100

Annual losses of petroleum and related products due to accidental spillage are tremendous. This spill prevention and fail-safe engineering manual was prepared to provide guideline information on spill prevention procedures for individuals who have the responsibility for maintaining a spill-free plant during the transportation, transfer, storage and processing of petroleum and related products.

An attempt has been made to cover every facet of spill prevention. Special emphasis is given to fail-safe engineering as an approach to preventing spills from the predominant cause—human failure. The book addresses state-of-the-art spill prevention practices and automation techniques that can reduce spills caused by human error. To emphasize the need for spill prevention measures, historic spills are briefly described, after which remedial action is defined in an appropriate section of the manual. The section on plant security goes into considerable depth, since few security guidelines have been provided for industrial facilities that transfer, store, and process petroleum and related products.

The condensed table of contents below lists **chapters and selected subtitles.**

ISBN 0-8155-0944-8(1983)

329 pages

Other Noyes Publications

REMEDIAL ACTION TECHNOLOGY FOR WASTE DISPOSAL SITES

by

P. Rogoshewski **H. Bryson** **K. Wagner**

JRB Associates, Inc.
for the U.S. Environmental Protection Agency

Pollution Technology Review No. 101

The remedial actions which can be applied to control, contain, treat, or remove contaminants from uncontrolled hazardous waste sites and the nature of contamination at waste disposal sites are described in this comprehensive handbook. Improper disposal of industrial, commercial, and municipal solid and hazardous wastes is one of the nation's most pressing environmental problems. The quantity of wastes generated and disposed of annually is tremendous and growing. Cases of improper waste management have resulted in contamination of local groundwater, surface water, land, air, and food and forage crops.

As a result of clean-up operations that have already been conducted, and in anticipation of site clean-up activities that will result from recent regulatory action, many technologies have and are being developed. Those technologies specifically designed for clean-up of waste disposal sites are called "remedial actions." Remedial actions include surface, groundwater, leachate and gas migration controls; direct treatment methods; techniques for contaminated water and sewer lines; and processes for contaminated sediment removal.

The book details available technologies and describes how they may be selected and applied for the clean-up of disposal sites, with particular emphasis on hazardous waste sites. Information on each remedial action includes a general description; applications; design, construction, and/or operating considerations; advantages and disadvantages; and installation and annual operating costs, with examples where possible. **Chapter titles and selected subtitles** are given below.

ISBN 0-8155-0947-2 **500 pages**

Other Noyes Publications

ENERGY EFFICIENT INDUSTRIAL TECHNOLOGY IN EUROPE AND JAPAN

by

A. G. Fassbender, M. J. McGee, and Y. Yanase
Battelle, Pacific Northwest Laboratories

Energy Technology Review No. 85

This book describes energy efficient industrial technology currently in use in Europe and Japan.

Energy prices in Europe and Japan have exceeded those in the U.S. for many years. These higher costs have created a demand for efficient industrial equipment. Information on this equipment has previously been available only in bits and pieces. This book collects information on these technologies into one reference volume.

The technologies described constitute most of the advanced, gas-related technologies available to European and Japanese industry. Some of the technologies are unique, and some are currently available in the U.S.

The book contains three parts. Part I concentrates on equipment in use in West Germany, France, and the United Kingdom. Part II stresses equipment used in the Japanese aluminum, steel, and cement industries. Part III describes those selected technologies considered to have potential for widespread use.

A condensed table of contents listing **parts, chapters, and selected subtitles** is listed below.

ISBN 0-8155-0958-8 (1983) **416 pages**

Other Noyes Publications

ENERGY AND RESOURCE RECOVERY FROM WASTE

by

Stephen C. Schwarz, P.E. and Calvin R. Brunner, P.E.
Malcolm Pirnie, Inc.
White Plains, New York

Energy Technology Review No. 86
Pollution Technology Review No. 102

This book provides a complete and coordinated guide to the available techniques for recovering energy and resources from waste. Selection, design, and financing are covered and the information given reflects current costs and up-to-date mechanisms for implementation of the facilities.

Natural resources do not exist in unlimited supply. As the population increases in size and in affluence, the demand for energy and raw materials increases at an alarming rate. No less a concern is the generation of waste and the need for its safe and economical disposal.

In the book, the relationship of these two trends is evaluated in a qualitative and quantitative manner. Current technology for the recovery of energy and other resources from solid waste is well proven. These technologies are presented and applied to various situations. Facilities or processes will not be implemented without an understanding of their economics. Thus, the book also presents cost data and financing alternatives.

Background material is presented which covers estimating solid waste generation, source separation techniques, evaluating the impact of a facility on the environment and a discussion of the legal aspects of resource recovery.

Listed below are **chapter titles and selected subtitles.**

ISBN 0-8155-0959-6 (1983) **272 pages**

Other Noyes Publications

PRESSURIZED FLUIDIZED BED COMBUSTION TECHNOLOGY

by

W. F. Podolski et al
Argonne National Laboratory

R. L. Graves et al
Oak Ridge National Laboratory

Energy Technology Review No. 87
Pollution Technology Review No. 103

The technology of pressurized fluidized bed combustion (PFBC), particularly in its application to a coal-burning combined-cycle (CC) plant, is evaluated in this book. Engineering systems analyses are performed for advanced concept plant components—boiler systems, including combustors, air-handling and air-injection equipment, and heat exchangers; solids handling, injection and ejection systems; hot-gas cleanup equipment; instrumentation and control systems; and turbines—and materials of construction are technically assessed.

Environmental performance as well as energy efficiency are examined, and economic considerations—cost and market potential—are briefly reviewed. A broad survey of the principal related research and development programs in the U.S. and other countries is included. Responses to a survey of PFBC developers are reviewed, with emphasis on high risk areas of the PFBC concept. Finally, a projection of the most likely technological developments and a summary of unresolved technical issues and problems are presented.

A condensed table of contents including **part, chapter, and selected subtitles** is given below.

ISBN 0-8155-0960-X (1983)

429 pages

Other Noyes Publications

ROBOTICS APPLICATIONS FOR INDUSTRY
A Practical Guide

by

L. L. Toepperwein, M. T. Blackmon, W. T. Park
General Dynamics Corp.
Fort Worth, Texas

W. R. Tanner
Productivity Systems Inc.
Farmington, Michigan

W. F. Adolfson
Advanced Technology, Inc.
McLean, Virginia

This book provides an in-depth introduction to the basic concepts and techniques pertaining to the use of robotics technology for manufacturing and industrial applications. A special section is devoted to robotics in motor vehicle manufacture.

In an era when cost-effective technology and increased productivity are of prime concern, these studies should be of value in many decision-making processes. There is little question that computer aided manufacturing is going to become the technology of choice in coming years, and one of its first major U.S. applications has been in motor vehicle manufacture.

The book contains two parts. Part I details the basic concepts of robotics technology—robot configuration, sensors, tooling, work station integration, current application information, and future trends. It provides an approach to the implementation of the technology, a means of determining implementation costs, and a review of documentation of available literature. Part II covers robotics use in motor vehicle manufacture. Patterns of usage and current robot populations are given for the worldwide auto industry. New technology development requirements are identified and current research efforts described.

A condensed table of contents listing **part, chapter, and selected subtitles** is given below. An outstanding glossary of terms is included in Part I.

ISBN 0-8155-0962-6 (1983)

326 pages